Large Deviation Techniques in Decision, Simulation, and Estimation

Large Deviation Techniques in Decision, Simulation, and Estimation

JAMES A. BUCKLEW

Department of Electrical and Computer Engineering
University of Wisconsin—Madison
Madison, Wisconsin

A Wiley-Interscience Publication
John Wiley & Sons, Inc.
New York / Chichester / Brisbane / Toronto / Singapore

Library of Congress Cataloging in Publication Data:

Bucklew, James A.

Large deviation techniques in decision, simulation, and estimation / James A. Bucklew.

p. cm. — (Wiley series in probability and mathematical statistics. Applied probability and statistics section)

"A Wiley-Interscience publication."

Includes bibliographical references.

ISBN 0-471-61856-X

1. Large deviations. I. Title. II. Series.

QA273.67.B86 1990

519.2—dc20 89-77144

CIP

Printed in the United States of America

10 9 8 7 6 5 4 3 2 1

To
Cary and the Pups

Preface

The idea for this book arose out of a seminar series in large deviation theory run by James Kuelbs and Peter Ney of the University of Wisconsin—Madison Mathematics Department. The seminars, which ran from January 1985 to May 1987, presented results of the major research currently being done in that area. As I sat in on and participated in the seminars, I came to several conclusions about the subject material. First, there is a huge potential for applying these ideas to the study, analysis, and design of stochastic systems—in particular, information systems. Second, the subject material is unusually technical. The proofs in large deviation theory tend to require great amounts of analytical machinery to establish even the most basic of lemmas. Third, the ideas behind the proofs are usually quite simple and intuitive.

There is some controversy these days about the level of mathematics that should be expected of engineering and applied statistics graduate students. The spectrum of opinion varies from mere formula presentation to complete mathematical rigor. My own expectations vary with the importance and difficulty of the subject material. If the material is simple and foundational, I believe that one should present a rigorous and precise version of it. If the needed material is very difficult or tedious, I think that the student should at least be given a heuristic sketch of the proof which provides the main ideas and possible groundwork for the full-blown treatment.

Such an approach is followed in this book. Chapter II contains rigorous presentations of Cramér's and Ellis's foundational theorems. Elsewhere, some other rigorous presentations of key results are given, but the general level of expectations is fairly relaxed. I'm content that the "handwaving" sketches of some results contain the essential ideas and might be made rigorous with sufficient amounts of real analysis and measure theory. Whenever a theorem is quoted at the end of one of these derivational arguments, I give the rigorous version of the theorem with its technical conditions. In the notes and comments sections, I indicate where to find the proof in the mathematics literature.

This book is intended to form the basis of an upper-level graduate course for students of information theory, communication systems, applied statistics, and statistical signal processing. The prerequisites would be a graduate course in engineering probability and random processes, the ability to follow and understand an ε–δ proof, and some motivation to learn more about detection and estimation theory. No measure theory is assumed or used in the text (although the key definitions of measure theory are given in a self-contained appendix and are used to define precisely some of the terms used in the text). Problems are provided at the end of each chapter which are intended to provide practice exercises utilizing the new concepts developed, to illuminate the material from different directions, to develop new extensions and results to the material in the chapter, and to provide further examples of the tools and techniques made available by large deviation theory. There is a tremendous range of difficulty to be found in these problems. However, complete solutions to all exercises are provided at the end of the text.

I have long felt a need for a second graduate course in probability for information system students. Others acknowledge this also, but we differ regarding what should comprise such a course. Traditionally, a second course would revolve around the stochastic calculus, Itô's formula, and the nonlinear filtering equations. I think that this is a fine choice, but useful primarily to students whose major focus is going to be the study of dynamical systems (particularly stochastic control). Nonlinear filtering via the stochastic calculus does not appear to have developed into a major application area. On the other hand, large deviation theory *is* one of the main branches of modern probability theory and research. It can be used as a justification to study many topics in probability theory (such as Markov processes, Gaussian processes, types of stochastic convergence, and stochastic differential equations), just as the nonlinear filtering equations were the reason to concentrate on Itô processes. I believe that eventually large deviation theory will in some form find its way into graduate, even undergraduate, information systems curriculae. This book is intended as a small contribution toward this goal.

ACKNOWLEDGMENTS

I would like to thank John Sadowsky of Purdue University, Randall Bahr of the University of Arizona, and John Roy of the University of Wisconsin for their critical comments of an early version of this manuscript. Paul Schuette and Shaun Katzenberger of the University of Wisconsin must be praised for proofreading far in excess of the call of duty. I would especially like to thank Peter Ney, James Kuelbs, and Thomas Kurtz of the University of Wisconsin Mathematics department, without whose patient explanations and proofs this book would never have been attempted. I also thank the NSF for the support that gave me the time to write it. Of course, all the errors, omissions, and other not inconsiderable deficiencies of the book are the author's sole responsibility.

JAMES A. BUCKLEW

Madison, Wisconsin

Contents

Large Deviation Techniques
in Decision, Simulation,
and Estimation

CHAPTER I

Introduction

Consider a sequence of random variables $\mathbf{S}_1, \mathbf{S}_2, \ldots$ converging (in probability) to a real constant c. By this we mean that $P\{|\mathbf{S}_n - c| > \varepsilon\} \to 0$ as $n \to \infty$ for all $\varepsilon > 0$. $\mathbf{S}_n$ could represent, for example, the nth partial sum average of another sequence of random variables, e.g.,

$$\mathbf{S}_n = \frac{1}{n} \sum_{j=1}^{n} \mathbf{y}_j$$

where the $\{\mathbf{y}_j\}$ sequence might be independent, identically distributed, or even perhaps the realization of an ergodic Markov chain. By the law of large numbers, we know that $\mathbf{S}_n$ should be converging to the mean (if it exists). Hence c is merely the expected value of the random process. Another possibility might be to let $\mathbf{S}_n$ be defined as $\mathbf{S}_1/\sqrt{n}$. Here, if one knows a single value of the sequence, all can be obtained by proper scaling. In this case the constant c is zero regardless of the original distribution of $\mathbf{S}_1$.

It is often the case that not only does $P\{|\mathbf{S}_n - c| > \varepsilon\}$ go to zero, but it does so exponentially fast. That is, $P\{|\mathbf{S}_n - c| > \varepsilon\} \approx K(\varepsilon, c, n) \exp[-nI(\varepsilon, c)]$, where $K(\cdot, \cdot, \cdot)$ is a slowly varying function of n (relative to the exponential) and $I(\cdot, \cdot)$ is a positive quantity. Loosely, if such a relationship is satisfied, we will say that the sequence $\{\mathbf{S}_n\}$ satisfies a large deviation principle. Large deviation theory is concerned primarily with determining the quantities $I(\cdot, \cdot)$ and (to a lesser extent) $K(\cdot, \cdot, \cdot)$. The reason for the nomenclature is that for a fixed $\varepsilon > 0$ and a large index n, a large deviation from the nominal value occurs if $|\mathbf{S}_n - c| > \varepsilon$. Large deviation theory can rightly be considered as a generalization or extension of the law of large numbers. The law of large numbers says that certain probabilities converge to zero. Large deviation theory is concerned with the rate of convergence.

Why should information system engineers and applied statisticians be interested in large deviation theory? In this book we concentrate on three major application areas: hypothesis testing, fast simulation methods, and estimation theory.

The statistical framework of hypothesis testing has long proved its worth for information system design and analysis. Every modern curriculum in communication systems includes courses in which the communications problem is discussed from this point of view. A problem with this framework is that whereas it is very general and consequently includes most systems of interest, if it often very difficult to implement in practice. The methodology for a system analysis proceeds by finding the statistics governing the test under various hypotheses and evaluating the different probabilities of error. For a test statistic that involves passing the data through nonlinearities, taking sums, multiplying by matrices, etc., it is a formidable task to take the approach of finding the final probability distribution of the statistic. In small-sample problems there is really no alternative; a researcher may be testing the efficacy of a control using 15 samples from some study. Every sample may have been very difficult and/or expensive to obtain. As a result, the probabilities of error will probably be very large by the standards of practical information systems. Conversely, statistical communications is primarily a large sample or small probability of error discipline. Frequently, minimizing system complexity is more desirable than system optimization. Hence we are often faced with the problem of analyzing suboptimal systems that perhaps involve nonlinearities and huge numbers of samples.

One *would* like to evaluate all digital communication systems on the basis of their probability of error. Usually, though, closed form expressions for the probability of error can be obtained only for a few special cases, e.g., if the receiver is linear or the noise is deterministic (e.g., continuous wave interference) or Gaussian. Because of the intractability of computing these error probabilities directly, many other fidelity criteria have been developed and considered for use in the analysis and design of digital signaling systems. For example, the statistical measures of asymptotic relative efficiency, Bhattacharya distance, Fisher information, etc., have all made many appearances in the engineering detection theory literature.

Usually, as the sample size or signal power gets large, the probability of error decreases as an exponential function of the sample size or signal power. This fact is necessary to understand a common basis of comparison in the evaluation of digital systems. For example, a statement that a communication systems analyst might make is "Phase shift keying is approximately 3 dB better than frequency shift keying." [Translation: The modulation method of phase shift keying uses one half as much power as the method known as frequency shift keying for approximately the same probability of error performance. A factor of 2 in power is equivalent to difference of 3 on the decibel (dB) scale.] Phase shift keying (PSK) is a digital signaling method whereby a "1" is transmitted as $A\cos(\omega_0 t)$ for T seconds and a "0" is transmitted as $-A\cos(\omega_0 t)$ for T seconds. Similarly, frequency shift keying (FSK) transmits a "1" by sending $A\cos(\omega_0 t)$ for T seconds and transmits a "0" by sending $A\cos(\omega_1 t)$ for T seconds. The additive white Gaussian noise probability of error for PSK is well known [Stremler, 1990] to be

$$P_{\mathrm{PSK}}(\mathrm{error}) = K\exp\{-c\,\mathrm{SNR}\}$$

and for FSK is

$$P_{\mathrm{FSK}}(\mathrm{error}) = K'\exp\left\{-c\frac{\mathrm{SNR}}{2}\right\}$$

where c is a constant and K and K' are known functions varying slowly with the signal to noise power ratio, SNR (asymptotically K and K' are proportional to $1/\sqrt{\mathrm{SNR}}$). Ignoring the "constants" in front of the exponentials, one sees that the signal power is divided by 2 in FSK relative to what it is in PSK. Hence PSK is 3 dB (or a factor of 2) better. *The point is that digital communication systems are commonly evaluated on the basis of the constants appearing in the argument of the exponential decay component of the probability of error expressions.*

The branch of probability known as large deviation theory is concerned precisely with calculating the asymptotic rates with which these probabilities approach zero. Using these theorems and techniques, some notoriously difficult (but quite common) communication receiver structures will be analyzed in this book.

In many system designs or analyses, an event of rare probability is a key parameter of the system's efficacy. For example, in a digital communications receiver, the probability of a bit error is the natural system performance parameter. This probability is typically no greater than 10^{-6}. As mentioned above, communication receivers are typically complex, with nonlinear components such as limiters, automatic gain control devices, and multipliers throughout. Analysis of these systems (even a large deviation theory analysis), while providing insight, must frequently make so many simplifying assumptions that the final answer is often in doubt. One is then faced with the task of attempting to simulate the system's operation on a computer. Direct simulation of a rare event is a formidable computational task. You need to "hit" the event several times to get a good idea of its probability. For example, suppose that you want no more than 20% error with 95% confidence on the estimation of a certain probability. This would imply (for a 10^{-6} probability) that for independent identically distributed (iid) Bernoulli random variables, we need a sample size of 10^8 (see chapter VIII). This, of course, imposes severe (if not impossible) demands on the random number generators.

Recently, some work in the problem of simulating large backlogs in queuing networks has shed some light on this problem. Recall the principle of importance sampling in simulation: Suppose that P is our original (or direct simulation) probability distribution with density $p(\cdot)$. A natural estimate of the probability of the set A carried out by n samples $\boldsymbol{\omega}_1, \boldsymbol{\omega}_2, \ldots, \boldsymbol{\omega}_n$ is

$$\frac{1}{n}\sum_{i=1}^{n} 1_A(\boldsymbol{\omega}_i)$$

where $1_A(x) = 1$ if $x \in A$ and is zero otherwise. One hopes that under some kind of ergodicity assumption this estimate converges to $P(A)$. Let P^* be another probability distribution, with density $p^*(\cdot)$, from which we will generate random numbers $\boldsymbol{\omega}_1^*, \boldsymbol{\omega}_2^*, \ldots, \boldsymbol{\omega}_n^*$. Under P^* the corresponding estimate of interest might be

$$\frac{1}{n}\sum_{i=1}^{n} 1_A(\boldsymbol{\omega}_i^*)\frac{p(\boldsymbol{\omega}_i^*)}{p^*(\boldsymbol{\omega}_i^*)}$$

Under some ergodicity assumption, this estimate converges to $P(A)$ also. If P^* is chosen correctly, this estimator can have a smaller variance than the original estimate. We will show in this book that by using large deviation theory, we can create estimators that have variance performance *orders of magnitude* better than the direct estimates.

A parameter estimator is said to be consistent if as the number of samples approaches infinity, the estimate (in probability) approaches the true value. Often, this convergence is exponentially fast and a large deviation principle is associated with it. An obvious criterion for estimator design would be to maximize the rate constant associated with the large deviation. In this book we discuss this problem and give a universal lower bound to the best performance possible of any consistent estimate. Several examples are given of estimates that actually attain this lower bound.

Organization of the Book

In chapter II we present two of the fundamental results of large deviation theory: Cramér's theorem and the Gärtner–Ellis theorem. Cramér's theorem is usually credited with being the first large deviation result. It gives the large deviation principle for sums of independent identically distributed random variables. The Gärtner–Ellis result is much more general. It establishes a large deviation principle of random variables based only upon convergence properties of the associated sequence of moment generating functions.

In chapter III we establish a large deviation principle due to Sanov for the empirical distributions of iid data. An empirical distribution is an estimate, formed from data observations, of the true underlying distribution that controls the observations. The rate function in this case turns out to be an information theoretic entropy known as the divergence. These results are fundamental to information theory and statistical mechanics. The last part of the chapter deals with a very useful tool called the contraction principle. Suppose that one has a sequence of random variables with an associated large deviation principle. Now consider transforming that sequence with a continuous function into another sequence. The contraction principle gives an explicit formula for translating the original large deviation principle into one for the transformed sequence.

In chapter IV we establish the large deviation principle for the important case of continuous-time Gaussian random processes. We then use this result and the contraction principle to establish the large deviation principle for general diffusion processes. A consideration of the asymptotics of the mean escape time of a diffusion process from an attracting domain follows. We then consider large deviation results for a class of Markov processes that include birth–death processes.

In the first part of chapter V we develop large deviation theory for sums involving Markov chains. In this section we establish, via the Gärtner–Ellis theorem, the large deviation principle for additive functionals of a Markov process. The second part is devoted to deriving a large deviation result for the empirical distributions for Markov data. Both of these topics are crucial to the understanding of how dependency in the random process models affect the large deviation rate functions.

In chapter VI we consider some applications of the foregoing theory to research in detection theory. In the first section we consider the asymptotic performance of Neyman–Pearson tests by proving an elegant result known as Stein's lemma. In the second section we establish a result on the asymptotic rate to zero of the probability of error in distinguishing between two Markov processes. This generalizes a well-known result of Chernoff for the iid case. In the third section we show how to choose the best detector over some set of constraints (in the Markov chain hypothesis testing problem) to minimize the asymptotic probability of error. These constrained design problems and solutions appear to be new. In the fourth section we consider rate results for Gaussian processes acted upon by quadratic form detectors. Specifically, we give closed form solutions for square law receivers and differential phase shift keyed signaling systems. These results appear to be new also. In the fifth section we consider some asymptotics of a particular multiple hypothesis testing problem called the "changepoint problem."

In chapter VII we return to fundamentals and establish bounds for large deviation probabilities that are sharper than just the exponential decay rate. There are also slowly varying functions in front of the exponential terms that get washed out with the logarithmic limit theorems considered up to this point. In this chapter we establish what these terms are for the cases of iid and Markov random variables. We call these results the first order asymptotic expansions.

In chapter VIII we discuss the utilization of large deviation theory for the purpose of speeding up the simulation of rare events. We give two separate methodologies to accomplish this task. We provide an analysis (using techniques from chapter VII) of the expected variance reduction when one employs these techniques instead of a direct simulation method. One of the methodologies and the variance analysis appear to be new. We then show that in some sense these techniques are optimal.

Chapter IX deals with establishing a theorem due to Bahadur on the best possible performance of consistent estimators. This is accomplished by establishing a powerful universal lower bound to large deviation probabilities. Optimality results are then discussed for maximum likelihood parameter estimators.

In chapter X we present new proofs of some key results in information theory via large deviation theory techniques. The emphasis in this chapter is on theoretical limits to data compression. It appears that large portions of classical information theory can rightly be viewed as a subdiscipline of large deviation theory. The material is intended to be self-contained but would be more meaningful if the student had some familiarity with the basic results and theorems of the field.

The appendixes are intended to provide a quick sketch of and reference to most of the major mathematical concepts and machinery used in the text. The key topics considered are point set topology, elementary analysis, diagonalization of covariances, Toeplitz matrices, Perron–Frobenius theory, perturbation theory of finite-dimensional matrices, probability inequalities, Markov chains, hypothesis testing, parameter estimation, and the calculus of variations. References are given to allow the student to pursue these topics in further depth.

Throughout the book, random quantities will be denoted by **boldface** type.

CHAPTER II

Cramér's Theorem and Extensions

A. CRAMÉR'S THEOREM

The simplest setting in which to obtain large deviation results is that of considering sums of independent identically distributed (iid) random variables on the real line. For example, we would like to consider the large excursion probabilities of sums such as the sample average

$$\mathbf{S}_n = \frac{\mathbf{x}_1 + \mathbf{x}_2 + \cdots + \mathbf{x}_n}{n}$$

where the $\{\mathbf{x}_i\}$, $i = 1, 2, \ldots$, are iid and n is approaching infinity. Suppose that $E\{\mathbf{x}_1\} = m$ exists and is finite. By the law of large numbers, we know that $\mathbf{S}_n$ is approaching m. Consider the function

$$I(x) \equiv \sup_{\theta} [\theta x - \log M(\theta)]$$

where θ is a real variable and

$$M(\theta) \equiv E\{\exp[\theta \mathbf{x}_1]\}$$

$M(\theta)$ is called the *moment generating function.* $I(x)$, known as the *large deviation rate function,* will play a crucial role in the development of the theory. Let us list some properties of the large deviation rate function and give a couple of examples.

***Property* 1.** $I(x)$ is convex. □

Proof: I is the pointwise supremum of a collection of affine (and hence convex) functions. Hence I is convex. More directly, let $\lambda \in [0, 1]$; then

$$\begin{aligned} I(\lambda x_1 + (1-\lambda)x_2) &= \sup_\theta \{\theta(\lambda x_1 + (1-\lambda)x_2) - \log M(\theta)\} \\ &= \sup_\theta \{\theta\lambda x_1 - \lambda \log M(\theta) + \theta(1-\lambda)x_2 - (1-\lambda)\log M(\theta)\} \\ &\leqslant \sup_\theta \{\lambda(\theta x_1 - \log M(\theta))\} + \sup_\theta \{(1-\lambda)(\theta x_2 - \log M(\theta))\} \\ &= \lambda I(x_1) + (1-\lambda)I(x_2) \end{aligned}$$ □

We remark that a convex function $I(\cdot)$ on the real line is continuous everywhere on $D_I \equiv \{x : I(x) < \infty\}$, the *domain* of $I(\cdot)$.

***Property* 2.** $I(x)$ has its minimum value at $x = E\{\mathbf{x}_1\} = m$. Furthermore, $I(m) = 0$. □

Proof: Since $M(0) = 1$, $I(x) \geqslant 0x - \log M(0) = 0$. By Jensen's inequality $M(\theta) \geqslant \exp[\theta m]$, hence $\theta m - \log M(\theta) \leqslant 0$ for all θ. Thus $I(m) = 0$ and $I(x) \geqslant I(m)$ for all x. □

See figure 1 for a graph of a typical rate function.

***Property* 3.** Assume that $M(\theta) < \infty$ for all θ. If $\mathbf{x}_1$ has no real upper or lower bound [i.e., the smallest interval that $\mathbf{x}_1$ appears in with probability 1 is $(-\infty, \infty) = \mathbb{R}$], then $I(x) = \theta_0 x - \log M(\theta_0)$ for some $\theta_0 = \theta_0(x) \in \mathbb{R}$. In other words, the supremum in the definition of $I(x)$ is attained for every value of x. □

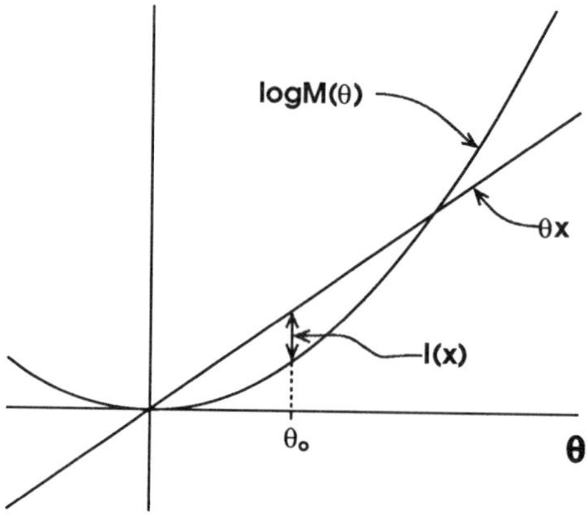

Figure 1. Graphical interpretation of the rate function $I(x)$.

Proof: Choose $K > 0$. Let $P_1(\cdot)$ be the distribution function of $\mathbf{x}_1$. Consider

$$\lim_{\theta\to\infty} \frac{\log M(\theta)}{\theta} = \lim_{\theta\to\infty} \frac{\log \int \exp[\theta x]\, dP_1(x)}{\theta}$$

$$\geqslant \lim_{\theta\to\infty} \frac{\log \int_{[K,\infty)} \exp[\theta x]\, dP_1(x)}{\theta}$$

$$\geqslant \lim_{\theta\to\infty} \frac{\theta K + \log P(\mathbf{x}_1 \in [K, \infty))}{\theta} = K$$

However, since K can be made arbitrarily large we must have that

$$\lim_{\theta\to\infty} \frac{\log M(\theta)}{\theta} = \infty$$

Similarly,

$$\lim_{\theta\to-\infty} \frac{\log M(\theta)}{-\theta} = \infty$$

Hence $\lim_{|\theta|\to\infty} \theta x - \log M(\theta) = -\infty$ for all x. So the $\sup\{\theta x - \log M(\theta)\}$ must be attained by some $\theta_0 = \theta_0(x) \in \mathbb{R}$. The fact that $M(\theta) < \infty$ for all θ implies (since all moments exist) that $M(\theta)$ (and hence $\log[M(\theta)]$) is differentiable for all values of θ. Differentiability implies that θ_0 must satisfy the equation $x = M'(\theta_0)/M(\theta_0)$. □

***Example* 1.** Suppose that $\mathbf{x}_1$ is a Gaussian random variable with mean zero and variance 1. The moment generating function is well known to be $M(\theta) = \exp\{\theta^2/2\}$. We then have $I(x) = \sup_\theta \{\theta x - \theta^2/2\} = x^2/2$. □

***Example* 2.** $\mathbf{x}_1$ takes the values 0 and 1 with probability $\frac{1}{2}$ each. Then $M(\theta) = [1 + \exp(\theta)]/2$ and calculation gives $I(x) = \log 2 + x \log x + (1 - x) \log(1 - x)$ for $0 \leqslant x \leqslant 1$ (take $0 \log 0 \equiv 0$ in this expression). $I(x) = \infty$ for x outside this range. □

Theorem (Cramér)
Assume that $M(\theta) < \infty$ for all θ. Then for every closed subset $F \subset \mathbb{R}$,

$$\overline{\lim_{n\to\infty}} \frac{1}{n} \log P\{\mathbf{S}_n \in F\} \leqslant -\inf_{x\in F} I(x)$$

and for every open subset $G \subset \mathbb{R}$,

$$\varliminf_{n\to\infty} \frac{1}{n} \log P\{\mathbf{S}_n \in G\} \geqslant -\inf_{x\in G} I(x)$$

Proof: We prove only a special case of the theorem. We assume that $\mathbf{x}_1$ has no real upper or lower bound. Furthermore, we only consider interval subsets of $\mathbb{R}$. In this setting it suffices to show for all $a < b$

$$\varliminf_{n\to\infty} \frac{1}{n} \log P\{\mathbf{S}_n \in (a,b)\} = \varlimsup_{n\to\infty} \frac{1}{n} \log P\{\mathbf{S}_n \in [a,b]\} = -\inf_{x\in(a,b)} I(x) = -\inf_{x\in[a,b]} I(x) \tag{1}$$

The last equality of (1) follows by the continuity of $I(\cdot)$. By properties 1 and 2, $I(x)$ is nonincreasing for $x \leqslant m$ and nondecreasing for $x \geqslant m$. Also, for $x \geqslant m$ and $\theta \leqslant 0$ we have that $\theta x - \log M(\theta) \leqslant \theta m - \log M(\theta) \leqslant 0$. Therefore, the supremum in the definition of $I(x)$ need only be taken over $\theta \geqslant 0$ for $x \geqslant m$. Similarly, we need consider only nonpositive θ for $x \leqslant m$. Since $\inf_{x\in(a,b)} I(x) = \inf_{x\in[a,b]} I(x)$, it suffices to show that

$$-\inf_{x\in[a,b]} I(x) \geqslant \varlimsup_{n\to\infty} \frac{1}{n} \log P\{\mathbf{S}_n \in [a,b]\} \geqslant \varliminf_{n\to\infty} \frac{1}{h} \log P\{\mathbf{S}_n \in (a,b)\} \geqslant -\inf_{x\in(a,b)} I(x)$$

Upper bound. If $a < m < b$, then by the law of large numbers and the fact that $I(m) = 0$, we have that all of the quantities above are zero and the statement is true. Therefore, without loss of generality, suppose that $a \geqslant m$ (the case $a < b \leqslant m$ is handled by replacing $\mathbf{x}_k$ with $-\mathbf{x}_k$). Then for $\theta \geqslant 0$,

$$\begin{aligned} P\{\mathbf{S}_n \in [a,b]\} &= \int_{[a,b]} dP_n(x) \leqslant \exp[-\theta a] \int_{[a,b]} \exp[\theta x]\, dP_n(x) \\ &\leqslant \exp[-\theta a] \int \exp[\theta x]\, dP_n(x) = \exp[-\theta a] M\left(\frac{\theta}{n}\right)^n \end{aligned}$$

where $P_n(\cdot)$ is the probability distribution function of $\mathbf{S}_n$. This implies that

$$\frac{1}{n} \log P\{\mathbf{S}_n \in [a,b]\} \leqslant -\frac{\theta a}{n} + \log M\left(\frac{\theta}{n}\right)$$

One may now replace θ by $n\theta$ to obtain

$$\frac{1}{n} \log P\{\mathbf{S}_n \in [a,b]\} \leqslant -[\theta a - \log M(\theta)]$$

Since this is true for all $\theta \geqslant 0$, we have

$$\frac{1}{n}\log P\{\mathbf{S}_n \in [a,b]\} \leqslant \inf_{\theta \geqslant 0} -[\theta a - \log M(\theta)] = -\sup_{\theta \geqslant 0}[\theta a - \log M(\theta)]$$

$$= -I(a) = -\inf_{x \in [a,b]} I(x)$$

One now merely takes the limit superior over n to finish the upper bound part of the proof.

Lower bound. Let $y \in (a,b)$. Let $\delta > 0$ be small enough so that $(y-\delta, y+\delta) \subset (a,b)$. It is then sufficient to show that

$$\varliminf_{n \to \infty} \frac{1}{n}\log P\{\mathbf{S}_n \in (y-\delta, y+\delta)\} \geqslant -I(y)$$

Let θ_0 be the point such that $I(y) = \theta_0 y - \log M(\theta_0)$. Without loss of generality, we can suppose that y is a number greater than or equal to the mean value m. This then implies that θ_0 is greater than or equal to zero by the arguments we made at the beginning of the proof. Define a new random variable $\mathbf{x}_{\theta_0}$ with probability distribution

$$P_{\theta_0}(z) \equiv P\{\mathbf{x}_{\theta_0} \leqslant z\} \equiv \frac{\int_{(-\infty,z]} \exp[\theta_0 x]\, dP_1(x)}{M(\theta_0)}$$

It is easy to see that the mean value of this new distribution is y (see figure 2).

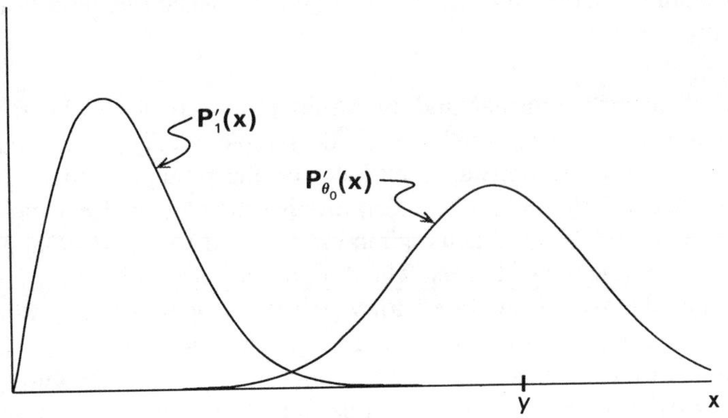

Figure 2

Therefore, by the law of large numbers, we must have for any $\delta_1 > 0$,

$$\lim_{n\to\infty} \int_{|(x_1+x_2+\cdots+x_n)/n-y|<\delta_1} dP_{\theta_0}(x_1)\,dP_{\theta_0}(x_2)\cdots dP_{\theta_0}(x_n) = 1$$

Now note that for $\delta_1 < \delta$ we have

$$\begin{aligned}
&P\{\mathbf{S}_n \in (y-\delta, y+\delta)\}\\
&= \int_{|(x_1+x_2+\cdots+x_n)/n-y|<\delta} dP_1(x_1)\,dP_1(x_2)\cdots dP_1(x_n)\\
&\geqslant \int_{|(x_1+x_2+\cdots+x_n)/n-y|<\delta_1} dP_1(x_1)\,dP_1(x_2)\cdots dP_1(x_n)\\
&\geqslant \exp[(-ny-\delta_1 n)\theta_0]\\
&\quad \int_{|(x_1+x_2+\cdots+x_n)/n-y|<\delta_1} \exp[\theta_0(x_1+x_2+\cdots+x_n)]\,dP_1(x_1)\,dP_1(x_2)\cdots dP_1(x_n)\\
&\geqslant \exp[(-ny-\delta_1 n)\theta_0]M(\theta_0)^n\\
&\quad \int_{|(x_1+x_2+\cdots+x_n)/n-y|<\delta_1} dP_{\theta_0}(x_1)\,dP_{\theta_0}(x_2)\cdots dP_{\theta_0}(x_n)
\end{aligned}$$

Therefore,

$$\varliminf_{n\to\infty} \frac{1}{n}\log P\{\mathbf{S}_n \in (y-\delta, y+\delta)\} \geqslant (-y-\delta_1)\theta_0 + \log M(\theta_0) = -I(y) - \delta_1\theta_0$$

Since δ_1 can be made arbitrarily small, we have finished the lower bound part of the proof. □

One of the more unusual (and nonintuitive) aspects of this theorem is that the asymptotics of the probability of the sample average lying in some set away from the mean depend essentially on one point, i.e., the logarithmic asymptotics of $P\{\mathbf{S}_n \in (m+\varepsilon, m+2\varepsilon)\}$ are the same as for $P\{\mathbf{S}_n \in (m+\varepsilon, \infty)\}$. The point where $I(x)$ attains its minimum is called the *dominating point* (see notes and comments) of the set. The dominating point of the set essentially determines the asymptotic rate with which the probability goes to zero.

One should point out that the asymptotic upper bound of Cramér's theorem actually holds for *every* n, not just for n large enough—i.e., $P\{\mathbf{S}_n \in [a,b]\} \leqslant \exp[-nI(a)]$ for all positive integers n (for the case $a > m$). This upper bound is usually called the *Chernoff bound* in communications theory. We consign further discussion of this important result to chapter VII, where more general results will be derived.

In the lower bound part of the proof of the theorem we defined a new probability distribution $P_{\theta_0}(\cdot)$. This distribution has a variety of names and uses. It is called variously the *twisted distribution, the tilted distribution, the exponential change of measure*, etc. The primary reason for this distribution is to shift the probability mass of the original random variables so that they have a specified mean value at the dominating point of the set under consideration. We will go into this point in more detail in chapter VIII.

The theorem that we have given is by no means the most general form that is known even for the iid case. Large deviation theorems are usually expressed as two separate limit theorems: an upper bound for closed sets and a lower bound for open sets. Only in the case of certain nice sets, such as intervals (or more generally in higher-dimensional spaces, convex sets), can one be guaranteed that the upper bound equals the lower bound. For the applications that we have in mind, the interval case will be sufficient.

We have imposed two other technical conditions that are not necessary but shorten the tedium of the proof considerably. First, we have required that $M(\theta)$, the moment generating function, be finite for all θ. Essentially, we thus demand that the probability density function of the $\mathbf{x}_i$'s (if it exists) has tails that decay to zero at least exponentially fast. In addition, we have required that $\mathbf{x}_1$ has no finite upper or lower bound. The primary reason for imposing these conditions was to be able to state property 3 of $I(x)$ which was employed heavily in the lower bound part of the proof. This property does not hold in general. Recall the definition $D_I \equiv \{x : I(x) < \infty\}$, the domain of $I(\cdot)$, and similarly, $D_M \equiv \{\theta : M(\theta) < \infty\}$. What is really needed to carry out the lower bound proof is that for every $y \in D_I$, we have that there exists a $\theta_0 = \theta_0(y)$ such that $I(y) = \{\theta_0 y - \log M(\theta_0)\}$. One condition that works quite well is to assume or require that $\log M(\theta)$ is steep. Steepness means simply that the magnitude of the derivative of $\log M(\theta)$ blows up or approaches infinity as it approaches the boundary of D_M, denoted ∂D_M. More precisely, a function $f: \mathbb{R}^d \to \mathbb{R}$ differentiable on its domain D_f is *steep* if $\{x_n\} \subset D_f$ and $x_n \to x_0 \in \partial D_f$ implies that $\|\nabla f(x_n)\| \to \infty$. This condition is enough to guarantee the existence of a maximizing θ_0 for all $y \in D_I$. With these technical considerations one can give a proof virtually identical to the proof above for this more general class of random variables. We consider some examples in exercises 1 and 2.

***Example* 3.** In this example we compare rate results obtained by careless use of the Central Limit Theorem with those obtained via Cramér's theorem. Let $\{\mathbf{x}_i\}$ be an iid collection of Gaussian, zero mean, variance 1, random variables. Define $\mathbf{S}_n = (1/n)\, \Sigma_{i=1}^n \mathbf{x}_i^2$. $\mathbf{S}_n$ converges to 1 by the law of large numbers. We are interested in the asymptotics of $P\{\mathbf{S}_n > \gamma\}$ for some γ greater than 1. Carelessly, one might say that $\mathbf{S}_n$ is approximately Gaussian with mean value 1 and variance $2/n$. Let $\mathbf{N}(m, \sigma^2)$ denote a Gaussian random variable of mean m and variance σ^2. Hence $P\{\mathbf{S}_n > \gamma\} \approx P\{\mathbf{N}(1,2/n) > \gamma\} = P\{\mathbf{N}(0, 1/n) > (\gamma - 1)/\sqrt{2}\} \approx k \exp[-n(\gamma - 1)^2/4]$ (where k is a slowly varying constant, see exercise 8). One would then conclude that $\lim_{n\to\infty} (1/n) \log P\{\mathbf{S}_n > \gamma\} = -(\gamma - 1)^2/4$. However, $\mathbf{x}_1^2$ has a chi-squared distribution. $\log[M(\theta)]$ is calculated (by completing a square) to be $-\frac{1}{2}\log[1 - 2\theta]$ for θ in the range $(-\infty, \frac{1}{2}) = D_M$. $I(\gamma)$ is then easily found to be $\frac{1}{2}(\gamma - \log \gamma - 1)$. It is trivial to check that $\log M(\theta)$ is steep. Hence we have that $\lim_{n\to\infty} (1/n) \log P\{\mathbf{S}_n > \gamma\} = I(\gamma)$. One makes more than a factor of 2 error in the rate when $\gamma = 3$ using the Central Limit Theory approximations. □

Note that in example 3, as γ approaches 1, the mean value of $\mathbf{x}_1^2$, the two rate calculations are asymptotically equivalent. Of course, they both go to zero as γ goes to 1; but also, and more important, they approach zero at the same rate; i.e., $\lim_{\gamma\to 1} [(\gamma - 1)^2/4]/I(\gamma) = 1$. This phenomenon is quite characteristic of large deviation calculations in settings more general than this example. The Central Limit Theorem approximations and the large deviation calculations coincide as the dominating points of the sets under consideration approach the mean value. We investigate this behavior more closely in chapter VII.

B. GÄRTNER–ELLIS THEOREM

Cramér [1938] first presented his theorem at a probability symposium in 1937. Since then, the theory has been broadened substantially. One of the most useful and surprising generalizations is the one due to Gärtner [1977] and, more recently, Ellis [1984]. These authors dispense with assumptions about the dependency structure of the random variable sequence and work entirely with assumptions on a sequence of moment generating functions. Their methods thus allow large deviation results to be derived for dependent random processes such as Markov chains, functionals of Gaussian random processes, etc. The main idea of their proof is to look closely at the proof of Cramér's theorem and see which aspects are really essential to the upper and lower bound derivations.

Suppose that we have some infinite sequence of random variables $\{\mathbf{Y}_n\}$. No assumptions are made about the dependency structure of this sequence (at least directly). Define

$$\varphi_n(\theta) \equiv \frac{1}{n} \log E\{\exp[\theta \mathbf{Y}_n]\}$$

Now let us list two hypotheses:

***Hypothesis* B1.** $\varphi(\theta) \equiv \lim_{n\to\infty} \varphi_n(\theta)$ exists for all $\theta \in \mathbb{R}$, where we allow ∞ both as a limit value and as an element of the sequence $\{\varphi_n(\theta)\}$. □

In other words, the limit exists and is defined to be $\varphi(\theta)$ for all $\theta \in \mathbb{R}$. Define $D_\varphi \equiv \{\theta : \varphi(\theta) < \infty\}$ and note that it is a convex set. $\varphi(\theta)$ automatically is convex since it is the limit of convex functions on a convex set.

***Hypothesis* B2.** φ is differentiable on D_φ. □

Now we may define

$$I(x) \equiv \sup_\theta [\theta x - \varphi(\theta)]$$

Furthermore, define

$$\varphi'(D_\varphi) \equiv \{\varphi'(\theta) : \theta \in D_\varphi\}$$

where φ' indicates the derivative of φ. Before proceeding to the main theorem, we first prove a simple but important lemma.

Lemma (exponential overbound)
Let $\mathbf{Z}$ be a real valued random variable. Then for $t \geqslant 0$, $z \in \mathbb{R}$, $P\{\mathbf{Z} > z\} \leqslant E\{\exp[t\mathbf{Z}]\} \exp[-tz]$.

Proof: The proof follows immediately from Chebyshev's inequality (see appendix D.2). More directly, let $F_z(\cdot)$ be the distribution function of $\mathbf{Z}$. $P\{\mathbf{Z} > z\} = \int_{(z,\infty)} 1\, dF_z(y)$. Now for $y \in [z, \infty)$, $1 \leqslant \exp[t(y - z)]$ for all $t \geqslant 0$. Hence

$$\begin{aligned} P\{\mathbf{Z} > z\} &\leqslant \int_{[z,\infty)} \exp[t(y - z)]\, dF_z(y) \leqslant \int_{(-\infty,\infty)} \exp[t(y - z)]\, dF_z(y) \\ &= E\{\exp[t\mathbf{Z}]\} \exp[-tz] \end{aligned}$$ □

We now state and prove a simple special case of a more general large deviations theorem due to Gärtner and Ellis. The general version of the theorem is given without proof in the next section, dealing with multidimensional results.

Theorem (Gärtner–Ellis)
Consider the interval with endpoints a and b ($a < b$). If hypothesis B1 holds and $[a, b] \cap D_I \neq \varnothing$ (the empty set), then

$$\overline{\lim_{n\to\infty}} \frac{1}{n} \log P\left\{\frac{\mathbf{Y}_n}{n} \in [a, b]\right\} \leqslant - \inf_{x \in [a,b]} I(x)$$

If hypotheses B1 and B2 hold and $(a, b) \subset \varphi'(D_\varphi)$, then

$$\underline{\lim}_{n\to\infty} \frac{1}{n} \log P\left\{\frac{\mathbf{Y}_n}{n} \in (a, b)\right\} \geqslant - \inf_{x \in (a,b)} I(x)$$

Proof: Upper bound. Fix some $\theta \in \mathbb{R}$. Define

$$H_\theta(c) = \{x : x\theta - \varphi(\theta) \leqslant c\}$$

Define $C = \bigcap_{i=1}^k H_{\theta_i}(c)$ for some set of fixed $\theta_1, \theta_2, \ldots, \theta_k$, $k < \infty$. Now

$$P\left\{\frac{\mathbf{Y}_n}{n} \notin C\right\} = P\left\{\frac{\mathbf{Y}_n}{n} \in \bigcup_{i=1}^k H^c_{\theta_i}(c)\right\} \qquad \text{(by DeMorgan's law)}$$

$$\leqslant \sum_{i=1}^k P\left\{\frac{\mathbf{Y}_n}{n}\theta_i - \varphi(\theta_i) > c\right\}$$

$$= \sum_{i=1}^k P\{\mathbf{Y}_n\theta_i > n\varphi(\theta_i) + cn\}$$

$$\leqslant \sum_{i=1}^k \exp[-n[c + \varphi(\theta_i)]]E\{\exp[\mathbf{Y}_n\theta_i]\}$$

(by the exponential overbound lemma with $t = 1$)

$$= \sum_{i=1}^k \exp[-n[c + \varphi(\theta_i)]]\exp[n\varphi_n(\theta_i)]$$

Hence $\overline{\lim}_{n\to\infty} (1/n) \log P\{\mathbf{Y}_n/n \notin C\} \leqslant -c$.

Consider now the interval $[a, b]$. Define $\bar{I} = \inf_{x\in[a,b]} I(x) < \infty$. It is clear that $[a, b] \subset \{x : I(x) > \bar{I} - \varepsilon\}$ for all $\varepsilon > 0$. Furthermore,

$$\{x : I(x) > \bar{I} - \varepsilon\} = \left\{x : \sup_\theta [\theta x - \varphi(\theta)] > \bar{I} - \varepsilon\right\} \subset \bigcup_\theta \{x : [\theta x - \varphi(\theta)] > \bar{I} - \varepsilon\}$$

Every set (i.e., every θ index) in this collection is an open set. Therefore, we have exhibited a collection of open sets (possibly uncountably infinite in number) that cover or include the set $[a, b]$. Such a collection of sets is called an open cover. Now the interval $[a, b]$ is closed and bounded and hence compact by the Heine–Borel theorem. Compact sets have the property that any open cover contains a finite subcover. Hence $[a, b] \subset \bigcup_{i=1}^k \{x : [\theta_i x - \varphi(\theta_i)] > \bar{I} - \varepsilon\}$ for some $k < \infty$ and some $\theta_1, \theta_2, \ldots, \theta_k$. Therefore,

$$\overline{\lim}_{n\to\infty} \frac{1}{n} \log P\left\{\frac{\mathbf{Y}_n}{n} \in [a, b]\right\} \leqslant \overline{\lim}_{n\to\infty} \frac{1}{n} \log P\left\{\frac{\mathbf{Y}_n}{n} \in \bigcup_{i=1}^k H^c_{\theta_i}(\bar{I} - \varepsilon)\right\} \leqslant -(\bar{I} - \varepsilon)$$

Since $\varepsilon > 0$ can be made arbitrarily small, we have

$$\overline{\lim_{n\to\infty}} \frac{1}{n} \log P\left\{\frac{\mathbf{Y}_n}{n} \in [a, b]\right\} \leqslant -\bar{I} = -\inf_{x\in[a,b]} I(x)$$

Lower bound. Take $v \in (a, b)$. Then $\varphi'(\theta) = v$ has a solution $\theta = \theta_v$ by assumption [since $(a, b) \subset \varphi'(D_\varphi)$]. This is enough to guarantee that $I(v) \equiv \sup_\theta [\theta v - \varphi(\theta)] = \theta_v v - \varphi(\theta_v)$. Let F_n denote the distribution function of $\mathbf{Y}_n$. Define

$$dF_n^{(\theta)}(\cdot) \equiv \frac{dF_n(\cdot)\exp[\theta x]}{\int \exp[\theta x]\, dF_n(x)} \equiv \frac{dF_n(\cdot)\exp[\theta x]}{\exp[n\varphi_n(\theta)]}$$

Let $\mathbf{Y}_n^{(\theta)}$ be a random variable with distribution $F_n^{(\theta)}(\cdot)$ and $B_\delta(v) \equiv \{x : |x - v| < \delta\}$. □

***Claim* 1.** $\lim_{n\to\infty} P\{\mathbf{Y}_n^{(\theta_v)}/n \notin B_\delta(v)\} = 0$. □

Proof of Claim 1: We will actually show the stronger result that this convergence is exponentially fast.

$$P\left\{\frac{\mathbf{Y}_n^{(\theta_v)}}{n} \geqslant v + \delta\right\} \leqslant E\{\exp[t(\mathbf{Y}_n^{(\theta_v)} - nv - n\delta)]\}$$

(for $t > 0$, by the exponential overbound lemma)

$$= \int \exp[t(y - nv - n\delta)]\, dF_n^{(\theta_v)}(y)$$

$$= \int \exp[y(\theta_v + t)]\, dF_n(y) \frac{\exp[-n(v + \delta)t]}{\exp[n\varphi_n(\theta_v)]}$$

$$= \exp[n\varphi_n(\theta_v + t)\exp[-n(v + \delta)t]\exp[-n\varphi_n(\theta_v)]$$

Therefore,

$$\overline{\lim_{n\to\infty}} \frac{1}{n} \log P\left\{\frac{\mathbf{Y}_n^{(\theta_v)}}{n} \geqslant v + \delta\right\} \leqslant \varphi(\theta_v + t) - \varphi(\theta_v) - (v + \delta)t \tag{2}$$

Since $(\varphi(\theta_v + t) - \varphi(\theta_v))/t \to \varphi'(\theta_v) = v$ as $t \to 0^+$, we have for sufficiently small t that the right-hand side of (2) can be made strictly negative. Hence $P\{\mathbf{Y}_n^{(\theta_v)} n \geqslant v + \delta\}$ converges to zero exponentially fast. Similarly,

$$P\left\{\frac{\mathbf{Y}_n^{(\theta_v)}}{n} \leqslant v - \delta\right\} \leqslant \exp[n\varphi_n(\theta_v - t) - n\varphi_n(\theta_v) + nt(v - \delta)]$$

for $t > 0$. Hence

$$\overline{\lim_{n\to\infty}} \frac{1}{n} \log P\left\{\frac{\mathbf{Y}_n^{(\theta_v)}}{n} \leqslant v - \delta\right\} \leqslant \varphi(\theta_v - t) - \varphi(\theta_v) + t(v - \delta) \tag{3}$$

Again, for t small enough, the upper bound of (3) becomes strictly negative, which implies that the probability goes to zero exponentially fast. Therefore, we have shown that $P\{\mathbf{Y}_n^{(\theta_v)}/n \notin B_\delta(v)\}$ not only goes to zero but does so exponentially fast, which completes the proof of the claim. □

Claim 2. $\underline{\lim}_{n\to\infty} (1/n) \log P\{\mathbf{Y}_n/n \in B_\delta(v)\} \geqslant -I(v) - \delta|\theta_v|$. □

Proof of Claim 2

$$P\left\{\frac{\mathbf{Y}_n}{n} \in B_\delta(v)\right\} = P\{\mathbf{Y}_n \in nB_\delta(v)\} \qquad [\text{note that } nB_\delta(v) \equiv B_{n\delta}(nv)]$$

$$= \int_{nB_\delta(v)} \exp[-\theta x + n\varphi_n(\theta)]\, dF_n^{(\theta)}(x)$$

$$= \exp[-[\theta v - \varphi_n(\theta)]n] \int_{nB_\delta(v)} \exp[-\theta[x - nv]]\, dF_n^{(\theta)}(x)$$

(now set $\theta = \theta_v$)

$$\geqslant \exp[-[\theta_v v - \varphi_n(\theta_v)]n] \exp[-n|\theta_v|\delta] F_n^{(\theta_v)}(nB_\delta(v))$$

where $F_n^{(\theta_v)}(A)$ denotes the probability of the set A under the probability distribution $F_n^{(\theta_v)}(\cdot)$. Note that $(1/n)\log(\cdot)$ of the first exponential product term converges to $-I(v)$ as n approaches infinity. $F_n^{(\theta_v)}(nB_\delta(v))$ converges to 1 by claim 1, and hence $(1/n)\log(\cdot)$ of it converges to zero. We have thus finished the proof of claim 2. □

Therefore,

$$\underline{\lim}_{n\to\infty} \frac{1}{n} \log P\left\{\frac{\mathbf{Y}_n}{n} \in (a, b)\right\} \geqslant \underline{\lim}_{n\to\infty} \frac{1}{n} \log P\left\{\frac{\mathbf{Y}_n}{n} \in B_\delta(v)\right\} \geqslant -I(v) - \delta|\theta_v|$$

for all sufficiently small $\delta > 0$ [since $v \in (a, b)$, (a, b) is an open set, and by claim 2]. Since δ can be made arbitrarily small, we have

$$\underline{\lim}_{n\to\infty} \frac{1}{n} \log P\left\{\frac{\mathbf{Y}_n}{n} \in (a, b)\right\} \geqslant -I(v)$$

Since v is any number in the interval (a, b), we have

$$\underline{\lim}_{n\to\infty} \frac{1}{n} \log P\left\{\frac{\mathbf{Y}_n}{n} \in (a,b)\right\} \geqslant - \inf_{x\in(a,b)} I(x)$$

This completes the proof of the theorem. □

Let us now consider an example.

***Example* 4.** Consider the autoregressive process $\mathbf{x}_n = \alpha\mathbf{x}_{n-1} + \mathbf{w}_n$, $n = 1, 2, \ldots; \alpha < 1$; and $\mathbf{x}_0 = 0$. $\{\mathbf{w}_n\}$, $n = 1, 2, \ldots$, is an iid sequence such that $P\{\mathbf{w}_1 \in [-\frac{1}{2}, \frac{1}{2}]\} = 1$, and $E\{\mathbf{w}_1\} = 0$. Define $\mathbf{Y}_n \equiv \Sigma_{i=1}^n \mathbf{x}_i$. We are interested in the logarithmic asymptotics of $P\{\mathbf{Y}_n/n > \beta\}$, where $\beta > 0$. Now $\mathbf{x}_1 = \mathbf{w}_1$, $\mathbf{x}_2 = \alpha\mathbf{w}_1 + \mathbf{w}_2, \ldots, \mathbf{x}_n = \Sigma_{i=1}^n \alpha^{n-i}\mathbf{w}_i$. Therefore,

$$\mathbf{Y}_n = \frac{\alpha^n - 1}{\alpha - 1}\mathbf{w}_1 + \frac{\alpha^{n-1} - 1}{\alpha - 1}\mathbf{w}_2 + \cdots + \mathbf{w}_n$$

Hence

$$\exp[n\varphi_n(\theta)] = \prod_{i=1}^n M_w\left(\frac{1 - \alpha^i}{1 - \alpha}\theta\right)$$

where $M_w(\cdot)$ is the moment generating function of the $\{\mathbf{w}_i\}$ random variables. This implies that

$$\varphi_n(\theta) = \frac{1}{n}\sum_{i=1}^n \log M_w\left(\frac{1 - \alpha^i}{1 - \alpha}\theta\right)$$

Therefore,

$$\varphi(\theta) \equiv \lim_{n\to\infty} \varphi_n(\theta) = \log M_w\left(\frac{\theta}{1 - \alpha}\right)$$

Hence the asymptotics of the sum $\mathbf{Y}_n$ are the same as the asymptotics of a sum of iid random variables with the same distribution as $\mathbf{w}_1/(1 - \alpha)$. To be more specific, consider $P\{\mathbf{w}_1 = \frac{1}{2}\} = \frac{1}{2} = P\{\mathbf{w}_1 = -\frac{1}{2}\}$. Then $M_w(\theta) = [\exp[\theta/2] + \exp[-\theta/2]]/2 = \cosh(\theta/2)$.

$$\begin{aligned} I(x) &= \sup_\theta\left\{\theta x - \log\left[\cosh\left(\frac{\theta}{2(1-\alpha)}\right)\right]\right\} \\ &= x(1-\alpha)\log\frac{m^+(x)}{m^-(x)} - \log\left[\frac{1}{2}\left(\sqrt{\frac{m^+(x)}{m^-(x)}} + \sqrt{\frac{m^-(x)}{m^+(x)}}\right)\right] \end{aligned}$$

where $m^+(x) = 1 + 2(1 - \alpha)x$ and $m^-(x) = 1 - 2(1 - \alpha)x$, for $0 < |x| < 1/[2(1 - \alpha)]$ and is infinity for x outside this range. Note that the assumptions that we have placed on the distribution of $\mathbf{w}_1$ guarantee that $\mathbb{R} \subset \varphi'(D_\varphi)$ and $\overline{\lim}_{n\to\infty} |\mathbf{Y}_n/n| \leqslant \frac{1}{2}/(1 - \alpha) < \infty$. □

It may be noted that we have imposed a few more technical conditions than Ellis [1984] does in order to make the proof more accessible. First, we have shown the upper bound only for bounded closed intervals. The proof technique that we have employed utilizes *only* the fact that $[a, b]$ is compact. Therefore, we may as well state our upper bound as holding for any compact set on $\mathbb{R}$. To show that the upper bound is true for any closed set (but not necessarily bounded), one needs an additional condition to avoid pathological cases. Ellis requires that $0 \in \mathring{D}_\varphi$—in other words, that zero be contained in the interior of the domain of $\varphi(\cdot)$. This guarantees an exponential decrease of certain probabilities as in claims 1 and 2 of the preceding theorem. This extra condition, it turns out, is very mild and will hold for the random variable sequence (or a shifted version of it) for all but the most pathological cases (which can be handled directly).

More serious technical difficulties lie with the lower bound. We essentially assumed that G is open and is wholly contained in $\varphi'(D_\varphi)$. With quite a lot of effort and the invocation of much convex analysis, Ellis weakens these requirements considerably and demands that $\varphi(\cdot)$ be steep on D_φ instead. Hypothesis B2 is necessary for the lower bound, as we see in the following example.

***Example* 5.** Suppose that $\mathbf{Y}_n$ is an independent sequence of random variables with $P\{\mathbf{Y}_n = n\} = \frac{1}{2} = P\{\mathbf{Y}_n = -n\}$. $\exp[n\varphi_n(\theta)] = \frac{1}{2}\exp[n\theta] + \frac{1}{2}\exp[-n\theta] = \cosh(n\theta)$. Therefore, $\lim_{n\to\infty} \varphi_n(\theta) = |\theta|$ and hypothesis B2 is violated. Hence $I(x) = 0$ for $-1 \leqslant x \leqslant 1$ and is equal to ∞ otherwise. However, $P\{\mathbf{Y}_n/n \in (-1, 1)\} = 0$, which implies that

$$\underline{\lim}_{n\to\infty} \frac{1}{n} \log P\left\{\frac{\mathbf{Y}_n}{n} \in (-1, 1)\right\} = -\infty \ngeqslant - \inf_{x\in(-1,1)} I(x) = 0$$ □

In example 5 the theorem failed because hypothesis B2 was violated at the origin. Other examples can be constructed in which the theorem fails because the hypothesis fails at other points. See exercises 4 and 5 at the end of the chapter for a sampling of other pathologies.

C. $\mathbb{R}^d$ VALUED RANDOM VARIABLES

So far, we have been discussing large deviation theory only as it pertains to one-dimensional random variables. Many of our proofs and techniques, however, extend to the more general setting of d-dimensional random vectors. We will merely state Ellis's theorem in this setting and indicate that the proof mirrors exactly what we have discussed in section B. Let $\{\mathbf{Y}_n\}$ be a sequence of $\mathbb{R}^d$ valued random variables. Define

$$\varphi_n(\theta) \equiv \frac{1}{n} \log E\{\exp[\langle \theta, \mathbf{Y}_n \rangle]\}$$

where $\theta \in \mathbb{R}^d$ and $\langle \cdot, \cdot \rangle$ denotes the usual inner or dot product between vectors. We define the relevant hypotheses as follows.

***Hypothesis* C1.** $\varphi(\theta) \equiv \lim_{n\to\infty} \varphi_n(\theta)$ exists for all $\theta \in \mathbb{R}$ where ∞ is allowed as both a limit value and as an element of the sequence. Furthermore, we assume that $\varphi(\theta)$ is a closed convex function. □

This hypothesis requires that $\varphi(\theta)$ be closed. A convex function is *closed* if and only if it is *lower semicontinuous*, that is, if for every sequence $\{\theta_n\}$ converging to $\theta \in D_\varphi$, we have $\underline{\lim}\, \varphi(\theta_n) \geqslant \varphi(\theta)$. We may now define

$$I(x) \equiv \sup_\theta [\langle \theta, x \rangle - \varphi(\theta)]$$

***Hypothesis* C2.** $0 \in \mathring{D}_\varphi$. □

***Hypothesis* C3.** $\varphi(\cdot)$ is steep on D_φ. □

Theorem (Ellis)
Suppose that hypothesis C1 is true; then

$$\overline{\lim_{n\to\infty}} \frac{1}{n} \log P\left\{\frac{\mathbf{Y}_n}{n} \in F\right\} \leqslant -\inf_{x\in F} I(x)$$

for all compact F. If hypotheses C1 and C2 are true, the limit above is true for all closed F. If hypotheses C1 and C3 are true, then

$$\underline{\lim}_{n\to\infty} \frac{1}{n} \log P\left\{\frac{\mathbf{Y}_n}{n} \in G\right\} \geqslant -\inf_{x\in G} I(x)$$

for all open G.

We will now work through some examples to show the types of calculation that one must go through in the d-dimensional case.

***Example* 6.** Let $\{\mathbf{x}_n\}$ be a d-dimensional iid sequence of Gaussian random vectors with mean value m and covariance matrix K. Assume that K is full rank (i.e., K is invertible). The moment generating function for a Gaussian random vector is well known to be $E\{\exp[\langle\theta, \mathbf{x}_1\rangle]\} = \exp[\langle\theta, m\rangle + \frac{1}{2}\theta^T K\theta]$, where the superscript T indicates transpose (for vectors and matrices). $I(x) = \sup_\theta\{\langle\theta, x\rangle - \langle\theta, m\rangle - \frac{1}{2}\theta^T K\theta\}$. Setting the gradient with respect to θ in the term in the curly brackets to zero results in $x - m - K\theta = 0$. This implies that $\theta_{opt} = K^{-1}(x - m)$. Substituting this value of θ back into the supremum expression yields $I(x) = \frac{1}{2}(x - m)^T K^{-1}(x - m)$. □

***Example* 7.** Let us consider the following generalization of example 4: $\mathbf{x}_n = A\mathbf{x}_{n-1} + \Gamma\mathbf{w}_n$, $n = 1, 2, \ldots$, and $\mathbf{x}_0 = 0$. We suppose that A is a $d \times d$ full rank matrix with the absolute value of all eigenvalues strictly less than 1. $\{\mathbf{w}_n\}$ is an iid sequence of random vectors with moment generating function $\varphi_w(\theta)$. This sort of model arises when one is considering a general autoregressive (AR) stochastic process. We are interested in the asymptotics of $\mathbf{Y}_n \equiv \Sigma_{i=1}^n \mathbf{x}_n$. Mimicking the arguments in example 4, it is easy to recast $\mathbf{Y}_n$ as $\mathbf{Y}_n = (I - A^n)(I - A)^{-1}\Gamma\mathbf{w}_1 + (I - A^{n-1})(I - A)^{-1}\Gamma\mathbf{w}_2 + \cdots + \Gamma\mathbf{w}_n$. Hence

$$
\begin{aligned}
E\{\exp[\langle\theta, \mathbf{Y}_n\rangle]\} &= \prod_{i=0}^{n-1} E\{\exp[\langle\theta, (I - A^{n-i})(I - A)^{-1}\Gamma w\rangle\} \\
&= \prod_{i=0}^{n-1} M_w([[I - A^{n-i}][I - A]^{-1}\Gamma]^T\theta)
\end{aligned}
$$

Therefore,

$$
\frac{1}{n}\log E\{\exp[\langle\theta, \mathbf{Y}_n\rangle]\} \xrightarrow[n\to\infty]{} \log M_w(([I - A]^{-1}\Gamma)^T\theta) \equiv \varphi(\theta)
$$

The final maximization to find $I(\cdot)$ of course will depend on the actual functional form of $\varphi_w(\theta)$ (see exercise 6 at the end of the chapter). □

Notes and Comments

Cramér [1938] is usually credited with presenting the first large deviation asymptotic expansion. He gave not only the exponential rate, but the constants in front of the exponential (see chapter VII) for the case of iid random variables with a continuous component and for the case of infinite intervals (such as $[a, \infty)$). Chernoff [1952] showed that whatever the type of random variable (discrete, continuous, or mixed) the large deviation rate function gives the correct exponential rate in the iid case. This original work was extended in various directions. Bahadur and Zabell [1979] and Groeneboom et al. [1979] give perhaps the most general form of Cramér's theorem for iid data. They allow the random variables to take on values in quite general topological vector spaces. More recently, Bolthansen [1986] gives full expansions in some quite general spaces.

The definition of a dominating point given in the text is different from that given originally by Ney [1983]. We call a dominating point any point that minimizes the large deviation function over a set. The original definition also requires that it be unique.

Gärtner [1977] considered looking only at the convergence properties of the moment generating function of a sequence of random variables. He assumed throughout that $M(\theta) < \infty$ for all θ. By extensive use of convexity theory, Ellis [1984] relaxes this fairly stringent condition. Ellis [1985] provides an excellent reference for the further pursuit of convexity theory and the interrelationships between it, large deviation theory, and statistical mechanics. Some readable but highly condensed lecture notes on Cramér's theorem and other topics for mathematically minded readers are found in Varadhan [1984]. An excellent but fairly abstract and even more highly mathematical treatment is found in Strook [1984] or Deuschel and Strook [1989].

There are several technical constructs and important considerations that have been avoided to simplify the presentation. The definition of a *large deviation principle* typically includes more than just an upper bound holding for closed sets and a lower bound holding for open sets. Usually, it is demanded that the rate function be lower semicontinuous. Furthermore, it is required that the level sets of the rate function, $\{x : I(x) \leqslant c\}$, be compact for each $c < \infty$. These are important technical conditions that one usually must show in order that the standard proofs of the upper and lower bounds will hold in very general settings. We will not follow this terminology for a large deviation principle in this book, as such technical considerations will rarely, if every, be necessary.

EXERCISES

1. Find $I(x)$ for iid averages of
   (a) Exponential random variables (density $p(x) = \lambda \exp[-\lambda x] x \geqslant 0$).
   (b) Bernoulli zero–one random variables ($P\{\mathbf{x}_1 = 1\} = p = 1 - P\{\mathbf{x}_1 = 0\}$).
   (c) Cauchy random variables (density $p(x) = [\pi(1 + x^2)]^{-1}$).

2. Are the moment generating functions in exercise 1 steep?

3. Let $\{\mathbf{x}_i\}$ be an iid sequence of random variables such that $M(\theta) < \infty$ for all θ, $E\{\mathbf{x}_1\} = 0$, and for some $b > 0$, $P\{x_1 \in [b, \infty)\} = 0$. Show that $I(x) = \infty$ for all $x \in (b, \infty)$. How does this problem illustrate the statement "The rate function blows up for the impossible values of the sample average"?

4. Let $\mathbf{S}_n^+$ and $\mathbf{S}_n^-$ be the number of 1's and -1's, respectively, in a sequence of n iid symmetric Bernoulli $\{-1, 1\}$ random variables. Define $\mathbf{Y}_n$ by $P\{\mathbf{Y}_n = \mathbf{S}_n^+\} = \frac{1}{2} = P\{\mathbf{Y}_n = -\mathbf{S}_n^-\}$. Find $M(\theta)$ for the $\{\mathbf{Y}_n\}$ sequence. Are the hypotheses of the Gärtner–Ellis theorem satisfied? Do the conclusions of that theorem hold in this case?

5. Let $\{\mathbf{x}_i\}$ be a sequence of iid $\{-1, 1\}$ symmetric Bernoulli random variables. $\mathbf{S}_n^+ \equiv n$ if *all* n of $\{\mathbf{x}_1, \mathbf{x}_2, \ldots, \mathbf{x}_n\}$ are $+1$; otherwise, $\mathbf{S}_n^+ \equiv 0$. Similarly, $\mathbf{S}_n^- \equiv -n$ if *all* $\{\mathbf{x}_1, \mathbf{x}_2, \ldots, \mathbf{x}_n\}$ are -1; otherwise, $\mathbf{S}_n^- \equiv 0$. $\mathbf{Y}_n$ is defined as $P\{\mathbf{Y}_n = \mathbf{S}_n^+\} = \frac{1}{2} = P\{\mathbf{Y}_n = \mathbf{S}_n^-\}$. What is $M(\theta)$ for the $\mathbf{Y}_n$ sequence? Compute $I(x)$. Are the upper and lower bounds correct? Do the conclusions of the Gärtner–Ellis theorem hold in this case?

6. What is $I(x)$ for example 7 when $\{\mathbf{w}_i\}$ is an iid sequence of zero mean Gaussian random vectors with covariance matrix K?

7. Stirling's approximation formula for the $\Gamma(\cdot)$ $[\Gamma(n+1) \equiv n!]$ function is $\Gamma(p) \approx \sqrt{2\pi/p}\,(p/e)^p$ for large p.

(a) Show that

$$\frac{1}{n}\log\binom{n}{k} \approx \frac{k}{n}\log\frac{k}{n} - \left(1 - \frac{k}{n}\right)\log\left(1 - \frac{k}{n}\right) \qquad \text{for large } n$$

(b) From this formula, verify directly the large deviation principle of example 2 (i.e., do not use Cramér's theorem).

8. Verify (for $w > 0$)

$$\left(\frac{1}{w} - \frac{1}{w^2}\right)\exp[-\tfrac{1}{2}w^2] \leqslant \int_w^\infty \exp[-\tfrac{1}{2}x^2]\,dx \leqslant \frac{1}{w}\exp[-\tfrac{1}{2}w^2]$$

(*Hint:* Show that

$$\int_w^\infty \left(1 - \frac{3}{x^4}\right)\exp[-\tfrac{1}{2}x^2]\,dx = \left(\frac{1}{w} - \frac{1}{w^3}\right)\exp[-\tfrac{1}{2}w^2])$$

Using this result, verify directly the large deviation principle of example 1.

CHAPTER III

Sanov's Theorem and the Contraction Principle

A. EMPIRICAL DISTRIBUTIONS AND SANOV'S THEOREM FOR THE MARGINAL DISTRIBUTIONS

Let $\{\mathbf{y}_n\}$ be an iid sequence of one-dimensional random variables. Suppose that the probability distribution function of this sequence is not known. Given the first n samples, one might try to estimate what the true probability distribution is by forming the *empirical distributions*,

$$\mathbf{L}_n \equiv \frac{1}{n}\sum_{j=1}^{n} \delta_{\mathbf{y}_j}$$

where δ_y is the unit point mass (a Dirac delta function) at point y. Note that $\mathbf{L}_n$ is a "random distribution" since it explicitly depends on the random variables $\mathbf{y}_1, \mathbf{y}_2, \ldots, \mathbf{y}_n$. As n approaches infinity we expect $\mathbf{L}_n$ to be converging in distribution (see appendix D.3) to the underlying unknown probability distribution. It is true, in many cases, that this convergence is exponentially fast.

To talk about a large deviation principle, in the way that we have seen it so far, one usually develops an upper bound for compact (or more generally closed) sets and a lower bound for the open sets. In this section we discuss convergence rate results in a space of probability measures. What do we mean by open and closed sets in such a space? The closed sets are easy. They will simply be defined as the complements of the open sets. Open sets will be defined as all possible unions of open balls. Hence all we need do is give a definition of an open ball in the space of probability measures.

If we had a distance measure (or metric) $d(\mu, \nu)$ between any two points μ and ν, say, we could define a radius r open ball with center μ as

$B_\mu(r) \equiv \{v : d(v, \mu) < r\}$. There are many choices of a metric on this space that we could choose. Different metrics can in general give us distinctly different classes of open and closed sets. We will be concerned with a specific metric, the Prohorov metric, that is closely associated with convergence in distribution.

The concept of an open set is closely tied to the concept of the convergence of sequences. An infinite sequence of points p_n in some space is said to *converge* to some point p in that space if every open set containing p also contains all but a finite number of the elements of the sequence. Different collections of open sets will then generate different notions of convergence (and vice versa). We already know that the empirical distributions should be converging in distribution to the true underlying distribution. Hence it is convergence in distribution that we want our metric to reflect.

One metric that generates the open sets associated with convergence in distribution is the *Prohorov metric*. We define the Prohorov metric for probability measures on $\mathbb{R}$ as follows. Let F be a closed set on the real line. Define $F^\varepsilon \equiv \{x \in \mathbb{R}: \inf_{y \in F} |x - y| < \varepsilon\}$. F^ε is said to be the "ε-fattening" of F. The Prohorov metric between two probability measures P and Q is then defined as $\rho(P, Q) \equiv \inf\{\varepsilon > 0 : P(F) \leqslant Q(F^\varepsilon) + \varepsilon$ for all closed F on $\mathbb{R}\}$. In exercise 1 you are asked to show that ρ has all the properties of a *metric*, i.e., $\rho(P, Q) = \rho(Q, P)$, $\rho(P, Q) = 0$ if and only if $P = Q$, and finally, the triangle inequality holds $\rho(P, Q) + \rho(Q, T) \geqslant \rho(P, T)$. We will denote the space of probability measures with the Prohorov metric as Ψ. We should note that there are other metrics that will also generate the same open sets as the Prohorov metric. The Prohorov metric is just convenient.

Let us consider a special simple case to see if we can derive the form of the large deviations theorem. Let $\{\mathbf{y}_n\}$ be an iid sequence of discrete random variables that take on only $d < \infty$ distinct values $\alpha_1, \alpha_2, \ldots, \alpha_d$. $P\{\mathbf{y}_1 = \alpha_i\} = \pi_i$, $i = 1, 2, \ldots, d$. We will call this the π distribution. We form the empirical distributions $\mathbf{L}_n$ as before. As n approaches infinity, we know that $\mathbf{L}_n$ is converging in distribution to the π distribution. A construct that should be equivalent to the above is to consider the standard basis vectors $e_1, e_2, \ldots, e_d$ along the coordinate axes of $\mathbb{R}^d$. Note that $e_1 = (1, 0, 0, \ldots, 0)$, $e_2 = (0, 1, 0, 0, \ldots, 0), \ldots, e_d = (0, 0, 0, \ldots, 0, 1)$. Let $\{\mathbf{y}'_n\}$ be a sequence of iid d-dimensional random vectors derived from the $\{\mathbf{y}_n\}$ sequence by defining $\mathbf{y}'_k = e_i$ when $\mathbf{y}_k = \alpha_i$ for $k = 1, 2, \ldots$. Then

$$\mathbf{S}_n \equiv \frac{1}{n} \sum_{i=1}^{n} \mathbf{y}'_i = \left(\frac{\mathbf{n}_1}{n}, \frac{\mathbf{n}_2}{n}, \ldots, \frac{\mathbf{n}_d}{n}\right) \in \mathbb{R}^d$$

where $\mathbf{n}_i \equiv$ number of times out of n trials that the random variables $\mathbf{y}_1, \mathbf{y}_2, \ldots, \mathbf{y}_n$ have equaled α_i. $\mathbf{S}_n$ can be viewed as a vector in $\mathbb{R}^d$ but also and more important, it can be viewed as a probability distribution for every value of n. It is simple to see that the distribution defined by $\mathbf{S}_n$ is the same distribution defined by $\mathbf{L}_n$. Convergence in distribution for $\mathbf{L}_n$ in this case corresponds to the usual pointwise convergence of $\mathbf{S}_n$ as a vector in $\mathbb{R}^d$ (see exercise 2). We can find the asymptotics of $\mathbf{S}_n$ viewed as a vector on $\mathbb{R}^d$ by an application of the multidimensional Gärtner–Ellis theorem from section C of chapter II.

Define

$$\varphi_n(\theta) \equiv \frac{1}{n}\log E\{\exp[\langle\theta, n\mathbf{S}_n\rangle]\} = \frac{1}{n}\log E\left\{\exp\left[\left\langle\theta, \sum_{i=1}^{n}\mathbf{y}'_i\right\rangle\right]\right\}$$

$$= \frac{1}{n}\log\prod_{i=1}^{n} E\{\exp[\langle\theta, \mathbf{y}'_i\rangle]\} = \log E\{\exp[\langle\theta, \mathbf{y}'_1\rangle]\} = \log\sum_{i=1}^{d}\pi_i\exp[\theta_i]$$

Note that $\varphi_n(\cdot)$ does not depend on n and hence trivially we have that

$$\varphi(\theta) \equiv \lim_{n\to\infty}\varphi_n(\theta) = \log\sum_{i=1}^{d}\pi_i\exp[\theta_i]$$

Now from chapter II we have that the rate function $I(x)$ for $\mathbf{S}_n$ exists and is given by (for $x\in\mathbb{R}^d$)

$$I(x) = \sup_\theta\{\langle\theta, x\rangle - \varphi(\theta)\} = \sup_\theta\left\{\sum_{i=1}^{d}\theta_i x_i - \log\sum_{i=1}^{d}\pi_i\exp[\theta_i]\right\} \tag{1}$$

Let us suppose that $x\in\mathbb{R}^d$ is such that $\Sigma_{i=1}^d x_i = 1$ and $x_i \geqslant 0$, $i = 1, \ldots, d$. To find the maximizing θ we simply take the partial with respect to θ_j of the expression in the curly brackets of (1) to obtain

$$x_j = \frac{\pi_j\exp[\theta_j]}{\Sigma_{i=1}^d\pi_i\exp[\theta_i]} = \frac{\pi_j\exp[\theta_j]}{K}$$

where $K \equiv \Sigma_{i=1}^d\pi_i\exp[\theta_i]$. This implies that

$$\theta_j = \log\frac{x_j}{\pi_j} + \log K \tag{2}$$

Hence from (1)–(2),

$$I(x) = \left\{\sum_{i=1}^{d}x_i\log\frac{x_i}{\pi_i} + \log K - \log K\right\}$$

$$= \sum_{i=1}^{d}x_i\log\frac{x_i}{\pi_i}\qquad\left(\text{for } x\in\mathbb{R}^d,\ \sum_{i=1}^{d}x_i = 1,\ x_i\geqslant 0\right) \tag{3}$$

Suppose that $x_m < 0$ for some m. Then let $\theta_i = 0$ for all $i \neq m$ and let $\theta_m \to -\infty$. It is then easy to see that $\langle \theta, x \rangle - \varphi(\theta)$ approaches infinity in this case. Suppose that $x_i \geqslant 0$ for every i but $\Sigma_{i=1}^d x_i = s \neq 1$. Suppose that $s > 0$. Find the optimum θ vector, θ^*, for the point $x^* \equiv (x_1/s, x_2/s, \ldots, x_d/s) \in \mathbb{R}^d$ (i.e., θ^* maximizes $\{\langle \theta, x^* \rangle - \varphi(\theta)\}$). Now consider

$$\sum_{i=1}^d \theta_i^* x_i - \log \sum_{i=1}^d \pi_i \exp[\theta_i^*] = \sum_{i=1}^d \theta_i^* \frac{x_i}{s} - \log \sum_{i=1}^d \pi_i \exp[\theta_i^*] + \sum_{i=1}^d \theta_i^* x_i \left(1 - \frac{1}{s}\right)$$

$$= I(x^*) + \sum_{i=1}^d \theta_i^* x_i \left(1 - \frac{1}{s}\right) \tag{4}$$

Notice that the $I(x^*)$ term does not change if you replace θ_i^* by $\theta_i^* + c$ for every i, where c is an arbitrary constant. By choosing a constant of sufficiently large magnitude [its sign depending on the sign of $(1 - 1/s)$], we can make the quantity given in (4) approach infinity also. Hence $I(x) = \infty$ in this case also. We leave it as an exercise to verify that $I(0) = \infty$. Therefore, to summarize,

$$I(x) = \begin{cases} \displaystyle\sum_{i=1}^d x_i \log \frac{x_i}{\pi_i} & \text{for } \{x_1, x_2, \ldots, x_d : x_i \geqslant 0 \text{ for all } i \text{ and } \sum_{i=1}^d x_i = 1\} \\ \infty & \text{otherwise} \end{cases} \tag{5}$$

Since convergence pointwise in $\mathbb{R}^d$, in this case, corresponds to convergence in distribution in the space of probability measures (see exercise 2), we know that the notions of open (and hence closed) sets must also coincide. We have thus proved the finite state space (or finite alphabet) case of the following theorem. First let us give a definition. Let P and Q be two probability measures. $Q \ll P$ (Q is said to be *absolutely continuous* with respect to P or P *dominates* Q) if $P(A) = 0$ implies that $Q(A) = 0$ for all (measurable) sets A. See appendix D.4 on likelihood ratios for results and consequences of this condition. In particular, the likelihood ratio dQ/dP exists and is finite between the two measures.

Theorem (Sanov)

Let $\mathbf{x}_1, \mathbf{x}_2, \ldots$ be a sequence of iid random variables taking values in $\mathbb{R}$. Let μ be the distribution of $\mathbf{x}_1$. For any other distribution ν on $\mathbb{R}$, define

$$I_\mu(\nu) \equiv \begin{cases} \displaystyle\int_{\mathbb{R}} \log \frac{d\nu(y)}{d\mu(y)}\, d\nu(y) & \text{for } \nu \ll \mu \text{ and } \displaystyle\int_{\mathbb{R}} \left|\log \frac{d\nu(y)}{d\mu(y)}\right| d\nu(y) < \infty \\ \infty & \text{otherwise} \end{cases} \tag{6}$$

Define the empirical distribution

$$\mathbf{L}_n \equiv \frac{1}{n} \sum_{i=1}^{n} \delta_{\mathbf{x}_i}$$

Then $\mathbf{L}_n$ obeys a large deviation principle for the open and closed sets on Ψ, the space of probability measures, with rate function $I_\mu(\cdot)$. Explicitly, we have for any open space set $G \subset \Psi$ and any closed set $F \subset \Psi$,

$$\underline{\lim}_{n\to\infty} \frac{1}{n} \log P\{\mathbf{L}_n \in G\} \geqslant -\inf_{r\in G} I_\mu(r)$$

and

$$\overline{\lim}_{n\to\infty} \frac{1}{n} \log P\{\mathbf{L}_n \in F\} \leqslant -\inf_{r\in F} I_\mu(r)$$

This theorem will extend to more general settings, as will be seen in succeeding sections. One of the nicest attributes of Sanov's theorem is that one immediately knows what the rate function is for the empirical distributions under consideration. The maximization in the definition of $I_\mu(\cdot)$ has been taken care of already. The difficulty with using Sanov's theorem is that finding the infimum over the set of probability measures is typically not very straightforward. The succeeding lemma is directed at this problem for a certain important class of sets in Ψ.

Let $g(\cdot)$ be some function mapping the state space of the random variables $\{\mathbf{y}_n\}$ into other points on the real line. We may be interested in the asymptotics of

$$\mathbf{S}_n \equiv \frac{1}{n} \sum_{i=1}^{n} g(\mathbf{y}_i)$$

Note that we may write $\mathbf{S}_n$ as

$$\mathbf{S}_n = \frac{1}{n} \sum_{i=1}^{n} \int g(y)\delta_{\mathbf{y}_i}(y)\,dy = \int \left(\frac{1}{n} \sum_{i=1}^{n} \delta_{\mathbf{y}_i}(y)\right) g(y)\,dy = \int g(y)\,d\mathbf{L}_n(y)$$

In other words, $\mathbf{S}_n$ is the expected value of $g(\cdot)$ under the distribution defined by $\mathbf{L}_n$. By the law of large numbers we know that $\mathbf{S}_n$ converges to m, the expected value of $g(\cdot)$ under the true distribution. Consider $P\{\mathbf{S}_n > m + \varepsilon\}$ for some $\varepsilon > 0$. This probability equals $P\{\int g(y)\,d\mathbf{L}_n(y) > m + \varepsilon\}$. We know by

Sanov's theorem what the rate functions of the empirical distributions are. Hence we must have (if the limit exists)

$$\lim_{n\to\infty} \frac{1}{n} \log P\left\{\int g(y)\, d\mathbf{L}_n(y) > m + \varepsilon\right\} = -\inf_{Q\in S} I(Q)$$

where $I(\cdot)$ is as given by (6) and

$$S \equiv \left\{Q \in \Psi : \int g(y)\, dQ(y) > m + \varepsilon\right\}$$

Now $I(\cdot)$ is a convex function of Q (see exercise 9). The minimization over a set of this form will then necessarily lie on the boundary, i.e., $-\inf_{Q\in S} I(Q) = -\inf_{Q\in \partial S} I(Q)$, where $\partial S \equiv$ boundary of $S = \{Q \in \Psi : \int g(y)\, dQ(y) = m + \varepsilon\}$. We can now perform this minimization with the aid of the following lemma. [We must caution here that in general settings, S will be guaranteed to be an open set if and only if $g(\cdot)$ is a bounded continuous function. When $g(\cdot)$ does not have these properties, there can be major technical difficulties to showing the existence of the limits.]

Lemma (Kullback)
Suppose that P is a given fixed probability measure with rate function $I(\cdot)$. Let

$$S_\beta \equiv \left\{Q \in \Psi : \int g(y)\, dQ = \beta\right\}$$

Define a new probability measure P_θ as

$$P_\theta(A) \equiv \frac{\int_A \exp[-\theta g(y)]\, dP(y)}{\varphi(\theta)}$$

where

$$\varphi(\theta) \equiv \int \exp[-\theta g(y)]\, dP(y)$$

and θ is the solution (it turns out to be unique) to the equation

$$\frac{\varphi'(\theta)}{\varphi(\theta)} = -\beta$$

Then

$$\inf_{Q\in S_\beta} I(Q) = I(P_\theta)$$

Proof: First note that (by formally differentiating under the integral),

$$\int g(y)\,dP_\theta(y) = \frac{\int g(y)\exp[-\theta g(y)]\,dP(y)}{\varphi(\theta)} = -\frac{\varphi'(\theta)}{\varphi(\theta)} = \beta$$

Therefore, $P_\theta \in S_\beta$. For a fixed $Q \ll P$ [if Q is not dominated by P, then $I(Q) = \infty$ and will not effect the infimum calculation] define

$$\rho(y) \equiv \frac{dQ(y)}{dP(y)}$$

and

$$\rho_\theta(y) \equiv \frac{dP_\theta(y)}{dP(y)} = \frac{\exp[-\theta g(y)]}{\varphi(\theta)}$$

Note also that

$$\begin{aligned} I(Q) &= \int \log\frac{dQ(y)}{dP(y)}\,dQ(y) = \int \frac{dQ(y)}{dP(y)}\log\frac{dQ(y)}{dP(y)}\,dP(y) \\ &= \int \rho(y)\log\rho(y)\,dP(y) \end{aligned} \tag{7}$$

Similarly,

$$I(P_\theta) = \int \rho_\theta(y)\log\rho_\theta(y)\,dP(y)$$

Let $Q \in S_\beta$. Since Q and P_θ are in the set S_β, we have the following identities:

$$\begin{aligned} I(P_\theta) &= \int \rho_\theta(y)\log\rho_\theta(y)\,dP(y) = \int \log\rho_\theta(y)\,dP_\theta(y) \\ &= -\int \theta g(y)\,dP_\theta(y) - \log\varphi(\theta) = -\theta\beta - \log\varphi(\theta) \end{aligned} \tag{8}$$

Similarly,

$$\int \rho(y) \log \rho_\theta(y)\, dP(y) = \int \log \rho_\theta(y)\, dQ(y) = -\int \theta g(y)\, dQ(y) - \log \varphi(\theta)$$

$$= -\theta\beta - \log \varphi(\theta) \tag{9}$$

Therefore, combining (8) and (9) yields

$$I(P_\theta) = \int \rho_\theta(y) \log \rho_\theta(y)\, dP(y) = \int \rho(y) \log \rho_\theta(y)\, dP(y) \tag{10}$$

Hence, from (7) and (10),

$$\begin{aligned} I(Q) - I(P_\theta) &= \int \rho(y) \log \rho(y)\, dP(y) - \int \rho_\theta(y) \log \rho_\theta(y)\, dP(y) \\ &= \int \rho(y) \log \rho(y)\, dP(y) - \int \rho(y) \log \rho_\theta(y)\, dP(y) \\ &= \int \rho(y) \log \frac{\rho(y)}{\rho_\theta(y)}\, dP(y) \geqslant \int \rho(y) \left[1 - \frac{\rho_\theta(y)}{\rho(y)}\right] dP(y) \end{aligned}$$

(since $\log x \geqslant 1 - 1/x$)

$$= \int \rho(y)\, dP(y) - \int \rho_\theta(y)\, dP(y) = \int dQ(y) - \int dP_\theta(y) = 1 - 1 = 0$$

This shows that P_θ is the minimizing point (dominating point) for the set S_β and hence the lemma is proved. □

Example* 1 *(Chernoff's Theorem). Suppose that we have one of two equally likely mutually absolutely continuous (see appendix D.4) probability distributions P_1 and P_2 controlling an observed iid sample sequence of discrete random variables $\{\mathbf{x}_n\}$. It is well known (see appendix E) that the minimum probability of error method to determine which distribution is controlling the data is the likelihood ratio test: Announce that P_1 is true if $\mathbf{T}_n \geqslant 0$; otherwise, announce P_2, where

$$\mathbf{T}_n \equiv \frac{1}{n} \sum_{i=1}^{n} \log \frac{dP_1(\mathbf{y}_i)}{dP_2(\mathbf{y}_i)}$$

is a function of the first n data samples only. There are two types of error that can be made; one can announce that P_1 is true when P_2 actually controls the data or, alternatively, announce that P_2 is true when P_1 actually controls the data. These errors are called type I and type II errors, respectively (also false-alarm and miss probabilities). The overall probability of error is an average of the probability of these two events. Hence $P\{\text{error given } n \text{ observations}\} = \frac{1}{2}P\{\text{error given } n \text{ observations} \mid P_1 \text{ is true}\} + \frac{1}{2}P\{\text{error given } n \text{ observations} \mid P_2 \text{ is true}\}$. Let us consider the type II error for now.

$$P\{\text{error for } n \text{ observations} \mid P_1 \text{ is true}\} = P_1\left\{\frac{1}{n}\sum_{i=1}^{n}\log\frac{dP_1(\mathbf{y}_i)}{dP_2(\mathbf{y}_i)} < 0\right\}$$

In this setup $\log[dP_1(\cdot)/dP_2(\cdot)]$ is taking the place of $g(\cdot)$ and 0 is taking the place of β in the lemma above.

$$\varphi(\theta) \equiv \int \exp[-\theta g(y)]\, dP_1(y)$$

Now

$$\log\frac{dP_1(\cdot)}{dP_2(\cdot)} = \log\frac{dP_1(\cdot)/dW(\cdot)}{dP_2(\cdot)/dW(\cdot)}$$

where W is any other probability distribution that dominates P_1 and P_2. $\phi(\theta)$ is then found to be H_θ, where for $\alpha \in \mathbb{R}$ we define

$$H_\alpha \equiv \int \left[\frac{dP_2(y)}{dW(y)}\right]^{\alpha}\left[\frac{dP_1(y)}{dW(y)}\right]^{1-\alpha} dW(y)$$

which is called the *Chernoff entropy* or the *α-entropy*. θ should be chosen so that $\varphi'(\theta)/\varphi(\theta) = 0$ or equivalently, $\varphi'(\theta) = 0$. This is equivalent to choosing θ to minimize $\varphi(\theta)$ or equivalently, $\log \varphi(\theta)$.

Therefore, from an application of Sanov's theorem we have (assuming limits exist)

$$\begin{aligned}
&\lim_{n\to\infty}\frac{1}{n}\log P\{\text{error for } n \text{ observations} \mid P_1 \text{ is true}\}\\
&\quad = \lim_{n\to\infty}\frac{1}{n}\log P_1\left\{\frac{1}{n}\sum_{i=1}^{n}\log\frac{dP_1(\mathbf{x}_i)}{dP_2(\mathbf{x}_i)} < 0\right\}\\
&\quad = \lim_{n\to\infty}\frac{1}{n}\log P_1\left\{E_{\mathbf{L}_n}\left\{\log\frac{dP_1(\cdot)}{dP_2(\cdot)}\right\} < 0\right\}\\
&\quad = -\inf_{Q\in S_0} I_1(Q)
\end{aligned}$$

where $I_1(\cdot)$ is the rate functional generated by the P_1 probability measure, $\mathbf{L}_n$ is the n-data-point empirical distribution, Q is a probability distribution, and $S_0 = \{Q \in \Psi : E_Q\{\log[dP_1(\cdot)/dP_2(\cdot)]\} = 0\}$. We now apply Kullback's lemma to find

$$-\inf_{Q \in S_0} I_1(Q) = -I_1(P_\theta)$$

where

$$\frac{dP_\theta(y)}{dP_1(y)} = \frac{\exp[-\theta g(y)]}{\varphi(\theta)}$$

and θ is the solution to $\varphi'(\theta) = 0$. Now

$$\begin{aligned} I_1(P_\theta) &= \int \log \frac{dP_\theta(y)}{dP_1(y)}\, dP_\theta(y) = -\theta \int g(y)\, dP_\theta - \log \varphi(\theta) \\ &= -\theta \int \log \frac{dP_1(y)}{dP_2(y)}\, dP_\theta - \log \varphi(\theta) \end{aligned} \tag{11}$$

However, $P_\theta \in S_0$, which implies that the integral in (11) is zero. Hence

$$\begin{aligned} \lim_{n\to\infty} \frac{1}{n} \log P\{\text{error for } n \text{ observations} \mid P_1 \text{ is true}\} &= \log \varphi(\theta) \\ &= \min_\alpha \log H_\alpha \\ &\equiv \log H_{\alpha_0} \end{aligned}$$

where α_0 is that value for which $\log H_\alpha$ attains its minimum. It is easy to check (and intuitively pleasing) that for an optimum test, one has that the asymptotic rate to zero of the type I error will equal that of the type II error. One can directly verify this statement from another application of Sanov's theorem to the type I error (which will be left as an exercise). Therefore, since the overall error is merely an average of these two errors, we have that the overall error rate is equal to the rate on just one of the errors. Hence

$$\lim_{n\to\infty} \frac{1}{n} \log P\{\text{error for } n \text{ observations}\} = \log H_{\alpha_0} \tag{12}$$

Chernoff [1952] first presented this important limit theorem employing a different method of proof. This result is usually credited (along with Cramér's theorem) as being one of the earliest large deviation results. □

***Example* 1a.** Consider the hypothesis testing problem between two exponential densities with mean values λ_1^{-1} and λ_2^{-1}. Without loss of generality assume that $\lambda_1 > \lambda_2$. We can then calculate

$$\begin{aligned} H_\alpha &\equiv \int_0^\infty (\lambda_1 \exp[-\lambda_1 x])^\alpha (\lambda_1 \exp[-\lambda_1 x])^{1-\alpha}\, dx \\ &= \lambda_1^\alpha \frac{\lambda_2^{1-\alpha}}{\lambda_1 \alpha + \lambda_2(1-\alpha)} \end{aligned}$$

We now find the solution to the equation $H'(\alpha_0) = 0$ is given by

$$\alpha_0 = \left(\log \frac{\lambda_1}{\lambda_2}\right)^{-1} - \frac{\lambda_2}{\lambda_1 - \lambda_2}$$

It is easy to check that α_0 is always in the interval $[0, 1]$ (see exercise 5b). Hence

$$\begin{aligned} \log H_{\alpha_0} = -\log(\lambda_1 - \lambda_2) + \log\left(\log \frac{\lambda_1}{\lambda_2}\right) + 1 \\ + [\lambda_1 \log \lambda_2 - \lambda_2 \log \lambda_1][\lambda_1 - \lambda_2]^{-1} \end{aligned}$$

which is the asymptotic rate to zero of the probability of error. □

***Example* 1b.** Now consider the hypothesis testing problem between two zero mean normal densities with variances σ_1^2, $\sigma_2^2(\sigma_2 > \sigma_1)$. We have

$$\begin{aligned} H_\alpha &\equiv \int_{-\infty}^\infty [\sqrt{2\pi}\,\sigma_1]^{-\alpha}[\sqrt{2\pi}\,\sigma_2]^{\alpha-1} \exp\left[-\frac{\alpha x^2}{2\sigma_1^2}\right] \exp\left[-\frac{(1-\alpha)x^2}{2\sigma_2^2}\right] dx \\ &= r^\alpha[\alpha(r^2 - 1) + 1]^{-1/2} \end{aligned}$$

where $r \equiv \sigma_2/\sigma_1$. A simple differentiation shows that

$$\alpha_0 = [\log(r^2)]^{-1} - [r^2 - 1]^{-1}$$

One may substitute this value back into the expression for $\log H_\alpha$ to obtain the asymptotic rate expression. □

B. MULTIDIMENSIONAL SANOV'S THEOREM

This section is intended to be a quick introduction to some of the extensions that have been made of Sanov's theorem to more general probability spaces and dependency structures. It is not a prerequisite to the rest of this chapter or the book.

Sanov's theorem has been extended in many different directions from the simple case above. One of the more useful extensions is to generalize the result to iid sequences of $\mathbb{R}^d$ valued random vectors. The following theorem states this important case.

Theorem (Sanov)
Let $\mathbf{x}_1, \mathbf{x}_2, \ldots$ be a sequence of iid random vectors taking values in $\mathbb{R}^d$. Let μ_d be the d-dimensional distribution of $\mathbf{x}_1$. For any other distribution ν_d on $\mathbb{R}^d$, define

$$I_{\mu_d} \equiv \begin{cases} \displaystyle\int_{\mathbb{R}^d} \log \frac{d\nu_d(y)}{d\mu_d(y)}\, d\nu_d(y) & \text{for } \nu_d \ll \mu_d \text{ and } \displaystyle\int_{\mathbb{R}^d} \left| \log \frac{d\nu_d(y)}{d\mu_d(y)} \right| d\nu_d(y) < \infty \\ \infty & \text{otherwise} \end{cases}$$

Define the empirical distribution

$$\mathbf{L}_n \equiv \frac{1}{n} \sum_{i=1}^{n} \delta_{\mathbf{x}_i}$$

Then $\mathbf{L}_n$ obeys a large deviation principle for the open and closed sets on Ψ, the space of probability measures, with rate function $I_{\mu_d}(\cdot)$.

The proof of this theorem requires many more technical details than the proof of the simpler one-dimensional finite state space case (i.e., random variables that take on only a finite number of values) that we have already shown; however, the essential ideas remain the same.

What we have been discussing so far is the empirical distribution of the one-dimensional or marginal data for iid sequences. We can also group our incoming samples into blocks or vectors of two, three, or higher dimensions. Then, in the same fashion, we can form an empirical estimate of the higher-order distributions. More specifically, let $\mathbf{y}_1, \mathbf{y}_2, \ldots$ be our incoming data. Define

$$\begin{aligned} \mathbf{Y}_1^{(n)} &\equiv (\mathbf{y}_1 \quad \mathbf{y}_2 \quad \cdots \quad \mathbf{y}_k) \\ \mathbf{Y}_2^{(n)} &\equiv (\mathbf{y}_2 \quad \mathbf{y}_3 \quad \cdots \quad \mathbf{y}_{k+1}) \\ &\vdots \\ \mathbf{Y}_m^{(n)} &\equiv (\mathbf{y}_m \quad \mathbf{y}_{m+1} \quad \cdots \quad \mathbf{y}_{k+m-1}) \end{aligned}$$

for $m \leqslant n - k + 1$ and the cyclic vector

$$\mathbf{Y}_{m'}^{(n)} = (\mathbf{y}_{m'} \quad \mathbf{y}_{m'+1} \quad \cdots \quad \mathbf{y}_n \quad \mathbf{y}_1 \quad \mathbf{y}_2 \quad \cdots \quad \mathbf{y}_{m'+k-1-n})$$

for $n \geqslant m' \geqslant n - k + 2$. We now define the empirical distribution

$$\mathbf{L}_n^k \equiv \frac{1}{n} \sum_{i=1}^{n} \delta_{\mathbf{Y}_i^{(n)}}$$

Note that the one-dimensional marginal distribution of $\mathbf{L}_n^k$ is merely $\mathbf{L}_n$, the empirical distribution we defined earlier. We will, without stating a theorem, merely note that there are various large deviation principles available under various hypotheses for $\mathbf{L}_n^k$ (see Notes and Comments at the end of the chapter).

More generally, one can consider trying to form an empirical distribution for an entire random process instead of just the k-dimensional marginal distributions. Consider a random process governed by a probability measure $\mu(\cdot)$. From observations of the incoming samples one can try to form an empirical measure for the whole random process at one time. Note that to do this, one would have to specify all finite-dimensional distributions from only a finite number of samples! One way that this is accomplished is by taking our data $(\mathbf{y}_1 \mathbf{y}_2 \quad \cdots \quad \mathbf{y}_n)$ and repeating it periodically in both directions. Call this infinitely repeated version of our data $\mathbf{y}^n$, which of course is an infinite periodic sequence (of period n) defined on the integers. From this "infinite" collection of data, we can then compute any desired k-dimensional marginal empirical distribution that we wish. More specifically, the empirical measure of the entire random process will be defined as

$$\mathbf{R}_n \equiv \frac{1}{n} \sum_{i=0}^{n-1} \delta_{T^k \mathbf{y}^n}$$

where T^k is the single shift operator T iterated k times, i.e., $T\mathbf{y} = T(\cdots \quad \mathbf{y}_{-1} \quad \mathbf{y}_0 \quad \mathbf{y}_1 \quad \mathbf{y}_2 \quad \cdots) = (\cdots \quad \mathbf{y}_{-2} \quad \mathbf{y}_{-1} \quad \mathbf{y}_0 \quad \mathbf{y}_1 \quad \cdots)$. Thus $\mathbf{R}_n(B)$ is the relative frequency with which $\mathbf{y}^n, T\mathbf{y}^n, \ldots, T^{n-1}\mathbf{y}^n$ is in B. Let $P_{\boldsymbol{\omega}}$ denote the conditional probability distribution with respect to P of $\mathbf{y}_1$ given $\boldsymbol{\omega} = \{\mathbf{y}_j;\ j \leqslant 0\}$ (i.e., $P_{\boldsymbol{\omega}}(A) \equiv P\{\mathbf{y}_1 \in A \mid \boldsymbol{\omega} = \{\mathbf{y}_j;\ j \leqslant 0\}\}$). Suppose that the empirical measure $\mathbf{R}_n$ is being generated by an iid sequence of the $\mathbf{y}_i$'s with distribution μ (denote the associated one-dimensional marginal distribution as μ_1). Define a function on the space of probability measures by

$$I_\mu(P) \equiv \begin{cases} \displaystyle\int \left(\int \log \frac{dP_\omega(y)}{d\mu_1(y)}\, dP_\omega(y) \right) dP(\omega) & \text{as long as the double integral exists absolutely and is finite} \\ \infty & \text{otherwise} \end{cases}$$

A theorem that generalizes everything that we have done so far is the following version of Sanov's theorem due to Bahadur.

Theorem (Bahadur)
The probability measures on infinite sequences associated with the $\mathbf{R}_n$ measures obeys a large deviation principle with rate function $I_\mu(\cdot)$.

We will not give a proof of this very general theorem, as the necessary measure theoretic technicalities would take us far afield. However, we note that it is usually the case that we can calculate I_μ by evaluating $\lim_{k\to\infty}(1/k)I_{\mu_k}(P_k)$, where the subscripts on the measures indicate that we are taking the k-dimensional marginals of the relevant stochastic process measures. Perez [1974] gives some technical conditions that are sufficient for these expressions to be equal to each other.

C. CONTRACTION PRINCIPLE

One of the most useful tools in large deviation theory is the "contraction principle." Basically, this tool gives a formula to compute the new rate function after the data have been transformed by a continuous (possibly nonlinear) transformation. Let $\{P_n\}$ be a family of probability measures on a space X satisfying a large deviation principle with rate function $I(\cdot)$. Let $F(\cdot)$ be a continuous function from X into another space Y. Then, for every n, $F(\cdot)$ and P_n induce a probability distribution Q_n on the space Y defined by $Q_n(\Gamma) \equiv P_n(F^{-1}(\Gamma))$. We then have the following theorem.

Theorem (Contraction Principle)
$\{Q_n\}$ obeys a large deviation principle with rate function $J(y)$, where

$$J(y) \equiv \begin{cases} \inf\limits_{x:F(x)=y} I(x) & \text{if } \{x:F(x)=y\} \neq \varnothing \\ \infty & \text{if } \{x:F(x)=y\} = \varnothing \end{cases}$$

Proof: Upper bound. Let $A \subset Y$ be a closed set. Let $C \equiv \{x: F(x) \in A\}$. Then $Q_n(A) = P_n(C)$. Note that C is also a closed set [since $F(\cdot)$ is continuous]. Then

$$\overline{\lim_{n\to\infty}} \frac{1}{n} \log Q_n(A) = \overline{\lim_{n\to\infty}} \frac{1}{n} \log P_n(C) \leqslant -\inf_{x\in C} I(x) = -\inf_{\substack{x:F(x)=y \\ \text{and } y\in A}} I(x)$$

$$= -\inf_{y\in A} \inf_{x:F(x)=y} I(x) = -\inf_{y\in A} J(y)$$

Lower bound. Let $G \subset Y$ be open. Let $y \in G$ and $x \in X$ such that $F(x) = y$. Since $F(\cdot)$ is continuous, we can find a neighborhood V of x such that

$F(V) \subset G$. Therefore,

$$\varliminf_{n\to\infty} \frac{1}{n} \log Q_n(G) \geqslant \varliminf_{n\to\infty} \frac{1}{n} \log P_n(V) \geqslant -I(x)$$

This is true for every x such that $F(x) \in G$, so we are finished. □

***Example* 2.** Let

$$\mathbf{S}_n \equiv \frac{1}{n} \sum_{i=1}^{n} \mathbf{x}_i$$

where the $\mathbf{x}_i$ are iid zero mean unit variance Gaussian random variables. Suppose that we are interested in the asymptotics of $P\{\mathbf{S}_n^2 \in (a,b)\}$. We know that $\mathbf{S}_n$ obeys a large deviation principle with rate function $I(x) = x^2/2$. We can apply the contraction principle then to obtain the desired asymptotics. $F(x) = x^2$ (in this case), which implies that

$$J(y) = \inf_{x:x^2=y} \frac{x^2}{2} = \frac{y}{2}$$

for $y > 0$. Therefore, we have

$$\lim_{n\to\infty} \frac{1}{n} \log P\{\mathbf{S}_n^2 \in (a,b)\} = -\inf_{y\in(a,b)} \frac{y}{2}$$ □

Notes and Comments

Much more general versions of the work of Sanov [1975] have been provided by various authors. Bahadur and Zabell [1979] give various results for iid Banach space (and even more general spaces) valued random variables. In Donsker and Varadhan [1975a, 1975b, 1976] and in Varadhan [1984] are given many results on Markov processes, including continuous-time results. Orey [1983] and Orey and Pelikan [1986] give some partial results on non-Markov data (essentially for the class of random processes that obey the Shannon–McMillan theorem of information theory).

While working on a new methodology to measure uncertainty for prior distributions, Kullback and Leibler [1951] presented the relative entropy as a criterion to measure distances between distributions. The theorem presented in the text [Kullback, 1959] evidently has tremendous application to statistical mechanics, as it characterizes the class of distributions known as *Gibbs distributions* to physicists [Ellis, 1985].

The contraction principle has far-ranging applications to both the theoretical development and the practical calculation of large deviation results. More general versions of this result are available. For example, Varadhan [1984] allows a sequence of continuous maps $F_n(\cdot)$ to be mapping X into Y. He requires that this sequence of maps converge uniformly on compact sets to a continuous function $F(\cdot)$. In this case $F_n(\cdot)$ and P_n induce a probability distribution Q_n on Y. He then shows that the same contraction principle holds.

EXERCISES

1. Show that the Prohorov metric has all the properties of a metric on the space of probability measures.

2. For iid random variables $\{\mathbf{x}_j\}$ with $P\{\mathbf{x}_1 = \alpha_i\} = \pi_i$, $i = 1, 2, \ldots, d < \infty$, show that for the construct given in the text $\mathbf{S}_n/n$ converging as a vector in $\mathbb{R}^d$ is the same as $\mathbf{L}_n$, the empirical distribution converging in distribution to π.

3. Show that convergence in Prohorov metric is equivalent to convergence in distribution.

4. Let $\{\mathbf{x}_i\}$ be a sequence of random variables converging in distribution to $\mathbf{x}$. If f is a bounded continuous function, show that $f(\mathbf{x}_n)$ converges in distribution to $f(\mathbf{x})$. If f is continuous but not bounded, can you give a counterexample? If f is bounded but not continuous, can you give a counterexample?

5. Show that $\log H_\alpha$ (the log of the Chernoff entropy)
   (a) Is convex as a function of α.
   (b) Has a minimizing α that lies in the interval $[0, 1]$.
   (c) Is independent of W the dominating measure.

6. Consider the hypothesis testing problem between iid variance 1 Gaussian data with H_0: zero mean and H_1: mean value $s > 0$. Find the asymptotic rate with which the probability of error approaches zero.

7. Let $I_\mu(\nu) \equiv \int_{\mathbb{R}} \log[d\nu(y)/d\mu(y)]\, d\nu(y)$. Consider the "twisted" measures ν defined by $d\nu(y)/d\mu(y) = \exp[\theta_0 + \theta_1 h_1(y) + \cdots + \theta_r h_r(y)]$ and which satisfy $\int_{\mathbb{R}} h_i(y)\, d\nu(y) = \beta_i$, $i = 1, \ldots, r$, where the θ's and the β's are real valued constants. Generalize Kullback's lemma to show that this set of measures equals the set of measures on which $\inf\{I_\mu(\nu), \int h_i(y)\, d\nu(y) = \beta_i\}$ is attained.

8. Using the notation of exercise 7, let $d\mu(y) = dy$. Then $d\nu(y)/dy$ is just the probability density function of the distribution $\nu(\cdot)$. Find functions h_1, $h_2, \ldots, h_r$ and constants $\beta_1, \beta_2, \ldots, \beta_r$ such that the following densities solve a minimum problem as in exercise 7.

(a) $f(x) = x^{a-1}(1-x)^{b-1}/(\Gamma(a)\Gamma(b)/\Gamma(a+b))\ 0 \leqslant x \leqslant 1$ (beta density).

(b) $f(x) = a\exp[-ax]\ 0 \leqslant x \leqslant \infty$ (exponential density).

(c) $f(x) = a^b x^{b-1}\exp[-ax]/\Gamma(b)\ 0 \leqslant x \leqslant \infty$, $a, b > 0$ (gamma density).

(d) Gaussian with mean m and variance σ^2.

(e) $f(x) = \frac{1}{2}a\exp[-a|x|] - \infty \leqslant x \leqslant \infty$, $a > 0$ (Laplace density).

9. Show that $I_\mu(\nu)$ is convex as a function of ν on the space of probability measures.

CHAPTER IV

Gaussian Processes and Wentzell–Freidlin Theory

A. GAUSSIAN PROCESSES

Consider a Gaussian process $\{\mathbf{x}(t) : 0 \leqslant t \leqslant T\}$ with mean value

$$m(t) \equiv E\{\mathbf{x}(t)\}$$

and autocovariance function

$$K(s, t) \equiv E\{(\mathbf{x}(s) - m(s))(\mathbf{x}(t) - m(t))\} \qquad \text{for all } 0 \leqslant s, t \leqslant T$$

Every realization of this random process generates a function on the interval $[0, T]$. Suppose that we generate a succession of independent realizations of the random process above and denote them as $\mathbf{x}_1, \mathbf{x}_2, \ldots$. We will in the sequel, whenever convenient and the context is clear, suppress the t dependence of functions (e.g., m will denote the mean value function). Every realization of

$$\mathbf{S}_n \equiv \frac{1}{n} \sum_{i=1}^{n} \mathbf{x}_i$$

is, of course, also a function on the interval $[0, T]$. However, we may also view $\mathbf{S}_n$ as a Gaussian random process in its own right with mean value $m(t)$ and autocovariance $(1/n)K(s, t)$. Note that $\mathbf{S}_n$ has the same distribution as $(\mathbf{x}_1 - m)/\sqrt{n} + m$. We would expect that $\mathbf{S}_n$ converges as n approaches infinity to the mean value m. The question that comes to mind (of course) is: Is there a large deviation principle here, and if so, what is its rate function?

To investigate large deviation principles in this setting, we first need to discuss the state space that we will be concerned with and the notions of open and closed sets in that space. A natural choice for the state space of Gaussian process realizations is the collection of all square integrable functions (usually denoted as $L_2[0, T]$). The open ball of radius r and center f is defined as

$$B_f(r) \equiv \left\{ g \in L_2[0, T] : \int_0^T (g(x) - f(x))^2 \, dx < r^2 \right\}$$

Open sets are defined to be all possible unions of these open balls. Closed sets are then defined (as usual) to be the complements of the open sets.

It is often desirable to find a nice set of functions $\{f_n(\cdot)\}$ with which we may expand the elements of $L_2[0, T]$. A particularly nice property for a collection $\{f_n(\cdot)\}$ to have is that it be a *complete orthonormal set*. By *orthonormal* we mean that

$$\int_0^T f_j(t) f_m(t) \, dt = \begin{cases} 1 & \text{if } j = m \\ 0 & \text{otherwise} \end{cases}$$

Completeness means that for *any* $g \in L_2[0, T]$ we have the representation (limits are understood to be taken in mean square)

$$g(t) = \sum_{k=0}^{\infty} g_k s_k(t) \qquad \text{where } g_k \equiv \int_0^T g(t) s_k(t) \, dt$$

The purpose of these constructions is to be able to work with the $\{g_k\}$ sequences instead of the actual functions $g(\cdot)$. It is simple to see from the orthonormality of the $\{f_k(\cdot)\}$ that

$$\sum_{k=0}^{\infty} g_k^2 = \int_0^T g(t)^2 \, dt$$

(this is called *Parseval's identity*). The space of all square summable sequences is usually denoted ℓ_2. There is then a 1–1 correspondence between the elements of ℓ_2 and $L_2[0, T]$ which preserves the sum of squares (or metric) relationships. We will freely go back and forth between these spaces in order to develop our large deviation principles. For an open or closed set $M \subset L_2[0, T]$, we denote the corresponding set in ℓ_2 as M^*. In general, we denote "starred" quantities as being the ℓ_2 counterparts to the unstarred quantities.

The Karhunen–Loève expansion of a Gaussian random process on the interval $[0, T]$ (see appendix C.2) allows us to specify a particular expansion set for $L_2[0, T]$ which has nice statistical properties for the expansion coefficients. Recall that subject to some technical conditions (see appendix C.2), our process $\mathbf{x}(t)$ can be represented in the following fashion:

$$\mathbf{x}(t) = \sum_{k=0}^{\infty} (\mathbf{a}_k + m_k)\varphi_k(t) \qquad \text{for } 0 \leqslant t \leqslant T$$

where the $\{\mathbf{a}_k\}$ are independent Gaussian random variables with $E\{\mathbf{a}_k^2\} = \lambda_k$, the $\varphi_k(\cdot)$ are eigenfunctions of the covariance operator, i.e.,

$$\int_0^T K(s, t)\varphi_k(s)\, ds = \lambda_k \varphi_k(t)$$

with $\{\lambda_k\}$ as the eigenvalues. Under certain technical conditions (see appendix C.2), the $\{\varphi_k(\cdot)\}$ form a complete orthonormal set for $L_2[0, T]$. $\{m_k\}$ and $\{\mathbf{a}_k\}$, the expansion coefficients, may be explicitly defined by the relations

$$\mathbf{a}_k \equiv \int_0^T (\mathbf{x}(t) - m(t))\varphi_k(t)\, dt$$

and

$$m_k \equiv \int_0^T m(t)\varphi_k(t)\, dt$$

Let us see if we can make a reasonable heuristic guess for the rate function of the Gaussian random process. Let M be an open or closed set in $L_2[0, T]$. Consider

$$P\{\mathbf{S}_n \in M\} = P\left\{\frac{\mathbf{x}_1 - m}{\sqrt{n}} + m \in M\right\}$$

$$= P\left\{\frac{1}{\sqrt{n}}(\mathbf{a}_1 \quad \mathbf{a}_2 \quad \cdots) + (m_1 \quad m_2 \quad \cdots) \in M^*\right\} \tag{1}$$

Now $\mathbf{a}^* = (\mathbf{a}_1 \quad \mathbf{a}_2 \quad \cdots)$ is an infinite-dimensional vector of independent Gaussian random variables. From example 6 of chapter II, we would know what the rate function for the probabilities of (1) would be if the vector were finite dimensional ($d < \infty$). That rate function would be

$$I(y^*) = \tfrac{1}{2}(y^* - m^*)K^{*-1}(y^* - m^*)^{\mathrm{T}} \tag{2}$$

where K^* is the d-dimensional covariance matrix and m^* is the mean vector. Let us see if we can interpret this result in the $d = \infty$ setting. y^* and m^* would merely be the infinite vectors $(y_1 \quad y_2 \quad \cdots)$ and $(m_1 \quad m_2 \quad \cdots)$, respectively. The $\{\mathbf{a}_k\}$ are independent Gaussian random variables, which means that their infinite rowed (and columned) covariance matrix is diagonal with $E\{\mathbf{a}_k^2\} = \lambda_k$ in the kth row and kth column. The inverse of the covariance matrix K^{*-1} is then an infinite diagonal matrix with $1/\lambda_k$ at the kth row and kth column position. Combining these arguments yields that (2) may be rewritten as

$$I(y^*) = \frac{1}{2} \sum_{k=0}^{\infty} \frac{(y_k - m_k)^2}{\lambda_k} \tag{3}$$

Using the fact (see appendix C.2) that

$$K(s, t) = \sum_{k=0}^{\infty} \lambda_k \varphi_k(s) \varphi_k(t)$$

gives the motivation to define an "inverse kernel" as

$$K^{-1}(s, t) = \sum_{k=0}^{\infty} \frac{1}{\lambda_k} \varphi_k(s) \varphi_k(t)$$

It is then easy to show (see exercise 1) that for $y \in L_2[0, T]$,

$$I(y) = \frac{1}{2} \int_0^T \int_0^T K^{-1}(s, t)(y(s) - m(s))(y(t) - m(t)\, ds)\, dt \tag{4}$$

For reason that are aside to our main interests, the square root of the integral quantity is sometimes called the *reproducing kernel Hilbert space norm* (usually denoted $\|\cdot\|_{\mathbf{RKHS}}$) of $y - m$. Consequently, in the literature one frequently sees the result above stated as follows.

Theorem (Shilder)
Let $\{\mathbf{x}(t): 0 \leqslant t \leqslant T\}$ be a Gaussian random process with mean value $m(t)$ and covariance function $K(s, t)$. For $y \in L_2[0, T]$, let

$$\|y\|^2_{\mathbf{RKHS}} \equiv \int_0^T \int_0^T K^{-1}(s, t) y(s) y(t)\, ds\, dt$$

Then $\{\mathbf{S}_n(t): 0 \leqslant t \leqslant T\}$ (or equivalently, $\{[\mathbf{x}(t) - m(t)]/\sqrt{n} + m(t): 0 \leqslant t \leqslant T\}$) obeys a large deviation principle with rate function $I(y) = \frac{1}{2}\|y - m\|^2_{\mathbf{RKHS}}$.

***Example* 1.** Suppose that $\{\mathbf{x}(t): 0 \leqslant t \leqslant T\}$ is a standard Brownian motion (or Wiener process) on the interval $[0, T]$. Then $m(t) = 0$ and $K(s, t) = \min(s, t)$. The eigenfunctions $\varphi_n(\cdot)$ and eigenvalues λ_n are (see Van Trees [1968])

$$\varphi_n(t) = \left(\frac{2}{T}\right)^{1/2} \sin \frac{(n - \frac{1}{2})\pi t}{T}$$

and

$$\lambda_n = \frac{T^2}{(n - \frac{1}{2})^2 \pi^2} \qquad \text{for } k = 1, 2, \ldots$$

Let y be a differentiable function on the interval $[0, T]$. Now

$$y(t) = \sum_{k=0}^{\infty} y_k \varphi_k(t)$$

which implies (formally) that

$$\begin{aligned} y'(t) &= \sum_{k=0}^{\infty} y_k \varphi_k'(t) = \sum_{k=0}^{\infty} y_k \left(\frac{2}{T}\right)^{1/2} (k - \tfrac{1}{2}) \frac{\pi}{T} \cos \frac{(k - \frac{1}{2})\pi t}{T} \\ &= \sum_{k=0}^{\infty} y_k \left(\frac{2}{T\lambda_k}\right)^{1/2} \cos \frac{(k - \frac{1}{2})\pi t}{T} \end{aligned}$$

Let us square this quantity and integrate it term by term. Note that

$$\int_0^T \cos \frac{(n - \frac{1}{2})\pi t}{T} \cos \frac{(m - \frac{1}{2})\pi t}{T} \, dt = \begin{cases} T/2 & \text{when } n = m \\ 0 & \text{otherwise} \end{cases}$$

Therefore,

$$\int_0^T y'(t)^2 \, dt = \sum_{k=0}^{\infty} \frac{y_k^2}{\lambda_k} = \|y\|_{\mathrm{RKHS}}^2$$

Hence one concludes that

$$I(y) = \begin{cases} \dfrac{1}{2} \displaystyle\int y'(t)^2 \, dt & \text{for differentiable } y \text{ such that } y(0) = 0 \\ \infty & \text{otherwise} \end{cases} \qquad \square$$

This is an interesting example for a number of reasons. Brownian motion noise drivers (the formal derivative of Brownian motion is white noise) are usually the noise sources in stochastic differential equations. To understand the large deviation principles associated with the solutions to these equations, one must have a clear understanding of the large deviation results for the Brownian motion. Extensive use of the results of this example is made in section B.

We also note that $I(\cdot)$ equals infinity for nondifferentiable functions in $L_2[0, T]$. It can be shown, however, that the sample paths of a Brownian motion are nowhere differentiable with probability 1. *In other words, the set of functions where $I(\cdot)$ does not equal infinity has zero probability of being a trajectory of the random process!* Suppose, then, that one asked: What is $P\{\mathbf{x}/\sqrt{n} \in \{\text{differentiable functions in } L_2[0, T]\}\}$? This probability for the Brownian motion must be zero from what has been stated above. Hence taking $(1/n)\log$ and the limit as n approaches infinity should formally give $-\infty$. However,

$$\inf_{y \in \{\text{differentiable functions}\}} I(y) = 0$$

This is since the all-zero constant function is in the set of differentiable functions whose value at the origin is zero (with corresponding RKHS norm of zero). What is wrong here (if anything)? The problem is that we have chosen a particularly bad set to which to apply our large deviation principle. The set of all differentiable functions is neither open nor closed (see exercise 2), so we have no large deviation principle defined for it. This illustrates that, especially in these infinite-dimensional spaces, one must be careful that all of the hypotheses of the theorems are satisfied.

B. WENTZELL–FREIDLIN THEORY

(1) Diffusion Processes

Consider the following real valued stochastic differential equation:

$$\begin{aligned} d\mathbf{X}_\varepsilon(t) &= \sqrt{\varepsilon}\,\sigma(\mathbf{X}_\varepsilon(t))\,d\boldsymbol{\beta}(t) + b(\mathbf{X}_\varepsilon(t))\,dt \qquad 0 \leqslant t \leqslant T \\ \mathbf{X}_\varepsilon(0) &= x \end{aligned} \tag{5}$$

where $\boldsymbol{\beta}(t)$ is a standard Brownian motion. This expression is shorthand for the integral equation

$$\mathbf{X}_\varepsilon(t) = x + \sqrt{\varepsilon}\int_0^t \sigma(\mathbf{X}_\varepsilon(\gamma))\,d\boldsymbol{\beta}(\gamma) + \int_0^t b(\mathbf{X}_\varepsilon(\gamma))\,d\gamma \qquad 0 \leqslant t \leqslant T \tag{6}$$

where the first integral is the standard Itô integral against Brownian motion (see appendix D.6). As $\varepsilon \to 0$ one expects that the contributions from the stochastic integral term will become smaller and the solution to the equation will converge to the solution of the deterministic equation,

$$\frac{dx(t)}{dt} = b(x(t)) \qquad 0 \leqslant t \leqslant T$$

$$x(0) = x \tag{7}$$

This is indeed the case and there is a large deviation principle associated with the convergence.

One way to view (6), the integral equation defining $\mathbf{X}_\varepsilon(t)$, is to observe that a continuous Brownian motion sample path $\sqrt{\varepsilon}\boldsymbol{\beta}(t)$, $0 \leqslant t \leqslant T$, is being transformed into another continuous path or trajectory by the integral equation. Let us denote the space of all continuous functions on the interval $[a, b]$ by $C[a, b]$. Then we may view the differential equation as performing a mapping from $C[0, T]$ into $C[0, T]$. Since we know that a large deviation principle exists for $\sqrt{\varepsilon}\boldsymbol{\beta}(t)$ and also know its rate function, we might hope that the mapping is continuous and hence that we can invoke the "contraction principle." This is close to being correct. We will now give a heuristic derivation (and indicate some technical difficulties) of the rate function for the solution to the stochastic differential equation.

The state space that we will use here is $C[0, T]$. A natural distance measure for this space is the *supremum metric* defined for all $f, g \in C[0, T]$ as

$$d(f, g) \equiv \sup\{|f(t) - g(t)| : 0 \leqslant t \leqslant t \leqslant T\}$$

Then the open ball of radius r and center f is $B_f(r) \equiv \{g \in C[0, T] : d(f, g) < r\}$. As always, the open sets will be all possible unions of open balls and closed sets are the complements of the open sets. We note that we derived the rate function for Brownian motion with the probability space being $L_2[0, T]$. Since the domain of the rate function is just the differentiable functions, we could have, for this example, used $C[0, T]$ (which includes the differentiable functions) as our state space and obtained the same results. (This last statement must really be justified on the basis of an embedding argument between the two spaces and their topologies as done in Freidlin and Wentzell [1984, p. 99].)

Let us consider a "δ" approximation to the problem. Consider approximating (5) as

$$d\mathbf{X}_{\varepsilon,\delta}(t) = \sqrt{\varepsilon}\sigma\left(\mathbf{X}_{\varepsilon,\delta}\left(\left[\frac{t}{\delta}\right]\delta\right)\right) d\boldsymbol{\beta}(t) + b\left(\mathbf{X}_{\varepsilon,\delta}\left(\left[\frac{t}{\delta}\right]\delta\right)\right) dt$$

$$\mathbf{X}_{\varepsilon,\delta}(0) = x \tag{8}$$

where $[y]$ denotes the greatest integer less than or equal to y. This approximation has the effect of freezing the coefficients for units of time δ in length, updating at the end of that time segment, and then freezing for another δ time unit. Choose δ to be of the form T/M, where M is an integer. Equation (8) may be easily solved to give

$$\begin{aligned}
\mathbf{X}_{\varepsilon,\delta}(t) &= x + \sqrt{\varepsilon}\sum_{i=0}^{[t/\delta]}(\sigma(\mathbf{X}_{\varepsilon,\delta}(i\delta))(\boldsymbol{\beta}((i+1)\delta) - \boldsymbol{\beta}(i\delta))) + \delta b(\mathbf{X}_{\varepsilon,\delta}(i\delta)) \\
&\quad + \sqrt{\varepsilon}\,\sigma\left(\mathbf{X}_{\varepsilon,\delta}\left(\left[\frac{t}{\delta}\right]\delta\right)\right)\left(\boldsymbol{\beta}(t) - \boldsymbol{\beta}\left(\left[\frac{t}{\delta}\right]\delta\right)\right) \\
&\quad + \left(t - \left[\frac{t}{\delta}\right]\delta\right) b\left(\mathbf{X}_{\varepsilon,\delta}\left(\left[\frac{t}{\delta}\right]\delta\right)\right) \qquad 0 \leqslant t \leqslant T \\
&= \mathbf{X}_{\varepsilon,\delta}\left(\left[\frac{t}{\delta}\right]\delta\right) + \sqrt{\varepsilon}\sigma\left(\mathbf{X}_{\varepsilon,\delta}\left(\left[\frac{t}{\delta}\right]\delta\right)\right)\left(\boldsymbol{\beta}(t) - \boldsymbol{\beta}\left(\left[\frac{t}{\delta}\right]\delta\right)\right) \\
&\quad + \left(t - \left[\frac{t}{\delta}\right]\delta\right) b\left(\mathbf{X}_{\varepsilon,\delta}\left(\left[\frac{t}{\delta}\right]\delta\right)\right) \qquad 0 \leqslant t \leqslant T
\end{aligned}$$

Consider the probability $P(\mathbf{X}_{\varepsilon,\delta}(\cdot) \in D)$, where D is an open set of the form

$$\begin{aligned}
D = \{f \in C[0, T] : f(\cdot) - x|_{[0,\delta]} \in O_1, f(\cdot) - f(\delta)|_{[\delta,2\delta]} \in O_2, \\
\ldots, f(\cdot) - f((M-1)\delta)|_{[(M-1)\delta,T]} \in O_M\}
\end{aligned}$$

and where the O_i are open sets in the space $C[(i-1)\delta, i\delta]$. Let $\mathscr{D}$ denote the class of all such sets. Define the event

$$E_{i+1} \equiv \{\mathbf{X}_{\varepsilon,\delta}(\cdot) - \mathbf{X}_{\varepsilon,\delta}(i\delta)|_{[i\delta,(i+1)\delta]} \in O_{i+1}\} \qquad i = 0, 1, \ldots, M-1$$

(Notationally, $f(\cdot)|_{[a,b]}$ denotes the function f restricted to the interval $[a, b]$.) We can then write

$$P(\mathbf{X}_{\varepsilon,\delta}(\cdot) \in D) = P(E_1 \cap E_2 \cap \cdots \cap E_M)$$

Consider the event E_1. There is a large deviation principle associated with this event as $\varepsilon \to 0$. For this first subinterval, we have

$$\mathbf{X}_{\varepsilon,\delta}(t) = x + \sqrt{\varepsilon}\,\sigma(x)\boldsymbol{\beta}(t) + tb(x) \qquad 0 \leqslant t \leqslant \delta$$

and hence

$$P(\mathbf{X}_{\varepsilon,\delta}(\cdot) - x|_{[0,\delta]} \in O_1) = P(\sqrt{\varepsilon}\sigma(x)\boldsymbol{\beta}(t) + tb(x)|_{t \in [0,\delta]} \in O_1)$$

Define the sets of functions $C_z[a,b] \equiv \{f \in C[a,b]: f(a) = z\}$ and $C_{1,z}[a,b] \equiv \{f \in C_z[a,b]: f$ is piecewise differentiable and $\sup\{|f'(t)|: t \in [a,b]\} < \infty\}$. Now by the contraction principle (assuming that the limits exist)

$$\begin{aligned}\lim_{\varepsilon\to 0} \varepsilon \log P(E_1) &= \lim_{\varepsilon\to 0} \varepsilon \log P(\mathbf{X}_{\varepsilon,\delta}(\cdot) - x|_{[0,\delta]} \in O_1)\\ &= -\inf_{y\in O_1} \inf_{\{f \in C_{1,0}[0,T]: \sigma(x)f(t) + tb(x)|_{t\in[0,\delta]} = y\}} \frac{1}{2}\int_0^\delta (\dot{f})^2\, ds\\ &= -\inf_{y\in O_1\cap C_{1,0}[0,\delta]} \frac{1}{2}\int_0^\delta \left[\frac{d}{ds}\frac{y - sb(x)}{\sigma(x)}\right]^2 ds\\ &= -\inf_{y\in O_1\cap C_{1,0}[0,\delta]} \frac{1}{2}\int_0^\delta \left[\frac{\dot{y} - b(x)}{\sigma(x)}\right]^2 ds\end{aligned}$$

This relation gives us a couple of pieces of information. Obviously, it tells us the rate that the probability of the event E_1 goes to zero, but it also tells us something else. Suppose that the infimum over the set $O_1 \cap C_{1,0}[0,\delta]$ is actually attained by a path $y_1 \in O_1$. Given that a deviation into O_1 occurs, one can show that with probability close to 1, the process is within an ε-tube of $y_1 + x$. In other words, if a large deviation does occur into O_1, then with great probability the path mapped out by $\mathbf{X}_{\varepsilon,\delta}(t) - x$, $0 \leqslant t \leqslant \delta$, will be within ε (in the supremum metric) of y_1. These statements should, of course, be proved (which we will not do), but they make heuristic sense. These infima over whole sets come down to looking at one path in the set (or on its boundary). That one path is, then, in some sense, the most probable excursion path of a large deviation. They correspond exactly to what we called dominating points in the Cramér theory.

Let us consider the event E_2. As $\varepsilon \to 0$, we know that with great probability $y_1 + x$ was the path chosen on the interval $[0,\delta]$. Therefore, $y_1(\delta) + x$ can be viewed as an initial condition for the stochastic differential equation on the interval $[\delta, 2\delta]$. Hence

$$\begin{aligned}&\lim_{\varepsilon\to 0} \varepsilon\, P(E_2 | \mathbf{X}_{\varepsilon,\delta}(\delta) = y_1(\delta) + x)\\ &\quad= \lim_{\varepsilon\to 0} \varepsilon\, P(\mathbf{X}_{\varepsilon,\delta}(\cdot) - \mathbf{X}_{\varepsilon,\delta}(\delta)|_{[\delta,2\delta]} \in O_2 | \mathbf{X}_{\varepsilon,\delta}(\delta) = y_1(\delta) + x)\\ &\quad= -\inf_{y\in O_2\cap C_{1,0}[\delta,2\delta]} \frac{1}{2}\int_\delta^{2\delta} \left[\frac{\dot{y} - b(y_1(\delta) + x)}{\sigma(y_1(\delta) + x)}\right]^2\end{aligned}$$

Suppose that this infimum is achieved at $y_2 \in O_2$. This again is the most likely path for the large deviation into the set O_2. Obviously, $y_2(\delta) = 0$. We can continue these arguments for the general events E_i, $i = 1, \ldots, M$. Hence making use of the fact that the solutions to equations (5) and (8) are Markov, we have

$$
\begin{aligned}
&\lim_{\varepsilon\to 0} \varepsilon \log P(\mathbf{X}_{\varepsilon,\delta}(\cdot)\in D)\\
&\quad = \lim_{\varepsilon\to 0} \varepsilon \log P(E_1 \cap E_2 \cap \cdots \cap E_M)\\
&\quad = \lim_{\varepsilon\to 0} \varepsilon \log P(E_2 \cap E_3 \cap \cdots \cap E_M | E_1) P(E_1)\\
&\quad \approx \lim_{\varepsilon\to 0} \varepsilon \log P(E_2 \cap E_3 \cap \cdots \cap E_M | \mathbf{X}_{\varepsilon,\delta}(\delta) = y_1(\delta) + x) P(E_1)\\
&\qquad\qquad \text{(by Markov property)}\\
&\quad = \lim_{\varepsilon\to 0} \varepsilon \log P(E_1)\\
&\qquad + \lim_{\varepsilon\to 0} \varepsilon \log P(E_3 \cap E_4 \cap \cdots \cap E_M | \mathbf{X}_{\varepsilon,\delta}(\delta) = y_1(\delta) + x, E_2)\\
&\qquad \times P(E_2 | \mathbf{X}_{\varepsilon,\delta}(\delta) = y_1(\delta) + x)\\
&\quad \approx \lim_{\varepsilon\to 0} \varepsilon \log P(E_1) + \lim_{\varepsilon\to 0} \varepsilon \log P(E_2 | \mathbf{X}_{\varepsilon,\delta} = y_1(\delta) + x)\\
&\qquad + \lim_{\varepsilon\to 0} \varepsilon \log P(E_3 \cap E_4 \cap \cdots \cap E_M | \mathbf{X}_{\varepsilon,\delta}(\delta) = y_1(\delta) + x,\\
&\qquad\qquad \mathbf{X}_{\varepsilon,\delta}(2\delta) = y_2(2\delta) + y_1(\delta) + x)\\
&\quad = \lim_{\varepsilon\to 0} \varepsilon \log P(E_1) + \lim_{\varepsilon\to 0} \varepsilon \log P(E_2 | \mathbf{X}_{\varepsilon,\delta}(\delta) = y_1(\delta) + x)\\
&\qquad + \lim_{\varepsilon\to 0} \varepsilon \log P(E_3 \cap E_4 \cap \cdots \cap E_M | \mathbf{X}_{\varepsilon,\delta}(2\delta) = y_2(2\delta) + y_1(\delta) + x)\\
&\quad \approx \lim_{\varepsilon\to 0} \varepsilon \log P(E_1)\\
&\qquad + \sum_{i=2}^{M} \lim_{\varepsilon\to 0} \varepsilon \log P(E_i | \mathbf{X}_{\varepsilon,\delta}((i-1)\delta) = y_{i-1}((i-1)\delta)\\
&\qquad\qquad + y_{i-2}((i-2)\delta) + \cdots + y_1(\delta) + x)\\
&\quad = -\frac{1}{2}\left[\int_0^{\delta} \left[\frac{\dot y_1 - b(x)}{\sigma(x)}\right]^2 - \int_{\delta}^{2\delta} \left[\frac{\dot y_2 - b(y_1(\delta)+x)}{\sigma(y_1(\delta)+x)}\right]^2 - \cdots\right.\\
&\qquad \left. - \int_{(M-1)\delta}^{M\delta} \left[\frac{\dot y_M - b(\Sigma_{i=1}^{M-1} y_i(i\delta)+x)}{\sigma(\Sigma_{i=1}^{M-1} y_i(i\delta)+x)}\right]^2\right]\\
&\quad = - \inf_{y\in D\cap C_{1,x}[0,T]} \frac{1}{2}\int_0^T \left[\frac{\dot y - b(y([t/\delta]\delta))}{\sigma(y([t/\delta]\delta))}\right]^2
\end{aligned}
$$

The minimizing y would be given by

$$y(t) = x + \sum_{i=1}^{[t/\delta]} y_i(i\delta) + y_{[t/\delta]+1}(t)$$

It is an exercise in real analysis to show that every open set in $C[0, T]$ contains an open set of the $\mathscr{D}$ class. This implies that open sets may be constructed entirely by unioning sets of the class $\mathscr{D}$. Closed sets will as always be the complements of the open sets. This indicates then that the rate function for $\{\mathbf{X}_{\varepsilon,\delta}(t)\ 0 \leqslant t \leqslant T\}$ defined by (8) is given by

$$I(y) = \begin{cases} \dfrac{1}{2}\displaystyle\int_0^T \left[\frac{\dot{y} - b(y([t/\delta]\delta))}{\sigma(y([t/\delta]\delta))}\right]^2 & \text{for } y \in C_{1,x}[0, T] \\ \infty & \text{otherwise} \end{cases}$$

The hope and expectation is that one can find the rate function for the original stochastic differential equation by letting $\delta \to 0$. This is indeed the case. We now consider a slightly more general setting before quoting the principal theorem of the section. Consider the following stochastic differential equation:

$$\begin{aligned} d\mathbf{X}_\varepsilon(t) &= \sqrt{\varepsilon}\,\sigma(\mathbf{X}_\varepsilon(t))\,d\boldsymbol{\beta}(t) + b(\mathbf{X}_\varepsilon(t))\,dt \qquad 0 \leqslant t \leqslant T \\ \mathbf{X}_\varepsilon(0) &= x \end{aligned} \tag{9}$$

where all quantities are defined on $\mathbb{R}^d$. We assume that $a(x) = \sigma(x)\sigma^*(x)$ is uniformly elliptic and bounded [in particular, $a^{-1}(\cdot)$ exists and is bounded] and $\sigma(\cdot)$ satisfies a global Lipshitz condition [$|\sigma(x) - \sigma(y)| \leqslant K|x - y|$]. Assume that $b(\cdot)$ is bounded and Lipshitz also. $\boldsymbol{\beta}(\cdot)$ is a standard d-dimensional Brownian motion. Then we have

Theorem (Wentzell–Freidlin)
The stochastic differential equation of (9) satisfies a large deviation principle with rate function given by

$$I_x(f, T) \equiv I_x(f) = \begin{cases} \dfrac{1}{2}\displaystyle\int_0^T < \dot{f}(t) - b(f(t)), a^{-1}(f(t))(\dot{f}(t) - b(f(t)) > dt \\ \qquad\qquad \text{for } f \in C_{1,x}[0, T] \text{ and } \displaystyle\int \dot{f}(t)^2\,dt < \infty \\ \infty \qquad\qquad \text{otherwise} \end{cases}$$

(2) Exit Times of Diffusion Processes

A problem of related interest for the diffusion equation theory is the question of the asymptotics of mean exit times. Consider the following system:

$$\dot{\mathbf{X}}(t, x) = b(\mathbf{X}(t, x))$$
$$\mathbf{X}(0, x) = x \tag{10}$$

where $b(\cdot)$ is bounded and Lipshitz. Let $s \in \mathbb{R}^d$ be an *asymptotically stable equilibrium point* of (10); i.e., for every neighborhood O' of s, let there exist a smaller neighborhood $O'' \subset O'$ such that for all $x \in O''$, $\mathbf{X}(t, x)$ converges to s as $t \to \infty$ without leaving O'. We say that $D \subset \mathbb{R}^d$ is *attracted* to s if the trajectories $\mathbf{X}(t, x)$, $x \in D$, converge to the equilibrium point s as $t \to \infty$ without leaving D.

Now suppose that we perturb (10) to

$$d\mathbf{X}_\varepsilon(t, x) = b(\mathbf{X}_\varepsilon(t, x))\, dt + \sqrt{\varepsilon}\, d\boldsymbol{\beta}(t)$$

In this setting, the rate function is given by

$$I_x(f, T) \equiv \begin{cases} \dfrac{1}{2}\displaystyle\int_0^T |\dot{f}(t) - b(f(t))|^2\, dt & \text{for } f \in C_{1,x}[0, T] \text{ and } \displaystyle\int \dot{f}(t)^2\, dt < \infty \\ \infty & \text{otherwise} \end{cases}$$

Define

$$V(s, y) \equiv \inf\{I_s(f, \tau) : f \in C_{1,s}[0, \tau],\ f(\tau) = y,\ \tau \in [0, \infty)\}$$

$V(s, y)$ is the infimum of the rate function over all paths that start at time zero from s and end at y at a later time.

Theorem (Wentzell–Freidlin)
Let s be an asymptotically stable equilibrium point and assume that $D \subset \mathbb{R}^d$ is attracted to s. Suppose that the boundary of D, ∂D, is a smooth manifold and $(b(y), n(y)) < 0$ for $y \in \partial D$, where $n(y)$ is the exterior normal at $y \in \partial D$. Then for $x \in D$ we have

$$\lim_{\varepsilon \to 0} \varepsilon \log E\{\tau^\varepsilon\} = \inf_{y \in \partial D} V(s, y)$$

where $\tau^\varepsilon = \inf\{t > 0 : X_\varepsilon(t, x) \notin D\}$ is the first exit of the process from D.

We argue that this is an intuitively reasonable result. Consider the typical behavior of a path as it evolves in time. Starting from a point $x \in D$, the path wanders to within a small neighborhood of s, the equilibrium point. The random perturbations of the Brownian motion term cause the path to make occasional sojourns away from the equilibrium point, but the restoring force tends to pull it back again to within a small neighborhood of s. Occasionally, however, a path will make a large deviation from its usual behavior and will wander outside the attracting set D. This is a rare event and we would like to obtain the mean time to occurrence asymptotics of it.

Let us denote our rare event by A: the event that starting from a small neighborhood of s, the path will leave D before returning to the neighborhood. From the theorem in the preceding subsection and the definitions above, we have that

$$\lim_{\varepsilon \to 0} \varepsilon \log P(A) = -\inf_{y \in \partial D} V(s, y) \equiv V$$

We know that large deviation rare events tend to "look like" exponentially distributed random variables, i.e., $P(A) \approx \exp[\varepsilon^{-1} V]$. Therefore, suppose that we start off our random process at a stable equilibrium point s and follow a path. With great probability $[1 - P(A)]$ it wanders around a bit and then returns to a close neighborhood of the equilibrium point. The average amount of time that it takes to do this sojourn from the equilibrium point and back again may be denoted T_{sojourn}. With probability $P(A)$ it escapes. Since the process is Markov we may calculate the average time to escape by calculating the average number of sojourns from and back to the equilibrium point and multiply by T_{sojourn}. We use the fact here that what happens on each sojourn is (approximately) independent. The average number of sojourns needed until escape (assuming independence) should then be like $1/P(A)$. Hence

$$E\{\tau^{\varepsilon}\} \approx E\{\#\text{ of sojourns}\} T_{\text{sojourn}} \approx \frac{T_{\text{sojourn}}}{P(A)}$$

Now as ε gets small, T_{sojourn} also becomes small. The reasoning then is that short sojourns (say those that do not take the path at least a δ amount away from equilibrium point) will not contribute much to the mean escape time. Call sojourns that travel at least δ away from the equilibrium point, δ-sojourns. Given that we have a δ-sojourn, the probability that it goes on and escapes should be approximately $P(A)$ also for small δ. Moreover, $T_{\delta\text{-sojourn}}$ should have a minimum value, as it does take a nonzero amount of time, say t_{δ}, to get δ away from the equilibrium. Furthermore, as ε goes to zero, the great majority of these paths sojourn times will be t_δ. Hence

$$E\{\tau^\varepsilon\} \approx E\{\#\text{ of }\delta\text{-sojourns}\} T_{\delta\text{-sojourn}} \approx \frac{T_{\delta\text{-sojourn}}}{P(A)} \approx \frac{t_\delta}{P(A)}$$

Therefore,

$$\lim_{\varepsilon\to 0} \varepsilon \log E\{\tau^\varepsilon\} = \lim_{\delta\to 0}\lim_{\varepsilon\to 0} \varepsilon \log \frac{t_\delta}{P(A)} = -\lim_{\varepsilon\to 0} \varepsilon \log[P(A)] = \min_{y\in\partial D} V(s, y)$$

C. SLOW MARKOV WALK

(1) Discrete-Time Case

Consider the following Markov chain model:

$$\mathbf{x}_{k+1}^n = \mathbf{x}_k^n + \frac{1}{n} V(\mathbf{x}_k^n, \xi_k) \qquad k \geqslant 0 \text{ and } \mathbf{x}_0^n = x$$

where $V(\cdot,\cdot)$ is a function mapping $\mathbb{R}^2$ into $\mathbb{R}$ and the $\{\xi_k\}$ are iid random variables. Define the quantities

$$\begin{aligned} m(y) &\equiv E\{V(y, \xi_1)\} \\ M_y(\theta) &\equiv E\{\exp[\theta V(y, \xi_1)]\} \\ I_y(u) &\equiv \sup_\theta \{\theta u - \log M_y(\theta)\} \end{aligned}$$

To deal with the asymptotics, we would like to construct continuous-time paths from the realizations of the $\{\mathbf{x}_k^n\}$. Hence define

$$\mathbf{x}^n(t) = \mathbf{x}_k^n \qquad \text{for } t = \frac{k}{n}$$

To define the random process for all values of t, one then interpolates linearly between these points (see figure 3). We see that $\mathbf{x}^n(t)$ jump sizes decrease but happen more frequently as n gets large. Now we would like to see if there is a limiting value of $\mathbf{x}^n(t)$ as $n \to \infty$. Note that

$$\mathbf{x}^n(t) \cong \mathbf{x}_{[tn]}^n = \sum_{j=0}^{[tn]-1} (\mathbf{x}_{j+1}^n - \mathbf{x}_j^n) + x = x + \frac{1}{n} \sum_{j=0}^{[tn]-1} V(\mathbf{x}_j^n, \xi_j)$$

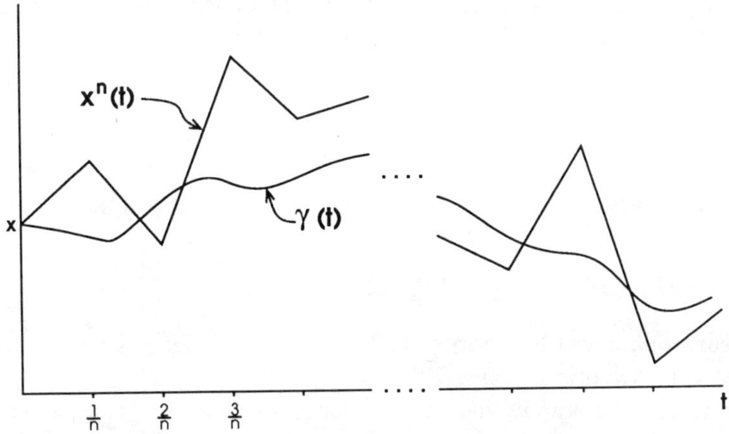

Figure 3

where $[y]$ is the greatest integer less than or equal to y. Let us consider the limiting behavior of the mean value. We have

$$\begin{aligned} E\{\mathbf{x}^n(t)\} &\cong x + \frac{1}{n}\sum_{j=0}^{[tn]-1} E\{V(\mathbf{x}_j^n, \xi_j)\} \\ &= x + \frac{1}{n}\sum_{j=0}^{[tn]-1} E\left\{m\left(\mathbf{x}^n\left(\frac{j}{n}\right)\right)\right\} \\ &\cong x + \int_0^t E\{m(\mathbf{x}^n(\sigma))\}\,d\sigma \qquad \text{(for large } n\text{)} \end{aligned}$$

If $\text{var}\{\mathbf{x}^n(\sigma)\}_{n\to\infty} \to 0$, we expect $E\{m(\mathbf{x}^n(\sigma))\} \approx m(E\{\mathbf{x}^n(\sigma)\}) \approx \mathbf{x}^n(\sigma)$. In that case, $\mathbf{x}^n(t)$ converges to some $\gamma(t)$, say, where $\gamma(t)$ satisfies the following integral equation:

$$\gamma(t) = x + \int_0^t m(\gamma(\sigma))\,d\sigma \tag{11}$$

This is, of course, equivalent to the differential equation

$$\gamma'(t) = m(\gamma(t))$$
$$\gamma(0) = x$$

We have still not determined if there is a limiting value of $\text{var}\{\mathbf{x}^n(t)\}$ as $n \to \infty$. We give a simple argument that there is for the special case of $V(x, \xi) \equiv V(\xi)$

(i.e., the jump does not depend on the position of the chain). For the jump-independent case the variance of $\mathbf{x}^n(t)$ is given by

$$\operatorname{var}\{\mathbf{x}^n(t)\} \cong \frac{1}{n^2} \sum_{j=0}^{[tn]-1} \operatorname{var}\{V(\xi_j)\} \xrightarrow[n\to\infty]{} 0 \tag{12}$$

[as long as $\operatorname{var}\{V(\xi_1)\}$ is bounded]. Equations (11) and (12) then imply that

$$\mathbf{x}^n(t) \xrightarrow[n\to\infty]{} \gamma(t)$$

in mean square and hence in probability. (The general position-dependent case is considered in exercise 3.)

Therefore, we know that the sequence of random processes $\mathbf{x}^n(t)$ is converging to some deterministic function $\gamma(t)$. The question, then, is: What is the underlying large deviation principle (if any)? We first need to determine what the underlying state space and associated metric should be. Note that the sample paths of $\mathbf{x}^n(t)$ are piecewise linear line segments. The limiting deterministic function $\gamma(t)$ is differentiable. We will choose our probability space to be $C_x[0, T]$ [recall that this is the space of all continuous functions, f, such that $f(0) = x$ and with the supremum metric $d(f, g) \equiv \sup_{t\in[0,T]} |f(t) - g(t)|$].

We now give a heuristic derivation for the large deviation principle in this setting. Consider a small ε-ball around some differentiable function $f \in C_x[0, T]$, i.e.,

$$B_f(\varepsilon) \equiv \{g \in C_x[0, T] : |f(t) - g(t)| < \varepsilon \quad \text{for all } 0 \leqslant t \leqslant T\}$$

Suppose that we are interested in $P\{\mathbf{x}^n(\cdot) \in B_f(\varepsilon)\}$. Let $t = M/n$ and $\Delta t = m/n$. Δt should be thought of as a fixed small number [as we take limits in n, then m will also have to get large to keep Δt the same (approximate) value]. Then

$$\begin{aligned}\frac{\mathbf{x}^n(t + \Delta t) - \mathbf{x}^n(t)}{\Delta t} &= \frac{\mathbf{x}^n_{M+m} - \mathbf{x}^n_M}{m/n} \\ &= \frac{n}{m} \sum_{j=0}^{m-1} (\mathbf{x}^n_{M+j+1} - \mathbf{x}^n_{M+j}) \\ &= \frac{1}{m} \sum_{j=0}^{m-1} V(\mathbf{x}^n_{M+j}, \xi_{M+j}) \end{aligned} \tag{13}$$

Now for values of $j = 0, 1, \ldots, m - 1$, $\mathbf{x}^n_{M+j}$ will not change by very much. Furthermore, as n approaches infinity, so does m, with the limit ratio

$\lim_{n\to\infty} m/n \equiv \Delta t$. Therefore, (13), when conditioned on $\mathbf{x}_M^n$, "looks like" a sample average of iid random variables. Hence we will be able to apply Cramér's theorem. From time t to $t + \Delta t$ we are interested in whether this sum is satisfying our condition of being in the ε-ball about f. For this to be true, we need $\mathbf{x}_M^n$ to be approximately $f(t)$ and the right-hand derivative should be approximately $f'(t)$. Therefore, from (13),

$$\begin{aligned} P\left\{\frac{1}{m}\sum_{j=0}^{m-1} V(\mathbf{x}_{M+j}^n, \xi_{M+j}) \approx f'(t) | \mathbf{x}_M^n \approx f(t)\right\} \\ \approx P\left\{\frac{1}{m}\sum_{j=0}^{m-1} V(\mathbf{x}_M^n, \xi_{M+j}) \approx f'(t) | \mathbf{x}_M^n \approx f(t)\right\} \\ \approx \exp[-mI_{f(t)}(f'(t))] = \exp[-n\,\Delta t\, I_{f(t)}(f'(t))] \end{aligned} \tag{14}$$

where the last line follows by invoking Cramér's theorem. Now define

$$E_i \equiv \{\mathbf{x}^n(i\,\Delta t) \approx f(i\,\Delta t) \quad \text{and} \quad \mathbf{x}^{n\prime}(i\,\Delta t) \approx f'(i\,\Delta t)\}$$

for $i = 0, 1, \ldots, L$, where $L \equiv [T/\Delta t]$. Hence

$$P\{\mathbf{x}^n(\cdot) \in B_f(\varepsilon)\} \approx P\{E_0 \cap E_1 \cap \cdots \cap E_L\} \tag{15}$$

Now, since $\mathbf{x}_k^n$ is Markov, we have

$$\begin{aligned} P\{\mathbf{x}^{n\prime}(L\,\Delta t) \approx f'(L\,\Delta t) | \mathbf{x}^n(L\,\Delta t) \approx f(L\,\Delta t), E_{L-1}, E_{L-2}, \ldots, E_0\} \\ = P\{\mathbf{x}^{n\prime}(L\,\Delta t) \approx f'(L\,\Delta t) | \mathbf{x}^n(L\,\Delta t) \approx f(L\,\Delta t)\} \\ \approx \exp[-n\,\Delta t\, I_{f(L\,\Delta t)}(f'(L\,\Delta t))] \end{aligned} \tag{16}$$

Also note that

$$\begin{aligned} P\{\mathbf{x}^n(L\,\Delta t) \approx f(L\,\Delta t) | E_{L-1}, E_{L-2}, \ldots, E_0\} \\ = P\{\mathbf{x}^n(L\,\Delta t) \approx f(L\,\Delta t) | E_{L-1}\} \\ = P\{\mathbf{x}^n(L\,\Delta t) \approx f(L\,\Delta t) | \mathbf{x}^n((L-1)\,\Delta t) \approx f((L-1)\,\Delta t) \\ \text{and } \mathbf{x}^{n\prime}((L-1)\,\Delta t) \approx f'((L-1)\,\Delta t)\} \approx 1 \end{aligned} \tag{17}$$

The idea in this last expression is that if we can fix the function value and its derivative at some point $y - \Delta t$ away from a point y, then we have essentially fixed the function value at y by a Taylor series approximation with error of the order $(\Delta t)^2$ at most. Hence, from (14)–(17),

$$
\begin{aligned}
&P\{\mathbf{x}^n(\cdot)\in B_f(\varepsilon)\} \approx P\{E_0 \cap E_1 \cap \cdots \cap E_L\} \\
&= P\{E_L | E_{L-1}, \ldots, E_0\} P\{E_{L-1} \cap \cdots \cap E_0\} \\
&= P\{\mathbf{x}^n(L\,\Delta t) \approx f(L\,\Delta t) \\
&\qquad \text{and } \mathbf{x}^{n\prime}(L\,\Delta t) \approx f'(L\,\Delta t) | E_{L-1}, \ldots, E_0\} P\{E_{L-1} \cap \cdots \cap E_0\} \\
&= P\{\mathbf{x}^{n\prime}(L\,\Delta t) \approx f'(L\,\Delta t) | \mathbf{x}^n(L\,\Delta t) \approx f(L\,\Delta t), E_{L-1}, \ldots, E_0\} \\
&\qquad \times P\{\mathbf{x}^n(L\,\Delta t) \approx f(L\,\Delta t) | E_{L-1}, E_{L-2}, \ldots, E_0\} P\{E_{L-1} \cap \cdots \cap E_0\} \\
&\approx \exp[-n\,\Delta t\, I_{f(L\,\Delta t)}(f'(L\,\Delta t))] 1 P\{E_{L-1} \cap \cdots \cap E_0\} \\
&= \exp[-n\,\Delta t\, I_{f(L\,\Delta t)}(f'(L\,\Delta t))] P\{E_{L-1} \cap \cdots \cap E_0\} \\
&\approx \prod_{i=0}^{[T/\Delta t]} \exp[-n\,\Delta t\, I_{f(i\Delta t)}(f'(i\,\Delta t))] \\
&\approx \exp\left[-n \int_0^T I_{f(t)}(f'(t))\, dt\right]
\end{aligned}
$$

This indicates that the functional

$$I(f) \equiv \int_0^T I_{f(t)}(f'(t))\, dt \tag{18}$$

should be our candidate for the rate function in this setting. The actual theorem is quite a bit more complicated. We now state a collection of conditions and give one available result. For a given function $\bar{g}(\theta)$, define $\underline{h}(u) \equiv \sup_\theta \{\theta u - \bar{g}(\theta)\}$.

***Condition* 1.** There exists a $\bar{g}(\theta) < \infty$ such that $\log[M_x(\theta)] \leqslant \bar{g}(\theta)$ for all x. □

***Condition* 2.** The set $D_{\underline{h}} \equiv \{u : \underline{h}(u) < \infty\}$ has at least one interior point u_0 and at that point $\sup_x I_x(u_0) < \infty$. □

***Condition* 3.** For any compact set U consisting entirely of interior points of the set $D_{\underline{h}}$, there are positive constants M and m such that for all x, and for all $u \in U$ we have

$$|I_x'(u)| \leqslant M \qquad \text{and} \qquad I_x''(u) \geqslant m$$ □

***Condition* 4.** The set of those u, in the closure of $D_{\underline{h}}$, for which $\underline{h}(u) < \infty$ is closed. Furthermore, for any compact U consisting entirely on points in $D_{\underline{h}}$, the function $I_x(u)$ is uniformly continuous on U for all x. □

Theorem (Wentzell)
Assume conditions 1–3. Then for all $f \in C_{1,x}[0, T]$ and for all $\varepsilon > 0$, we have

$$\varliminf_{n\to\infty} \frac{1}{n} \log P\left\{ \sup_{0\leqslant t\leqslant T} |\mathbf{x}^n(t) - f(t)| < \varepsilon \right\} \geqslant -I(f)$$

Define $\Upsilon(k) \equiv \{f \in C_{1,x}[0, T] : I(f) \leqslant k\}$. Assume conditions 1, 2, and 4. Then for $\varepsilon > 0$,

$$\varlimsup_{n\to\infty} \frac{1}{n} \log P\left\{ \inf_{f\in\Upsilon(k)} \sup_{0\leqslant t\leqslant T} |\mathbf{x}^n(t) - f(t)| \geqslant \varepsilon \right\} \leqslant -k \qquad (19)$$

The major impediment to using this result (besides verifying the conditions) is to perform the necessary minimization of the rate function over the set of interest in order to find the rate with which the probability of that set approaches zero. For an important class of sets, the level crossing sets, this minimization may be done straightforwardly. Recall that $C_x[0, T]$ is the space of all continuous functions on the interval $[0, T]$ whose value at the origin is x. For $f \in C_x[0, \infty)$ define for $r \in \mathbb{R}$, $\tau_{f,r} \equiv \inf\{t \in [0, \infty) : f(t) = r\}$. If f never equals r, then $\tau_{f,r} \equiv \infty$. A set $A_{r,x}$ is a *level crossing set with level crossing threshold* r if it can be represented as follows:

$$A_{r,x} \equiv \{f \in C_x[0, \infty) : \tau_{f,r} < \infty\}$$

We wish to solve the problem of minimizing the rate function over the level crossing set $A_{r,0}$. First consider the cost of minimizing the rate function over $A_{r,x}$,

$$E(x) \equiv \inf\left\{ \int_0^{\tau_{g,r}} I_{g(t)}(g'(t))\, dt : g \in A_{r,x} \cap C_{1,x}[0, \infty) \right\}$$

Note that $E(0)$ is the rate parameter that we are interested in, but considering $E(x)$ will help us to compute it. We proceed by a perturbation argument. We assume that there is an optimum g for every value of x to minimize $E(x)$. Starting from x, that g will (in a Δt time interval) take you to a new point $x + v\,\Delta t$. $E(x)$ is then a combination of the cost of going from x to $x + v\,\Delta t$ and the cost of $E(x + v\,\Delta t)$. Note that we are tacitly making use of the fact that the rate function is time invariant; i.e., $I_f(f')$ depends on t only through f

and f'. This allows us to "splice" together the optimal path going from 0 to x with the optimal path of going from x to r. The direction v that is chosen by the optimal g is, of course, optimum, so we must have the following relationship:

$$E(x) = \inf_v \{I_x(v)\,\Delta t + E(x + v\,\Delta t)\} = \inf_v \{I_x(v)\,\Delta t + E(x) + E'(x)v\,\Delta t\}$$

Cancel $E(x)$ from both sides, divide by Δt, and let $\Delta t \to 0$. One then obtains

$$\inf_v \{I_x(v) + E'(x)v\} = 0$$

or, equivalently,

$$\sup_v \{-E'(x)v - I_x(v)\} = 0 \tag{20}$$

We now need to note that $I_x(u) = \sup_\theta \{\theta u - \log M_x(\theta)\}$ is the convex dual (see appendix B.3) to $\log M_x(\theta)$, i.e., $\log M_x(\theta) = \sup_u \{u\theta - I_x(u)\}$ (see exercise 4). Therefore, (20) becomes

$$\log M_x(-E'(x)) = 0$$

Now integrate $-E'(x)$ from 0 to r, the threshold crossing level. Note that $E(r)$ should equal zero since there is no cost once we have arrived at the crossing level. Therefore,

$$\int_0^r -E'(x)\,dx = E(0)$$

Hence we have the following result:

$$\lim_{n\to\infty} \frac{1}{n} \log P\{\mathbf{x}^n \in A_{r,0}\} = -E(0) = \int_0^r E'(x)\,dx \tag{21}$$

where $-E'(x)$ is a solution to

$$\log M_x(-E'(x)) = 0 \tag{22}$$

Now suppose that the $\sup_v \{-E'(x)v - I_x(v)\}$ is attained at point $\underline{v}$. Then we have that $-E'(x) = I_x'(\underline{v})$. Furthermore, it is true (see exercise 5) that

$$\left.\frac{dI_x(u)}{du}\right|_{u=M_x'(\theta)/M_x(\theta)} \equiv I_x'([\log M_x(\theta)]') = \theta$$

and

$$\left.\frac{M'_x(\theta)}{M_x(\theta)}\right|_{\theta=I'_x(u)} \equiv [\log M_x(I'_x(u))]' = u$$

Therefore, from the derivative relationships, we have that

$$[\log M_x(-E'(x))]' = [\log M_x(I'_x(\underline{v}))]' = \underline{v}$$

Now $\underline{v}$ is, for the optimum path, playing the part of the slope of the path as a function of time. On the other hand, x is playing the part of position. If $g_{\text{opt}}(t)$ is the optimum path, then, as t varies, it gives the values of $x = g_{\text{opt}}(t)$ in a parametric fashion. $\underline{v}$ is similarly given parametrically by $\underline{v} = g'_{\text{opt}}(t)$. Therefore, one can obtain $g'_{\text{opt}}(t)$ by finding the set of $\underline{v}$ that come out of the equation $[\log M_x(-E'(x))]' = \underline{v}$, where $-E'(x)$ is a solution to $\log M_x(-E'(x)) = 0$.

***Example* 2.** Suppose that we are interested in the following jump process model:

$$\mathbf{x}_{k+1} = \mathbf{x}_k + V(\mathbf{x}_k, \xi_k) \qquad \mathbf{x}_0 \equiv 0$$

and for $x \geqslant 0$,

$$P\{\xi_k = 1\} = \lambda < \tfrac{1}{2}$$
$$P\{\xi_k = -1\} = 1 - \lambda \equiv \bar{\lambda}$$
$$V(x, e) = \begin{cases} e & \text{for } x > 0 \\ 1 & \text{for } x = 0 \end{cases}$$

which means that automatically $\mathbf{x}_1 = 1$. This is a Markov chain that starts off at zero, jumps up to some positive integer, and eventually will return to zero. (We note that the process will violate some of the conditions of the theorem, due to the behavior at the zero state [$V(x, e)$ is not differentiable at $x = 0$]. These "boundary" problems are of a technical nature and can be taken care of by successive approximation techniques. Essentially, one starts off the process at a fixed nonnegative integer other than zero and shows that after scaling the results are the same as if zero was the starting point.) Given that the chain is not at zero, it is more likely that it will jump down than up since $\lambda < \frac{1}{2}$. Let A be the set of trajectories of this chain that starting from 0, reach N, before hitting 0 again. We are interested, say, in the asymptotic behavior of $P\{A\}$ as N gets large. Let us consider the equivalent situation of

$$\mathbf{x}^N_{k+1} = \mathbf{x}^N_k + \frac{1}{N} V(\mathbf{x}^N_k, \xi_k) \qquad \mathbf{x}^N_0 \equiv 0$$

with the same definition of the distributions of $V(\mathbf{x}, \xi_k)$. This chain starting from 0, reaches 1, before hitting 0 also with probability $P\{A\}$. Let us define the continuous-time process $\mathbf{x}_k^N(\cdot)$ by

$$\mathbf{x}^N\left(\frac{k}{N}\right) \equiv \mathbf{x}_k^N \qquad k = 0, 1, \dots$$

and we linearly interpolate between these points to define the process at all points on the interval $[0, \infty)$. Therefore,

$$P\{A\} = P\{\mathbf{x}^N(\cdot) \text{ hits 1 before 0 given that } \mathbf{x}^N(0) = 0\}$$

and

$$M_x(\theta) = \lambda \exp[\theta] + \bar{\lambda} \exp[-\theta] \qquad \text{for } x > 0$$

We need to consider solutions to the equation

$$\log M_x(-E'(x)) = 0$$

or, equivalently,

$$M_x(-E'(x)) = 1$$

It is simple to verify that the only nonzero solution of this equation is

$$-E'(x) = \log\frac{\bar{\lambda}}{\lambda} \qquad x > 0$$

Now it must be true that the optimal trajectory starts off at 0 and wanders up to 1. This implies that in our previous notation $r = 1$. Therefore,

$$-E(0) = \int_0^r E'(x)\,dx = \int_0^1 -\log\frac{\bar{\lambda}}{\lambda}\,dx = -\log\frac{\bar{\lambda}}{\lambda}$$

Hence, by (21)–(22),

$$\lim_{N\to\infty} \frac{1}{N} \log P\{A\} = -\log\frac{\bar{\lambda}}{\lambda}$$

Note that

$$\frac{d}{du}[\log M_x(u)]|_{u=-E'(x)} = \frac{M_x'(-E'(x))}{M_x(-E'(x))} = M_x'(-E'(x)) = 1 - 2\lambda = \underline{v}$$

This is the only possible value for $\underline{v}$, which means that there is only one value of slope for the optimal trajectory. Hence the optimal trajectory is a line with slope $1 - 2\lambda$ that starts off at the origin and ends at 1 at time $\tau = 1/(1 - 2\lambda)$. □

(2) Continuous-Time Case

Consider a continuous-time birth–death random process $\mathbf{x}(t)$. $\mathbf{x}(t)$ jumps to $\mathbf{x}(t) + e_j$ with Poisson rate $\lambda_j(\mathbf{x}(t))$, where $e_1 = 1$ and $e_{-1} = -1$; i.e., suppose that at time t, the process is at $\mathbf{x}(t) = x$; then consider starting off two Poisson random processes with rates $\lambda_1(x)$ and $\lambda_{-1}(x)$. If the $\lambda_1(x)$ rate process jumps first at time t_1, then define $\mathbf{x}(t + \tau) \equiv x$ for $0 < \tau < t_1$ and $\mathbf{x}(t + t_1) \equiv x + 1$. Similarly, if the $\lambda_{-1}(x)$ rate process jumps first, say, at time t_1', then define $\mathbf{x}(t + \tau) \equiv x$ for $0 < \tau < t_1'$ and $\mathbf{x}(t + t_1') \equiv x - 1$. $\lambda_1(x)$ and $\lambda_{-1}(x)$ are called the birth and death rates, respectively, when the population is at size x.

The waiting time until a Poisson rate λ random process jumps is well known to be an exponential random variable with mean $1/\lambda$. An exponential random variable can be approximated in discrete time by a geometric random variable. Recall the number of times one flips a coin until the first heads appears has a geometric distribution:

$$P(k \text{ flips until heads appears}) = (1 - p)^{k-1} p$$

where p is the probability of heads. We will use such an approximation to determine the rate function for the continuous-time birth–death process via the results for the discrete-time case considered in the preceding subsection.

Suppose that $\mathbf{x}_n(t)$ is a birth–death process with jump rates $n\lambda_j(x)$ and corresponding jump directions e_j/n. Let $\mathbf{x}_n(0) = x_0$ for the initial condition. That is, $\mathbf{x}_n(t)$ jumps quickly and in small steps as $n \to \infty$. One can show (see exercise 3) that as $n \to \infty$, $\mathbf{x}_n(t)$ collapses to a deterministic trajectory defined by

$$\frac{d}{dt} x_\infty(t) = \lambda_1(x_\infty(t)) - \lambda_{-1}(x_\infty(t)) \qquad x_\infty(0) = x_0$$

We, of course, seek the large deviation principle associated with this convergence.

Let us consider a sequence (indexed by N) of discrete-time approximations to the foregoing continuous-time process. Suppose that the $\lambda_j(\cdot)$ are bounded (by B, say). Then for all $N > B$ we may define

$$\mathbf{x}_{k+1}^{(n,N)} = \mathbf{x}_k^{(n,N)} + \frac{1}{nN} V^{(N)}(\mathbf{x}_k^{(n,N)}, \xi_k) \qquad \mathbf{x}_0^{(n,N)} = x_0$$

where

$$P(V^{(N)}(x, \xi_1) = N) = \frac{\lambda_1(x)}{N}$$

$$P(V^{(N)}(x, \xi_1) = -N) = \frac{\lambda_{-1}(x)}{N}$$

and

$$P(V^{(N)}(x, \xi_1) = 0) = 1 - \frac{\lambda_1(x) + \lambda_{-1}(x)}{N}$$

where $\{\xi_k\}_{k=1}^{\infty}$ is an iid sequence of random variables. Obviously, we may turn this discrete-time process into continuous trajectories via the construction of the preceding subsection ($\mathbf{x}^{n,N}(k/nN) \equiv \mathbf{x}_k^{(n,N)}$; $k = 0, 1, \ldots, [nNT]$) and write the large deviation principle for it. To see that this is the correct approximation, notice that this process has jumps up and down of height $1/n$ [as $\mathbf{x}_n(t)$ does] and the process has an average jump rate of $nN\lambda_j(x)/N = n\lambda_j(x)$ events per unit time interval. Therefore, by (21)–(22), we have for some set of functions $A \subset C_{x_0}[0, T]$ (assuming that limits exist)

$$\lim_{n\to\infty} \frac{1}{nN} \log P(\mathbf{x}^{n,N}(t) \in A) = -\inf_{f\in A} \int_0^T I_f^{(N)}(f'(t))\, dt$$

or

$$\lim_{n\to\infty} \frac{1}{n} \log P(\mathbf{x}^{n,N}(t) \in A) = -\inf_{f\in A} \int_0^T NI_f^{(N)}(f'(t))\, dt \tag{23}$$

where

$$I_x^{(N)}(u) \equiv \sup_\theta \{\theta u - \log E\{\exp[\theta V^{(N)}(x, \xi_1)]\}\}$$

Now as $N \to \infty$, we should approach the desired rate result for the continuous-time birth–death process. Hence we must first calculate $I_x^{(N)}(u)$. First

$$\begin{aligned} M_x^{(N)}(\theta) &\equiv E\{\exp[\theta V^{(N)}(x, \xi_1)]\} \\ &= \exp[N\theta]\frac{\lambda_1(x)}{N} + \exp[-N\theta]\frac{\lambda_{-1}(x)}{N} + \left(1 - \frac{\lambda_1(x)}{N} - \frac{\lambda_{-1}(x)}{N}\right) \end{aligned}$$

Therefore,

$$\begin{aligned} I_x^{(N)}(u) = \sup_\theta \Bigg\{\theta u - \log\Bigg\{&\exp[N\theta]\frac{\lambda_1(x)}{N} + \exp[-N\theta]\frac{\lambda_{-1}(x)}{N} \\ &+ \left(1 - \frac{\lambda_1(x)}{N} - \frac{\lambda_{-1}(x)}{N}\right)\Bigg\}\Bigg\} \end{aligned}$$

Maximize this over θ by taking the derivative to find

$$\theta_{\text{opt}}^{(N)} = \frac{1}{N}\log[A + B]$$

where

$$A \equiv \frac{u}{N}\left(1 - \frac{\lambda_1(x)}{N} - \frac{\lambda_{-1}(x)}{N}\right)\Big/ D$$

$$B \equiv \frac{[A^2D^2 - 4(u\lambda_1(x)/N^2 - \lambda_1(x)/N)(u\lambda_{-1}(x)/N^2 - \lambda_{-1}(x)/N)]^{1/2}}{D}$$

$$D \equiv 2\left(\frac{u\lambda_1(x)}{N^2} - \frac{\lambda_1(x)}{N}\right)$$

One then has that

$$\theta_0 \equiv \lim_{N\to\infty} \theta_{\text{opt}}^{(N)} N = \log\frac{u + \sqrt{u^2 + 4\lambda_1(x)\lambda_{-1}(x)}}{2\lambda_1(x)}$$

Therefore,

$$\begin{aligned}
\lim_{N\to\infty} NI_x^{(N)}(u) &= \lim_{N\to\infty} \sup_\theta \{\theta Nu - N\log(M_x(\theta))\} \\
&= \lim_{N\to\infty} \sup_\theta \Bigg\{\theta Nu - N\log\Bigg(\exp[N\theta]\frac{\lambda_1(x)}{N} \\
&\quad + \exp[-N\theta]\frac{\lambda_{-1}(x)}{N} + \left(1 - \frac{\lambda_1(x)}{N} - \frac{\lambda_{-1}(x)}{N}\right)\Bigg)\Bigg\} \\
&= \lim_{N\to\infty} \Bigg[\theta_{\text{opt}}^{(N)} Nu - N\log\Bigg(\exp[\theta_{\text{opt}}^{(N)} N]\frac{\lambda_1(x)}{N} \\
&\quad + \exp[-\theta_{\text{opt}}^{(N)} N]\frac{\lambda_{-1}(x)}{N} + \left(1 - \frac{\lambda_1(x)}{N} - \frac{\lambda_{-1}(x)}{N}\right)\Bigg)\Bigg] \\
&= \theta_0 u - [\lambda_1(x)\exp[\theta_0] \\
&\quad + \lambda_{-1}(x)\exp[-\theta_0] - \lambda_1(x) - \lambda_{-1}(x)] \\
&= \theta_0 u - \sum_j \lambda_j(x)(\exp[\theta_0 e_j] - 1) \\
&= \sup_\theta \left\{\theta u - \sum_j \lambda_j(x)(\exp[\theta e_j] - 1)\right\} \qquad (24)
\end{aligned}$$

where the last line is just an algebraic identity and may be verified directly by differentiation. For $f \in C_{1,x_0}[0, T]$ define the functional

$$I(f) \equiv \int_0^T \sup_\theta \left\{\theta f'(t) - \sum_j \lambda_j(f(t))(\exp[\theta e_j] - 1)\right\} dt$$

Our arguments embodied in (23)–(24) would make this our candidate for the rate function of a continuous birth–death process. The following theorem makes this relationship more precise.

Theorem (Wentzell–Weiss)
If the rate functions $\lambda_j(r)$ are uniformly bounded, Lipshitz continuous, and uniformly positive, then for any $f \in C_{1,x_0}[0, T]$ we have

$$\overline{\lim_{\varepsilon \to 0^+}}\ \overline{\lim_{n \to \infty}} \frac{1}{n} \log P\left\{\sup_{0 \leq t \leq T} |\mathbf{x}_n(t) - f(t)| < \varepsilon\right\} \geq -I(f)$$

Define $\Upsilon(k) \equiv \{f \in C_{1,x_0}[0, T] : I(f) \leq k\}$. Then

$$\underline{\lim_{\varepsilon \to 0^+}}\ \underline{\lim_{n \to \infty}} \frac{1}{n} \log P\left\{\inf_{f \in \Upsilon(k)} \sup_{0 \leq t \leq T} |\mathbf{x}_n(t) - f(t)| \geq \varepsilon\right\} \leq -k$$

This theorem implies that the rate function is $I(f)$. (See Freidlin and Wentzell [1984] for a discussion on the equivalence of various rate functional definitions.) With little or no additional work we can consider a multidimensional (d-dimensional) birth–death process with the jump directions $e_j \equiv (0, \ldots, 0, 1, 0, \ldots, 0)$ and $e_{-j} \equiv (0, \ldots, 0, -1, 0, \ldots, 0)$, where all the elements of these vectors are zero except for the jth position ($j = 1, 2, \ldots, d$). The jump rates at $x \in \mathbb{R}^d$ in the e_j and e_{-j} directions are $\lambda_j(x)$ and $\lambda_{-j}(x)$, respectively. Let $f(t) \in \mathbb{R}^d (f(t) = (f_1(t), f_2(t), \ldots, f_d(t))$ such that $f_i \in C_{1,x_{0i}}[0, T]$, $i = 1, 2, \ldots, d$, where $x_0 = (x_{01}, x_{02}, \ldots, x_{0d})$ is the initial condition for the multidimensional birth–death process. Define

$$I(f) \equiv \int_0^T \sup_\theta \left\{\langle \theta, f'(t) \rangle - \sum_{j=1}^d [\lambda_j(f(t))(\exp[\langle \theta, e_j \rangle] - 1) + \lambda_{-j}(f(t))(\exp[\langle \theta, e_j \rangle] - 1)]\right\} dt$$

The last theorem holds with this rate function in this more general setting.

Again (as usual) the hard part of making use of these results is to perform the minimization of the rate function over the path set of interest. The class of the one-dimensional level crossing problems, however, can always be solved quite easily as we did in the preceding subsection. We will again solve for the large deviations of these types of sets but this time by a classical calculus of variations approach (see appendix G).

Consider the general problem of finding

$$\inf_{f \in A_{r,x}} \int_0^{\tau_{f,r}} I_f(f')\, dt$$

where

$$\tau_{f,r} \equiv \inf\{t : f(t) = r\}$$
$$A_{r,x} = \{f \subset C_x[0, \infty) : \tau_{f,r} < \infty\}$$

and

$$I_x(u) = \sup_\theta \{\theta u - \log M_x(\theta)\}$$

Euler's equation (see appendix G) is then

$$\frac{\partial}{\partial f} I_f(f') - \frac{d}{dt}\frac{\partial}{\partial f'} I_f(f') = 0$$

Since $I_f(f')$ does not depend on t directly, Euler's equation can be recast into the first order differential equation (see appendix G)

$$I_f(f') - f' \frac{\partial}{\partial f'} I_f(f') = c \equiv \text{constant}$$

The stopping endpoint $\tau_{f,r}$ is free. (The starting endpoint, of course, is fixed at $t = 0$.) The transversality condition then becomes

$$I_{f(\tau_{f,r})}(f'(\tau_{f,r})) - f'(\tau_{f,r}) \frac{\partial}{\partial f'} I_{f(\tau_{f,r})}(f'(\tau_{f,r})) = 0$$

The transversality condition implies that the constant c must equal zero. Hence

$$I_f(f') - f' \frac{\partial}{\partial f'} I_f(f') = 0 \tag{25}$$

Therefore,

$$\int_0^{\tau_{f,r}} I_f(f')\, dt = \int_0^{\tau_{f,r}} f' \frac{\partial}{\partial f'} I_f(f')\, dt$$

Now

$$I_f(f') = \sup_\theta \{\theta f' - \log M_f(\theta)\} = \theta^* f' - \log M_f(\theta^*)$$

which easily implies that

$$\frac{\partial}{\partial f'} I_f(f') = \theta^*(f) \equiv \theta^*$$

Equation (25) implies that

$$I_f(f') = \theta^* f'$$

or

$$\log M_f(\theta^*) = 0 \tag{26}$$

Hence

$$\inf_{f \in A_{r,x}} \int_0^{\tau_{f,r}} I_f(f')\,dt = \int_0^{\tau_{f,r}} \theta^* f'(t)\,dt = \int_0^{\tau_{f,r}} \theta^*(f(t)) f'(t)\,dt = \int_x^r \theta^*(f)\,df \tag{27}$$

where θ^* is the solution to (26).

Note that if we compare this case with the preceding subsection θ^* is playing the same role that $-E'(\cdot)$ was in the discrete-time case. The two variational problems are actually identical. The optimum trajectory could certainly have been found via the techniques derived and explained in the preceding subsection. The calculus of variations is, however, perhaps the more direct and powerful approach to use in these problems.

***Example* 3.** An M/M/1 queue consists of a Poisson arrival stream (at rate λ customers/second), a single server, and a waiting room for those customers queued up for service. The service time for each customer is an independent exponentially distributed random variable with mean value $1/\mu$. The queue is stable (in the sense of average queue length) if and only if $\lambda < \mu$. We are interested in the large deviations of the event that the queue length exceeds n people. We first scale so that the arrival rate is $n\lambda$ and service rate is $n\mu$ (this corresponds to speeding up time by a factor of n). The jumps are constrained to be $\pm 1/n$, which means that we are interested in the set of trajectories $A_{1,0}$ (reaching n for the unscaled queue is like reaching 1 for the scaled process). [The fact that the service rate must be zero when there are no customers in the system violates the rate continuity assumption of the theorem (just as in example 2). We are interested in a large excursion away from zero. We may as well just ask what is the probability that a trajectory in the scaled process reaches n before returning to zero given that we start out at some $\varepsilon > 0$. We can then let ε approach zero to obtain the desired rate result. We will not actually go through this formal procedure, but it is implicit in the justification for the following derivation.] Now

$$\log M_f(\theta) = \sum_j \lambda_j(f)(\exp[\theta e_i] - 1) = \lambda(\exp[\theta] - 1) + \mu(\exp[-\theta] - 1)$$

Hence the only nonzero solution to the equation $\log M_f(\theta^*) = 0$ is $\theta^* = \log(\mu/\lambda)$. Hence (27) becomes in this example,

$$I \equiv \inf_{f \in A_{1,0}} \int_0^{\tau_{f,1}} I_f(f')\,dt = \int_0^1 \theta^*\,df = \log\frac{\mu}{\lambda}[1 - 0] = \log\frac{\mu}{\lambda}$$

Therefore,

$$\lim_{n\to\infty} \frac{1}{n} \log P(\text{M/M/1 queue exceeds } n \text{ before returning to } 0) = -\log\frac{\mu}{\lambda}$$

It is easy to verify that the optimum trajectory is given by

$$f_{\text{opt}}(t) = (\mu - \lambda)t \qquad 0 \leqslant t \leqslant \frac{1}{\mu - \lambda} \equiv \tau_{f_{\text{opt}},1}$$

That is, the way the queue achieves a large level n is for it to build up linearly at the rate of $(\mu - \lambda)$ customers/second. See exercises 6–8 for ramifications and extensions of this observation. □

***Example* 4.** Suppose that n statistically independent sources feed one communication channel. The sources are modeled as two-state Markov processes: spending an exponentially (mean value $1/\lambda$) time off and exponentially (mean value $1/\mu$) time on. When "on" a source transmits at rate 1 message per second. The average number of transmitting sources is then

$$\mathbf{z}_n(t) \equiv \frac{1}{n}(\#\text{ of "on" sources})$$

which is a birth–death process with birth rate $\lambda(1 - \mathbf{z}_n(t))$, death rate $\mu\mathbf{z}_n(t)$, and jumps of $\pm(1/n)$. We assume that the channel has capacity nC, where $C < 1$. When more than nC sources are on, the excess traffic is shunted to a buffer. The buffer fills at rate $n(\mathbf{z}_n(t) - C)$, and when $\mathbf{z}_n(t)$ dips below level C, the buffer empties through the channel at the maximum possible rate of

$n(C - \mathbf{z}_n(t))$. Let $n\mathbf{b}(t)$ be the number of messages in buffer at time t. (We note that this example suffers from the same drawbacks in the properties of the process near zero as the previous examples. They can be taken care of in a similar fashion.)

Note that a source is "on" approximately $\lambda/(\lambda + \mu)$ amount of the time. For this system to be stable, then obviously $C > \lambda/(\lambda + \mu)$. Given that the system is stable, what are the asymptotics of the event $\{\mathbf{z}_n(t) > C\}$ (which turns out to have the same asymptotics as the event $\{b(t) > 0\}$)? Therefore, we wish to calculate the probability of a large deviation from the equilibrium of $\mathbf{z}_n(t)$ $[E\{\mathbf{z}_n(t)\} = \lambda/(\lambda + \mu) \equiv p]$. First we must calculate

$$I \equiv \inf_{f \in A_{C,p}} \int_0^{\tau_{f,C}} I_f(f')\, dt$$

Now

$$\log M_f(\theta) = \lambda(1 - x)(\exp[\theta] - 1) + \mu x(\exp[-\theta] - 1)$$

which means that the only nonzero solution to $\log M_r(\theta^*) = 0$ is

$$\theta^* = \log \frac{\mu x}{\lambda(1 - x)}$$

Hence (27) becomes in this example

$$I = \int_p^C \log \frac{\mu f}{\lambda(1 - f)}\, df = C \log \frac{C}{p} + (1 - C) \log \frac{1 - C}{1 - p}$$

and

$$\lim_{n \to \infty} \frac{1}{n} \log P(\mathbf{z}_n(t) > C) = -\left[C \log \frac{C}{p} + (1 - C) \log \frac{1 - C}{1 - p}\right] \qquad \square$$

Notes and Comments

It turns out that many of the ideas presented in section A such as the RKHS norm and the inverse kernel can be discussed without necessarily having the stringent conditions of the Karhunen–Loève (KL) expansion holding. The KL expansion does make many operations formally transparent that would appear only as abstract manipulations with a less direct (and more general) approach. See the textbook by Strook [1984] for a rigorous derivation along these lines. Schilder [1966] gives a more general result than what is quoted in the text. That paper gives full asymptotic expansions (somewhat as in chapter

VII) for these Wiener integrals. Pincus [1968] extends many of these ideas to more general Gaussian processes than the Wiener process.

Azencott [1980] gives a more general version of the first theorem in section B, where he relaxes the condition that $a(\cdot)$ be everywhere invertible. In later papers Azencott [1983, 1984] derives asymptotic expansions (as in chapter VII) for some of these probabilities. Freidlin and Wentzell [1984] extend their theorem in another direction by allowing an arbitrary Gaussian process to be the driver of the Itô equation instead of a simple Brownian motion. It is easy to see how our heuristic arguments could be modified to take care of this case since we already know the rate function for an arbitrary Gaussian process on the interval. In the book we explicitly consider only one-dimensional stochastic differential equations. Most of the foregoing references state their results more generally in $\mathbb{R}^d$.

For the second theorem in section B, the more general case when $\sigma(x) \neq 1$ is considered in Freidlin and Wentzell [1984]. This case is much more difficult and only partial results are available. The basic form of the theorem still holds but with different technical conditions. Kushner [1984] has also considered many extensions and applications of this result.

The theorems in section C were originally derived by Wentzell [1976a, 1976b]. Other versions of these results are given by Azencott and Ruget [1977] and Borokov [1967]. The heuristic derivation and optimization given in subsection 1 is due to Parekh and Walrand [1986]. The theorem stated in subsection 2 was stated and proved in that form by Weiss [1987]. However, the heuristic derivation given in the text is new. Examples 3 and 4 and exercises 6–8 are due to Weiss [1987]. More recently, Dupuis [1988] has extended the slow Markov walk model to allow the possibility of non-iid drivers for the $\{\xi_i\}$ in the defining expression of the walk. The work involves estimates of the large deviations by considering the limits of the moment-generating functions via a Gärtner–Ellis theorem type of formulation.

EXERCISES

1. Verify that $\int_0^T\int_0^T K^{-1}(s, t)(y(s)-m(s))(y(t)-m(t))\,ds\,dt = \sum_{k=0}^{\infty}(y_k - m_k)^2/\lambda_k$.

2. Give an example of a sequence of real valued functions f_n where:

(a) f_n is differentiable everywhere for every n and $\lim_{n\to\infty}\int_0^T(f_n(x) - f(x))^2\,dx = 0$, but $f(x)$ is not differentiable everywhere.

(b) f_n is not differentiable everywhere for any n and $\lim_{n\to\infty}\int_0^T(f_n(x) - f(x))^2\,dx = 0$, but f is differentiable everywhere.

(c) How do these two examples show that the set of all differentiable functions is neither open nor closed in $L_2[0, T]$?

3. Show that $\lim_{n\to\infty} P\{\sup_{0\leq t\leq 1} |\mathbf{x}^n(t) - \gamma(t)| > \varepsilon\} = 0$ for all $\varepsilon > 0$, where $\mathbf{x}^n(t)$ is the continuous-time interpolation of the slow Markov walk of section C and $\gamma(t)$ is the solution to the differential equation $\gamma'(t) = m(\gamma(t))$ and $\gamma(0) = x$.

4. Suppose that $f(x)$ is a real valued convex function finite everywhere on the real line. Let f^* be its convex dual, i.e., $f^*(\theta) \equiv \sup_x \{\theta x - f(x)\}$.. Show that $f^{**}(x) = f(x)$. Suppose for simplicity that $f(x)$ is differentiable and the derivative has a unique inverse function $f'^{-1}(\cdot)$.

5. Show that the derivative relationships are true:

$$\left.\frac{dI_x(u)}{du}\right|_{u = M_x'(\theta)/M_x(\theta)} \equiv I_x'([\log M_x(\theta)]') = \theta$$

and

$$\left.\frac{M_x'(\theta)}{M_x(\theta)}\right|_{\theta = I_x'(u)} \equiv [\log M_x(I_x'(u))]' = u$$

6. We know from example 3 that for an M/M/1 queue to build up to a high level, it does so in a linear fashion with $(\mu - \lambda)$ customers/second. There are many different ways this rate can be achieved; e.g., the arrival rate can be $(\mu - \lambda)$ and the service rate 0 for a while. Argue that this rate is actually achieved by the service and arrival rates interchanging.

7. Consider two M/M/1 queues fed by a single arrival stream with rate v. Each arrival generates one customer for each queue. The service rates in each queue are μ_1 and μ_2, respectively. Let $\mathbf{x}_i$ be the number of customers in queue i. Show that if $\mu_1 > \mu_2$, then

$$\lim_{n\to\infty} \frac{1}{n} E\{\mathbf{x}_2 | \mathbf{x}_1 = n\} = \frac{\mu_1 - \mu_2}{\mu_1 - v}$$

Show that if $\mu_1 < \mu_2$, then

$$\lim_{n\to\infty} E\{\mathbf{x}_2 | \mathbf{x}_1 = n\} = \frac{\mu_1}{\mu_2 - \mu_1}$$

8. Suppose that exercise 7 is changed by queue 1 receiving a Poisson (λ_1) stream in addition to the v stream and queue 2 receiving a Poisson (λ_2) stream in addition to the v stream. Develop the asymptotics of $E\{\mathbf{x}_2 | \mathbf{x}_1 = n\}$ for this situation.

CHAPTER V

Large Deviations for Markov Processes

In this chapter we consider large deviation theory in the setting of Markov processes. In section A our method is to invoke the Gärtner–Ellis theorem of chapter II directly to determine the rate function for the sum of a function of successive samples taken from a finite state space Markov chain. The main effort here is to compute the limit of the relevant moment generating function sequence. Section B is concerned with large deviations of the empirical distributions taken from a Markov chain. Our method here will be quite different and less direct. We first prove a large deviation asymptotic integral formula in which the rate function appears explicitly. We then invoke a result from convexity theory to determine what the form of the rate function must be.

A. FUNCTIONALS OF A FINITE STATE SPACE MARKOV PROCESS

In this section we consider only finite state space ($k < \infty$ elements) Markov chains. Functions on this state space may be thought of in two ways. They may be thought of as functions mapping from the finite state space into $\mathbb{R}$. Alternatively, the function values may be arranged into a vector with dimension k, the cardinality of the state space. Similarly, linear operators may be thought of as mapping functions of the state space into other functions on the state space or, alternatively, as $k \times k$ matrices operating on vectors. We use these equivalent viewpoints freely in the following derivation.

Let

$$p(y|x) \equiv P\{\mathbf{x}_{n+1} = y|\mathbf{x}_n = x\}$$

be the transition probabilities of a Markov chain with finite state space **S**. Let f be a fixed function that maps from **S** into $\mathbb{R}$. Let us suppose that the chain is irreducible (and hence ergodic). Then

$$\mathbf{S}_n \equiv \frac{1}{n} \sum_{i=1}^{n} f(\mathbf{x}_i)$$

converges to

$$E_\mu\{f\} \equiv \sum_{y \in \mathbf{S}} f(y)\mu(y)$$

where $\mu(\cdot)$ is the stationary probability vector (function) of the chain (see appendix D.5). In such a setting, one would certainly expect a large deviation principle to hold for $\mathbf{S}_n$. Let's try to deduce the form of the rate functional.

Define

$$\varphi_n(\theta) \equiv \frac{1}{n} \log E\{\exp[\theta n \mathbf{S}_n]\} \tag{1}$$

and hence

$$E\{\exp[\theta n \mathbf{S}_n]\} = \sum_{x_1} \sum_{x_2} \cdots \sum_{x_n} \exp\left[\theta \sum_{i=1}^{n} f(x_i)\right] p(x_1, x_2, \ldots, x_n) \tag{2}$$

where $p(x_1, x_2, \ldots, x_n) \equiv P\{\mathbf{x}_1 = x_1, \mathbf{x}_2 = x_2, \ldots, \mathbf{x}_n = x_n\}$ is the multidimensional probability function of the first n steps of the chain. This probability function (by the Markov property) can be written as

$$p(x_1, x_2, \ldots, x_n) = \left(\prod_{i=1}^{n-1} p(x_{i+1} | x_i)\right) p(x_1)$$

where $p(x_1) \equiv P\{\mathbf{x}_1 = x_1\}$ is the initial distribution of the starting value of the chain. We then have, from (1)–(2),

$$\exp[n\varphi_n(\theta)] = \sum_{x_1} \sum_{x_2} \cdots \sum_{x_n} \exp\left[\theta \sum_{i=1}^{n} f(x_i)\right]\left(\prod_{i=1}^{n-1} p(x_{i+1} | x_i)\right) p(x_1) \tag{3}$$

Let $\mathscr{G}$ be the collection of all functions mapping from S into $\mathbb{R}$, i.e., $\mathscr{G} \equiv \{g \,|\, g: S \to \mathbb{R}\}$. Define the linear operator T_θ, $T_\theta: \mathscr{G} \to \mathscr{G}$, by

$$T_\theta(g)(x) \equiv \sum_{y \in S} \exp[\theta f(y)] g(y) p(y \,|\, x) \tag{4}$$

We now would like to rewrite (3) in terms of this operator. Let us perform the summation over x_n first in (3). That summation is

$$\sum_{x_n} \exp[\theta f(x_n)] p(x_n \mid x_{n-1}) = T_\theta(1)(x_{n-1})$$

One then needs to perform the summation over x_{n-1}, which is seen to be

$$\sum_{x_{n-1}} \exp[\theta f(x_{n-1})] T_\theta(1)(x_{n-1}) p(x_{n-1} \mid x_{n-2}) = T_\theta(T_\theta(1))(x_{n-2}) \equiv T_\theta^2(1)(x_{n-2})$$

It is not hard to see by continuing this argument that (3) may be written as

$$\exp[n\varphi_n(\theta)] = \sum_{x_1} T_\theta^{n-1}(1)(x_1) \exp[\theta f(x_1)] p(x_1) \tag{5}$$

Now $T_\theta^{n-1}(1)(x)$ is the result of a particular iterated operator operating on the constant function 1. In this finite-dimensional setting, it corresponds exactly to multiplying the matrix representation of T_θ by itself $n-1$ times and then operating that result on the vector consisting of ones at every position. Define

$$\|T_\theta\|_s \equiv \sup_{g:\|g\| \leqslant 1} \|T_\theta(g)(\cdot)\|$$

where $\|\cdot\|_s$ is called the (strong) matrix norm and $\|\cdot\|$ denotes the usual Euclidean distance for k-dimensional space. (Recall that k is the number of states that the chain can attain.) For these types of positive operators (vectors with all components strictly positive are mapped into vectors of the same type), the Perron–Frobenius theorem (see appendix C) states that T_θ has a largest eigenvalue $\lambda(\theta)$ such that $\|T_\theta\|_s = \lambda(\theta)$ and associated unique (up to scaling) eigenvector ψ. This then implies that a first order expansion exists for the iterates of T_θ as follows:

$$T_\theta^{n-1}(g)(x) \approx c\lambda(\theta)^{n-1}\psi(x)$$

with other terms in the expansion being exponentially insignificant as n approaches infinity, and $c > 0$ depends on the function (vector) g. Therefore,

$$\begin{aligned} \lim_{n\to\infty} \varphi_n(\theta) &= \lim_{n\to\infty} \frac{1}{n} \log \sum_{x_1} c\lambda(\theta)^{n-1}\psi(x_1) \exp[\theta f(x_1)] p(x_1) \\ &= \log \lambda(\theta) + \lim_{n\to\infty} \frac{1}{n} \log \sum_{x_1} c\psi(x_1) \exp[\theta f(x_1)] p(x_1) \\ &= \log \lambda(\theta) \end{aligned} \tag{6}$$

Let us define

$$I(x) \equiv \sup_{\theta} \{\theta x - \log \lambda(\theta)\} \tag{7}$$

For this finite state space case, it is easy to verify that the domain (as a function of θ) of $\log \lambda(\theta)$ is all of $\mathbb{R}$, i.e., $\lambda(\theta) < \infty$ for all $\theta \in \mathbb{R}$. Hence $\log \lambda(\theta)$ is a closed, convex, and steep function with 0 in the interior of its domain. This means that the conditions of the Gärtner–Ellis theorem hold and we have the following.

Theorem (Miller)
Let $\{\mathbf{x}_n\}$ be an irreducible Markov chain with finite state space S. Let f be a function mapping S into $\mathbb{R}$. Then $S_n \equiv (1/n)\,\Sigma_{i=1}^{n} f(\mathbf{x}_i)$ obeys a large deviation principle with rate function $I(x)$ as in (7).

Let us define another operator, T_θ^*, $T_\theta^*\colon \mathscr{G} \to \mathscr{G}$, by

$$T_\theta^*(g)(x) \equiv \exp[\theta f(x)] \sum_y g(y) p(x \mid y)$$

This operator when viewed as a matrix is the transpose (or matrix adjoint) of T_θ. In (3), if we perform the summation over x_1 first, one must consider

$$\exp[\theta f(x_2)] \sum_{x_1} \exp[\theta f(x_1)] p(x_2 \mid x_1) p(x_1) = T_\theta^*(p)(x_2)$$

The x_2 summation then requires that we find

$$\exp[\theta f(x_3)] \sum_{x_2} T_\theta^*(p)(x_2) p(x_3 \mid x_2) = T_\theta^*(T_\theta^*(p))(x_3) \equiv T_\theta^{*2}(p)(x_3)$$

One may continue this process to obtain

$$\exp[n\varphi_n(\theta)] = \sum_{x_n} T_\theta^{*n-1}(p)(x_n) \tag{8}$$

A matrix and its adjoint share exactly the same set of eigenvalues with each other (but not eigenvectors; see exercise 1). It is then easy to verify that

$$\begin{aligned} \lim_{n\to\infty} \varphi_n(\theta) &= \lim_{n\to\infty} \frac{1}{n} \log \sum_{x_n} T_\theta^{*n-1}(p)(x_n) = \lim_{n\to\infty} \frac{1}{n} \log \sum_{x_n} c^* \lambda(\theta)^{n-1} \psi^*(x_n) \\ &= \log \lambda(\theta) \end{aligned} \tag{9}$$

which is the same result that we obtained previously, where ψ^* is the eigenvector associated with the largest positive eigenvalue of T_θ^* (and has no relationship, necessarily, to ψ defined above). There can be (depending on the application) computational advantages to employing one operator or the other; thus it is useful to be familiar with both formulations.

One can mimic these arguments for sums of the form

$$S_n \equiv \frac{1}{n} \sum_{i=1}^{n-1} f(x_{i+1}, x_i)$$

In this case we define $T_\theta, T_\theta: \mathscr{G} \to \mathscr{G}$, by

$$T_\theta(g)(x) \equiv \sum_y \exp[\theta f(x, y)] g(y) p(y \mid x)$$

and one may quickly verify that

$$\exp[n\varphi_n(\theta)] = \sum_{x_1} T_\theta^{n-1}(1)(x_1)p(x_1)$$

and

$$\lim_{n \to \infty} \varphi_n(\theta) = \log \lambda(\theta) \tag{10}$$

where $\lambda(\theta)$ is the largest positive eigenvalue of T_θ. The arguments from this point onward proceed in an identical fashion to those of the previous case. However, the methodology breaks down for functions that depend on three or more past values of the chain.

***Example* 1.** Let us compute the rate function of a two-state $\{0, 1\}$ Markov chain that stays where it is with probability $p \leqslant \frac{1}{2}$ and flips state with probability $1 - p \equiv q$. Let us define the function

$$f(x) \equiv 1 - x$$

where x is the state of the chain. Then, in order to find $\lambda(\theta)$, we must find the largest eigenvalue of the matrix operator T_θ:

$$\begin{bmatrix} \exp[\theta]p & q \\ \exp[\theta]q & p \end{bmatrix}$$

Therefore, expanding the determinant of the matrix minus λ above times the identity matrix gives the equation

$$\lambda^2 + \lambda(-p\exp[\theta] - p) + \exp[\theta](p^2 - q^2) = 0$$

The largest root of this equation is found via the quadratic formula to be

$$\lambda(\theta) = \tfrac{1}{2}[p(\exp[\theta] + 1) + \sqrt{p^2(\exp[\theta] + 1)^2 + 4\exp[\theta](1 - 2p)}]$$

One then can solve for $I(x) = \sup_\theta\{\theta x - \log \lambda(\theta)\}$ to obtain the rate function. □

By this simple example and the resulting messy algebraic expression, one can appreciate that it is usually desirable to automate the calculations of the largest eigenvalue by standard numerical techniques. Note that when $p = \frac{1}{2}$, we are back to summing independent Bernoulli random variables and $\lambda(\theta) = (\exp[\theta] + 1)/2$, which checks with example 2 of chapter II.

We can also handle Markov processes that evolve in continuous time by essentially the same techniques. Let $\{\mathbf{x}(t)\}$ be a continuous-time finite state space S Markov process. We would then wish to consider the large deviations of functionals of the form

$$\mathbf{S}_t \equiv \frac{1}{t}\int_0^t f(\mathbf{x}(s))\,ds$$

Let us determine heuristically the form of the rate function. We first want to calculate (for large t)

$$\varphi_t(\theta) \equiv \frac{1}{t}\log E\{\exp[\theta t \mathbf{S}_t]\} = \frac{1}{t}\log E\left\{\exp\left[\theta\int_0^t f(\mathbf{x}(s))\,ds\right]\right\} \tag{11}$$

Let Δ denote a small time interval. $\varphi_t(\theta)$ may be closely approximated by

$$\varphi_t(\theta) \approx \frac{1}{t}\log E\left\{\exp\left[\theta\sum_{i=1}^{[t/\Delta]} f(\mathbf{x}(i\Delta))\Delta\right]\right\} \tag{12}$$

From the work we just completed on discrete-time Markov chains, we know that (12) may be rewritten as

$$\varphi_t(\theta) \approx \frac{1}{t}\log\sum_x T_{\theta,\Delta}^{[t/\Delta]}(1)(x)\exp[\theta f(x)\Delta]P\{\mathbf{x}(\Delta) = x\} \tag{13}$$

where

$$T_{\theta,\Delta}(g)(x) \equiv \sum_{y} \exp[\theta f(y)\Delta] g(y) P\{\mathbf{x}(i\Delta) = y \mid \mathbf{x}((i-1)\Delta) = x\} \tag{14}$$

Let $\lambda(\theta, \Delta)$ denote the maximum eigenvalue of this operator. Then from the Perron–Frobenius theory, we have for large t

$$\begin{aligned}\varphi_t(\theta) &\approx \frac{1}{t} \log \sum_{x} T_{\theta,\Delta}^{[t/\Delta]}(1)(x) \exp[\theta f(x)\Delta] P\{\mathbf{x}(\Delta) = x\} \\ &\approx \frac{1}{t} \log \lambda^{[t/\Delta]}(\theta, \Delta) \approx \frac{1}{\Delta} \log \lambda(\theta, \Delta)\end{aligned}$$

Let us investigate this last quantity as Δ approaches zero. By considering (14), it is clear that, as Δ approaches zero, $\lambda(\theta, \Delta)$ approaches 1. Hence

$$\frac{1}{\Delta} \log \lambda(\theta, \Delta) \xrightarrow[\Delta \to 0]{} \left.\frac{d \log \lambda(\theta, \Delta)}{d\Delta}\right|_{\Delta=0} = \left.\frac{d\lambda(\theta, \Delta)}{d\Delta}\right|_{\Delta=0} \equiv \lambda(\theta) \tag{15}$$

Now consider the behavior of the operator $T_{\theta,\Delta}$ for small positive Δ:

$$\begin{aligned}T_{\theta,\Delta}(g)(x) &= \sum_{y} \exp[\theta f(y)\Delta] g(y) P\{\mathbf{x}(i\Delta) = y \mid \mathbf{x}((i-1)\Delta) = x\} \\ &= \sum_{y} (1 + \theta f(y)\Delta) g(y) P\{\mathbf{x}(i\Delta) = y \mid \mathbf{x}((i-1)\Delta) = x\} + O(\Delta^2) \\ &= g(x) + \Delta\theta g(x) f(x) + \Delta G(g + \theta f g \Delta)(x) + O(\Delta^2)\end{aligned} \tag{16}$$

where

$$G(g)(x) \equiv \lim_{\delta \to 0} \left[\frac{\Sigma_y\, g(y) P\{\mathbf{x}(i\delta) = y \mid \mathbf{x}((i-1)\delta) = x\} - g(x)}{\delta}\right]$$

$G(\cdot)$ is called the *infinitesmal generator* of the continuous-time Markov process. It plays a key role in the understanding of these types of probability models. Note that $G(\cdot)$ is a linear operator. From (16) and the linearity of $G(\cdot)$, we have

$$\begin{aligned}T_{\theta,\Delta}(g)(x) &= \sum_{y} \exp[\theta f(y)\Delta] g(y) P\{\mathbf{x}(i\Delta) = y \mid \mathbf{x}((i-1)\Delta) = x\} \\ &\approx g(x) + \Delta\theta g(x) f(x) + \Delta G(g)(x)\end{aligned} \tag{17}$$

where we have suppressed all terms involving squared and higher powers of Δ. Let $\psi_\Delta(x)$ denote the eigenvector that goes with the eigenvalue $\lambda(\theta, \Delta)$. It seems reasonable to assume that ψ_Δ would have a limiting value as $\Delta \to 0$. Let us define this limit as

$$\lim_{\Delta \to 0} \psi_\Delta(x) \equiv \psi(x)$$

Then we have from (15),

$$\frac{\lambda(\theta, \Delta)\psi_\Delta(x) - \psi_\Delta(x)}{\Delta} \xrightarrow[\Delta \to 0]{} \left.\frac{d\lambda(\theta, \Delta)}{d\Delta}\right|_{\Delta=0} \psi(x) = \lambda(\theta)\psi(x) \tag{18}$$

From (17), the eigenvalue/eigenfunction relation may be written (for small Δ) as

$$T_{\theta,\Delta}(\psi_\Delta)(x) = \lambda(\theta, \Delta)\psi_\Delta(x) = \psi_\Delta(x) + \Delta\theta\psi_\Delta(x)f(x) + \Delta G(\psi_\Delta)(x)$$

or

$$\theta\psi_\Delta(x)f(x) + G(\psi_\Delta)(x) = \frac{\lambda(\theta, \Delta)\psi_\Delta(x) - \psi_\Delta(x)}{\Delta} \tag{19}$$

Using (18), we may take the limit as Δ approaches zero in (19) to obtain

$$\theta f(x)\psi(x) + G(\psi)(x) = \lambda(\theta)\psi(x) \tag{20}$$

Equation (20) is the eigenvalue/eigenfunction problem that can be solved to obtain $\lambda(\theta)$. From (15) and (20), $\lambda(\theta)$ has the dual interpretation of being the largest eigenvalue of the operator $\theta f(\cdot) + G(\cdot)$ *and*

$$\lambda(\theta) = \left.\frac{d\lambda(\theta, \Delta)}{d\Delta}\right|_{\Delta=0}$$

***Example* 2.** Let $\{\mathbf{x}(t)\}$ be the "random telegraph"; i.e., $\mathbf{x}(t)$ remains at state x for an exponentially distributed (with mean value $1/\mu$) amount of time and then flips to state $1-x$, $x = 0, 1$. By the memoryless property of the exponential distribution (see exercise 2), it is easy to verify that this is a Markov process. Let us compute the infinitesmal generator and large deviation rate function for this random process. One may derive the transition probabilities for the sampled process to be

$$P\{\mathbf{x}(i\Delta) = y | \mathbf{x}((i-1)\Delta) = x\} = \begin{cases} \exp[-\Delta\mu] + o(\Delta) & \text{if } y = x \\ 1 - \exp[-\Delta\mu] + o(\Delta) & \text{if } y = 1 - x \end{cases}$$

Therefore,

$$\frac{\Sigma_y g(y)P\{\mathbf{x}(i\Delta) = y \mid \mathbf{x}((i-1)\Delta) = x\} - g(x)}{\Delta}$$

$$= \frac{g(x)\exp[-\Delta\mu] + g(1-x)(1-\exp[-\Delta\mu]) - g(x) + o(\Delta)}{\Delta}$$

$$= \frac{g(x)(-\Delta\mu) + g(1-x)(\Delta\mu) + o(\Delta)}{\Delta}$$

Now we take the limit as Δ goes to zero to find that

$$G(g)(x) = [g(1-x) - g(x)]\mu$$

is the infinitesmal generator of the process. Suppose that we are interested in the asymptotic behavior of $(1/t)\int_0^t \mathbf{x}(s)\,ds$. We search for the solution to the following eigenvalue problem:

$$\theta x\psi(x) + [\psi(1-x) - \psi(x)]\mu = \lambda(\theta)\psi(x)$$

We may choose $\psi(0) = 1$ since eigenfunctions may be scaled arbitrarily. The two values of x will give us two more equations. We thus have three equations for the three quantities $\psi(0)$, $\psi(1)$, and $\lambda(\theta)$. One may check that

$$\psi(1) = \frac{\theta/\mu + \sqrt{(\theta/\mu)^2 + 4}}{2}$$

and

$$\lambda(\theta) = \frac{\theta}{2} + \sqrt{\left(\frac{\theta}{2}\right)^2 + \mu^2} - \mu$$

Entering this result into the expression for the rate function, we obtain

$$I(x) = \begin{cases} (2x-1)r(x)^{1/2} + (r(x) + \mu^2)^{1/2} - \mu & \text{for } 0 \leqslant x \leqslant 1 \\ \infty & \text{otherwise} \end{cases}$$

where

$$r(x) \equiv \mu^2 \frac{(1-2x)^2}{1-(1-2x)^2} \qquad \square$$

B. RATE FUNCTIONS FOR THE EMPIRICAL DISTRIBUTIONS

Knowing how the asymptotics of functionals of the Markov chain behave is enough for most of the applications considered in this book. However, much work has been devoted to developing large deviation results for the empirical distributions of Markov processes. As in the Sanov theory, one develops rate functions whose arguments are probability distributions in the space of probability measures topologized by convergence in distribution. In this section of the chapter we present a very simple formulation of some of the key ideas for the finite state space case. Our formulation will follow a heuristic discussion by Strook [1984], who discussed some of the key theorems first presented by Donsker and Varadhan [1975a, 1975b, 1976]. We first prove a very general large deviation result due to Varadhan [1984].

Theorem (Varadhan)
Let $\{P_n\}$ be a sequence of probability distributions on a metric space X that satisfy a large deviation principle with rate function $I(\cdot)$. Suppose that the level sets of $I(\cdot)$ are compact, i.e., $\{x \in X : I(x) \leqslant \alpha\}$ is a compact set for all $\alpha \in \mathbb{R}$. Then for any real bounded continuous function $F(x)$ on X, we have

$$\lim_{n \to \infty} \frac{1}{n} \log \left[\int_X \exp[nF(x)]\, dP_n(x) \right] = \sup_{x \in X} [F(x) - I(x)] \tag{21}$$

Proof: Upper bound. Given a $\delta > 0$, we can find a finite number of closed sets $C_1, C_2, \ldots, C_m$ that cover X such that $F(\cdot)$ does not vary by more than δ on each one of these sets. To see this, just divide the abcissa up into a series of closed intervals of length δ. You will need only a finite number of these intervals since $F(\cdot)$ is bounded. The inverse image of these intervals under $F(\cdot)$ will produce the C_i's, since the inverse image of a closed set is closed under a continuous map. Then

$$\begin{aligned} \int_X \exp[nF(x)]\, dP_n &\leqslant \sum_{i=1}^{m} \int_{C_i} \exp[nF(x)]\, dP_n \\ &\leqslant \sum_{i=1}^{m} \int_{C_i} \exp[n(F_i + \delta)]\, dP_n = \sum_{i=1}^{m} \exp[n(F_i + \delta)] P_n(C_i) \end{aligned}$$

where F_i is the minimum of $F(x)$ on C_i. Therefore, since $\overline{\lim}_{n \to \infty} (1/n) \log P_n(C_i) \leqslant -\inf_{x \in C_i} I(x)$, we have

$$\overline{\lim_{n\to\infty}} \frac{1}{n} \log\left[\int_X \exp[nF(x)]\, dP_n(x)\right] \leqslant \sup_{1\leqslant i\leqslant m}\left[(F_i+\delta) - \inf_{x\in C_i} I(x)\right]$$
$$\leqslant \sup_{1\leqslant i\leqslant m}\ \sup_{x\in C_i}\,[F(x)-I(x)]+\delta$$
$$= \sup_x\,[F(x)-I(x)]+\delta$$

Letting δ approach zero (since it is arbitrary) finishes the proof of the upper bound.

Lower bound. Given a $\delta > 0$, there exists a point $y \in X$ such that

$$F(y) - I(y) \geqslant \sup_x\,[F(x) - I(x)] - \frac{\delta}{2}$$

By continuity, we can find an open set U containing y such that $F(x) \geqslant F(y) - \delta/2$ for all $x \in U$. Therefore, we have

$$\underline{\lim}_{n\to\infty} \frac{1}{n} \log\left[\int_X \exp[nF(x)]\, dP_n(x)\right]$$
$$\geqslant \underline{\lim}_{n\to\infty} \frac{1}{n} \log\left[\int_U \exp[nF(x)]\, dP_n(x)\right] \geqslant F(y) - \frac{\delta}{2} - \inf_{x\in U} I(x)$$
$$\geqslant F(y) - I(y) - \frac{\delta}{2} \geqslant \sup_x\,[F(x) - I(x)] - \delta$$

Again, since δ is arbitrary we are finished. □

We now would like to use this theorem to guess the form of the rate function for the empirical distributions of a Markov chain. We first need to define the quantities of interest.

$$\mathbf{L}_n \equiv \frac{1}{n}\sum_{i=1}^{n} \delta_{\mathbf{x}_i}$$

where $\{\mathbf{x}_i\}$ is an irreducible, finite state space S Markov chain with stationary transition probabilities

$$p(y\,|\,x) \equiv P\{\mathbf{x}_{i+1} = y\,|\,\mathbf{x}_i = x\}$$

For every value of n, $\mathbf{L}_n$ is a probability distribution. We would expect that $\mathbf{L}_n$ should be converging (in distribution) to μ, the stationary probability distribution of the chain. We in fact expect this convergence to be exponentially fast in the space of probability measures Ψ with the Prohorov metric as the distance measure. Let us heuristically derive the form of the rate function.

Let $F(\cdot)$ be a bounded continuous function mapping from the space of probability measures Ψ into $\mathbb{R}$. Then from (21) we would have

$$\lim_{n\to\infty} \frac{1}{n} \log E\{\exp[nF(\mathbf{L}_n)]\} = \sup_{\nu\in\Psi} [F(\nu) - I(\nu)]$$

A class of bounded continuous $F(\cdot)$ mapping from Ψ into $\mathbb{R}$ may be defined by

$$F(\nu) \equiv \int f(y)\, d\nu(y)$$

where $f(\cdot)$ is a bounded continuous function mapping $\mathbb{R}$ into $\mathbb{R}$. In other words, $F(\nu)$ gives the expected value of f with respect to the probability distribution ν. In this case (21) would have the form

$$\lambda(f) \equiv \lim_{n\to\infty} \frac{1}{n} \log E\left\{\exp\left[n \int f(y)\, d\mathbf{L}_n(y)\right]\right\} = \sup_{\nu\in\Psi}\left[\int f(y)\, d\nu(y) - I(\nu)\right] \quad (22)$$

$\lambda(f)$ is then the convex conjugate of $I(\nu)$, the rate function for which we are searching. Now, if $I(\cdot)$ is a closed proper convex function, it is conversely true that $I(\nu)$ is the convex conjugate of $\lambda(f)$. (See appendix B.3 or exercise 3 of chapter IV for a simpler case.) In that case

$$I(\nu) = \sup_{f\in C_b}\left(\int f(y)\, d\nu(y) - \lambda(f)\right) \quad (23)$$

where C_b is the set of all bounded continuous functions.

Let us see if we can find an alternative form for $\lambda(f)$. Define the operator

$$T_f(g)(x) \equiv \exp[f(x)] \sum_y g(y) P\{\mathbf{x}_i = y \mid \mathbf{x}_{i-1} = x\}$$

Note that this operator is different from (but related to) the other operators defined in section A. Consider

$$T_f^2(g)(x) \equiv T_f(T_f(g))(x)$$

$$= \exp[f(x)] \sum_y \left(\exp[f(y)] \sum_{y'} g(y') P\{\mathbf{x}_i = y' \mid \mathbf{x}_{i-1} = y\}\right) P\{\mathbf{x}_i = y \mid \mathbf{x}_{i-1} = x\}$$

$$= \exp[f(x)] \sum_{y} \left(\exp[f(y)] \sum_{y'} g(y') P\{\mathbf{x}_3 = y' \mid \mathbf{x}_2 = y\} \right) P\{\mathbf{x}_2 = y \mid \mathbf{x}_1 = x\}$$

$$= \sum_{y} \sum_{y'} \exp[(f(x) + f(y)] g(y') P\{\mathbf{x}_3 = y' \mid \mathbf{x}_2 = y\} P\{\mathbf{x}_2 = y \mid \mathbf{x}_1 = x\}$$

$P\{\mathbf{x}_3 = y' \mid \mathbf{x}_2 = y\} P\{\mathbf{x}_2 = y \mid \mathbf{x}_1 = x\}$ is the joint probability distribution of the first three samples of the chain given that $\{\mathbf{x}_1 = x\}$. Hence, letting $E_x\{\cdot\}$ denote the conditional expectation operator given that $\{\mathbf{x}_1 = x\}$, we have

$$T_f^2(g)(x) = E_x \left\{ \exp\left[\sum_{i=1}^{2} f(\mathbf{x}_i) \right] g(\mathbf{x}_3) \right\}$$

Containing this argument, we may show that, in general,

$$T_f^n(g)(x) = E_x \left\{ \exp\left[\sum_{i=1}^{n} f(\mathbf{x}_i) \right] g(\mathbf{x}_{n+1}) \right\}$$

Therefore,

$$\lambda(f) \equiv \lim_{n \to \infty} \frac{1}{n} \log E \left\{ \exp\left[n \int f(y) \, d\mathbf{L}_n(y) \right] \right\}$$

$$= \lim_{n \to \infty} \frac{1}{n} \log E \left\{ \exp\left[\sum_{i=1}^{n} f(\mathbf{x}_i) \right] \right\} = \lim_{n \to \infty} \frac{1}{n} \log \sum_{x \in S} T_f^n(1)(x) p(x) \tag{24}$$

where $p(x)$ is the initial distribution chosen for $\mathbf{x}_1$. In a fashion similar to the arguments presented in section A, we now invoke the Perron–Frobenius theorem to assert that $\lambda(f) = \log \gamma(f)$, where $\gamma(f)$ is the largest eigenvalue of T_f. Let u_f denote the eigenfunction (eigenvector) associated with this eigenvalue. Then

$$f - \lambda(f) = f - \log \gamma(f) = f - \log \frac{T_f(u_f)}{u_f} \tag{25}$$

Now define the *transition operator* of the chain as

$$T(g)(x) \equiv \sum_{y} g(y) P\{\mathbf{x}_i = y \mid \mathbf{x}_{i-1} = x\} = \exp[-f(x)] T_f(g)(x)$$

Therefore, (25) becomes

$$f - \lambda(f) = -\log \frac{T(u_f)}{u_f} \tag{26}$$

Then note that, from (23),

$$
\begin{aligned}
I(v) &= \sup_{f\in C_b}\left(\int f(y)\,dv(y) - \lambda(f)\right) \\
&= \sup_{f\in C_b}\left(\int f(y)\,dv(y) - \int \lambda(f)\,dv(y)\right) \\
&= \sup_{f\in C_b}\left(\int (f(y) - \lambda(f))\,dv(y)\right) \\
&= \sup_{f\in C_b}\left(\int -\log\frac{T(u_f)(y)}{u_f(y)}\,dv(y)\right) \\
&= -\inf_{f\in C_b}\int \log\frac{T(u_f)(y)}{u_f(y)}\,dv(y) \qquad (27)
\end{aligned}
$$

Now consider a positive function $u \in C_b$. Define the function

$$
f_u \equiv -\log\frac{T(u)}{u}
$$

Consider

$$
\begin{aligned}
T_{f_u}(u)(x) &= \exp[f_u(x)]\sum_y u(y)P\{\mathbf{x}_i = y \mid \mathbf{x}_{i-1} = x\} \\
&= \frac{u(x)}{T(u)(x)}\,T(u)(x) = u(x)
\end{aligned}
$$

Therefore, u is an eigenfunction of T_{f_u} with eigenvalue 1. Since there is only one positive eigenvalue, we have that $u = u_{f_u}$ in our notation for the eigenfunctions of the operators under consideration. Hence as $f \in C_b$ is varied, we will sweep out every possible positive function $u \in C_b$. Since the eigenfunctions of these operators are nonnegative, the converse holds also. Therefore, we may write

$$
I(v) = -\inf_{u\in C_b \text{ and } u>0}\left[\int \log\frac{T(u)(y)}{u(y)}\right]dv(y) \qquad (28)
$$

Hence, if a large deviation principle does hold, this is our candidate for the rate function. A simple form of a more general available result is given immediately below.

Theorem (Donsker–Varadhan)
Let $\{\mathbf{L}_n\}$ be the empirical distributions of an irreducible finite state space Markov chain. Then $\mathbf{L}_n$ obeys a large deviation principle with rate function $I(\cdot)$ given in (28).

Example **3.** Let us consider the degenerate Markov case of iid random variables with probability function α as our Markov chain. We desire to compute the form of the rate function in this simple setting. This special case should reduce to the results of Sanov's theorem of chapter III. Let $u = \exp[f]$; then

$$T(u)(x) = T(\exp[f])(x) = \sum_y \exp[f(y)]P\{\mathbf{x}_i = y \mid \mathbf{x}_{i-1} = x\} = \sum_y \exp[f(y)]\alpha(y)$$

Hence

$$\begin{aligned} I(v) &= -\inf_{f\in C_b}\left[\log\sum_z \exp[f(z)]\alpha(z) - \int \log[\exp[f(y)]\,dv(y)\right] \\ &= \sup_{f\in C_b}\left[\int f(y)\,dv(y) - \log\sum_z \exp[f(z)]\alpha(z)\right] \\ &= \sup_{f\in C_b}\left[\int f(y)\,dv(y) - \log E\{\exp[f(\mathbf{x}_1)]\}\right] \end{aligned}$$

This case should match up exactly with the rate function given by Sanov's theorem. We leave it to exercise 3 to show that the rate function above is indeed equal to the differential entropy that was derived in chapter III. □

Notes and Comments
Miller [1961] is usually credited as having done the first work on deriving large deviation results for finite-dimensional Markov chains. However, he quotes previous results in the Soviet literature. The technical conditions quoted in the text are actually looser (and the theorem weaker) than that actually appearing in Miller [1961]. Subject to various technical conditions (by no means trivial), Iscoe et al. [1985] gave results for additive $\mathbb{R}^d$ valued functionals of discrete- and continuous-time Markov processes. Later, Ney and Nummelin [1987a, 1987b] relaxed various technical conditions and provided more general results for the Markov additive processes but again for $\mathbb{R}^d$ valued functionals of the process. de Acosta [1985] gave upper bounds only, but in the more general setting of functionals of a Banach space valued Markov chain. More recently, de Acosta [1988] gives lower bounds for some very general Markov chains on general topological spaces.

The material in section B is essentially a simplified version of the work of

Donsker and Varadhan [1975a, 1975b, 1976] in a now classic series of papers. In principle, their methodology allows one to handle more general continuous functionals of abstract state space Markov chains than the $\mathbb{R}^d$ valued ones of the section A theory. The computational aspects, however, of using entropic formulations are in general (in the author's opinion) more difficult.

EXERCISES

1. Consider the matrix $\begin{bmatrix} 1 & c \\ 1 & 0 \end{bmatrix}$, $c > 0$. What is its largest eigenvalue and an associated eigenvector? Compute the same quantities for its adjoint.

2. Show that the "random telegraph" is a Markov process.

3. Show that the rate function calculated in example 3 is the same as in Sanov's theorem.

4. Let F denote the set of all bivariate distributions $f(\cdot, \cdot)$ such that $\Sigma_y f(x, y) = \Sigma_x f(x, y)$. For a given Markov transition function $p(x \mid y)$ define $I(f, p) \equiv \Sigma_x \Sigma_y \log[f(x \mid y)/p(x \mid y)] f(x, y)$. Furthermore define F_a as the subset of F such that $\Sigma_x \Sigma_y f(x, y) a(x, y) \geqslant 0$. Consider the probability $P_n \equiv P\{(1/n) \Sigma_{i=1}^n a(\mathbf{x}_i, \mathbf{x}_{i-1}) > 0\}$, where $\{\mathbf{x}_i\}$ is the realization of a Markov chain with transition probability $p(x \mid y)$. Find the asymptotics of this probability in terms of $I(f, p)$.

5. Consider the following discrete-time finite state space Markov chain: At time n you flip a coin with success probability p; if successful, you choose an independent sample from some distribution $v(\cdot)$ defined on the state space. At time $n + 1$ you then move to that value. If the coin flip is unsuccessful, at time $n + 1$, you stay where you are. Consider, for a given $f(\cdot)$, computing the asymptotics of $P((1/n) \Sigma_{i=1}^n f(\mathbf{x}_i) > 0)$. What is the eigenvalue problem for this particular case? Solve it as far as possible and indicate how you would automate the final calculations on a computer.

6. Repeat exercise 5 but in continuous time. Here you wait an exponentially distributed (mean value μ^{-1}) amount of time, then choose a new state from the distribution $v(\cdot)$. The exponential is the continuous analog of the geometric waiting time of exercise 5.

7. Consider an arbitrary Markov jump process. Here you wait an exponential amount of time in one state x, but the jump distribution $v(\cdot, x)$ for the next state is state dependent. What is the infinitesmal generator of this process? What is the associated eigenvalue problem when computing the asymptotic probability of a functional of the process?

CHAPTER VI

Applications to Detection Theory

A. ASYMPTOTIC THEORY OF NEYMAN–PEARSON TESTS

In many hypothesis testing situations, the prior probabilities of the hypotheses being true are unknown. The Neyman–Pearson formulation allows one to set the false-alarm rate (size or α) and maximize the detection rate (power or β) (see appendix E). A likelihood ratio test (by the Neyman–Pearson lemma) is the optimal detector in this setting. It is the same mathematical formulation if alternatively, we set the detection rate and minimize the false alarm rate. In this section we consider the asymptotics of the false-alarm rate if the detection rate is fixed to the value $0 < \beta < 1$.

Consider the hypothesis testing problem

$$H_0: \text{ iid data with probability density } q(x)$$

$$H_1: \text{ iid data with probability density } p(x)$$

where for simplicity we assume that densities $q(\cdot)$ and $p(\cdot)$ exist. Let $0 < \beta < 1$ be the fixed detection probability [P(decide $H_1 \mid H_1$ is true)] of the test. The Neyman–Pearson lemma gives that the optimal test to minimize the false-alarm rate is

$$\frac{1}{n}\sum_{i=1}^{n} \log \frac{p(\mathbf{x}_i)}{q(\mathbf{x}_i)} \underset{H_0}{\overset{H_1}{\gtrless}} u_n \tag{1}$$

where u_n is chosen so that detection probability β is maintained. (The notation means decide H_1 if test is greater than u_n; otherwise, decide H_0.) Let $P_i\{A\}$ denote the probability of the set A given that hypothesis i ($i \in \{0, 1\}$) is true. Hence we have that

$$P_1\left(\frac{1}{n}\sum_{i=1}^{n} \log \frac{p(\mathbf{x}_i)}{q(\mathbf{x}_i)} > u_n\right) = \beta \tag{2}$$

However, under H_1 being true,

$$\frac{1}{n}\sum_{i=1}^{n}\log\frac{p(\mathbf{x}_i)}{q(\mathbf{x}_i)} \tag{3}$$

is converging by the strong law of large numbers to

$$K(q, p) \equiv \int \log\frac{p(x)}{q(x)}\, p(x)\, dx \tag{4}$$

the Sanov theorem rate function (sometimes called the divergence or the Kullback–Leibler information number). Since $0 < \beta < 1$, (2)–(4) imply that u_n must be converging to $K(q, p)$ also.

Now consider the false-alarm rate of the test as a function of n, i.e.,

$$\alpha_n \equiv P_0\left(\frac{1}{n}\sum_{i=1}^{n}\log\frac{p(\mathbf{x}_i)}{q(\mathbf{x}_i)} < u_n\right) = P_0\left(\frac{1}{n}\sum_{i=1}^{n}\left(\log\frac{p(\mathbf{x}_i)}{q(\mathbf{x}_i)} - u_n\right) < 0\right)$$

To uncover the large deviation asymptotics of α_n, we propose to use the Gärtner–Ellis theorem.

Consider

$$\varphi_n(\theta) \equiv \frac{1}{n}\log E_0\left\{\exp\left[\theta\left[\sum_{i=1}^{n}\left(\log\frac{p(\mathbf{x}_i)}{q(\mathbf{x}_i)} - u_n\right)\right]\right]\right\} \tag{5}$$

where $E_i\{\cdot\}$ denote the expectation operator assuming that hypothesis H_i is true. Then

$$\begin{aligned}\exp[n\varphi_n(\theta)] &= \exp[-u_n n\theta]E_0\left\{\prod_{i=1}^{n}\left[\frac{p(\mathbf{x}_i)}{q(\mathbf{x}_i)}\right]^{\theta}\right\}\\ &= \exp[-u_n n\theta]\int\left(\prod_{i=1}^{n}p(x_i)\right)^{\theta}\left(\prod_{i=1}^{n}q(x_i)\right)^{1-\theta} dx_1\, dx_2\cdots dx_n\\ &= \exp[-u_n n\theta]\left[\int p(x)^{\theta}q(x)^{1-\theta}\, dx\right]^n\end{aligned}$$

Hence

$$\begin{aligned}\varphi(\theta) &\equiv \lim_{n\to\infty}\varphi_n(\theta)\\ &= \lim_{n\to\infty}\frac{1}{n}\log E_0\left\{\exp\left[\theta\left[\sum_{i=1}^{n}\log\frac{p(\mathbf{x}_i)}{q(\mathbf{x}_i)} - u_n\right]\right]\right\}\end{aligned}$$

$$= \lim_{n\to\infty}\left[-u_n\theta + \log\int p(x)^\theta q(x)^{1-\theta}\,dx\right]$$

$$= -K(q, p)\theta + \log\int p(x)^\theta q(x)^{1-\theta}\,dx \tag{6}$$

Therefore, invoking the Gärtner–Ellis theorem, we have, from (6),

$$\lim_{n\to\infty}\frac{1}{n}\log\alpha_n = -\sup_\theta\left\{K(q, p)\theta - \log\int p(x)^\theta q(x)^{1-\theta}\,dx\right\} \tag{7}$$

The optimum θ can be found by differentiation or

$$\begin{aligned}K(p, q) &= \frac{\partial}{\partial\theta}\log\int p(x)^\theta q(x)^{1-\theta}\,dx \\ &= \int p(x)^\theta q(x)^{1-\theta}[\log p(x) - \log q(x)]\,dx \bigg/ \int p(x)^\theta q(x)^{1-\theta}\,dx \\ &= \int p(x)^\theta q(x)^{1-\theta}\left[\log\frac{p(x)}{q(x)}\right]dx \bigg/ \int p(x)^\theta q(x)^{1-\theta}\,dx\end{aligned}$$

Obviously, when $\theta = 1$, we have a solution. By a convexity argument, the solution will be unique. Hence $\theta_{\text{opt}} = 1$.

Therefore, from (7) and $\theta_{\text{opt}} = 1$, we have the following theorem.

Theorem (Stein's Lemma)
Fix the detection probability of a Neyman–Pearson detector operating on iid data to be $\beta, 0 < \beta < 1$. Then the exponential rate of the probability of false alarm as a function of sample size is given by

$$\lim_{n\to\infty}\frac{1}{n}\log\alpha_n = -K(q, p)$$

Note that this result is independent of β as long as β is not zero or 1.

B. ASYMPTOTIC DISCERNIBILITY OF TWO MARKOV CHAINS

Chernoff's theorem (see example 1 of chapter III) on the asymptotic discernibility of two iid sequences is credited with being one of the oldest large deviation theorems. Quite a bit of work has been expended to extend this result to more general random processes. In this section we discuss the problem of discriminating between two finite state space Markov chains.

Consider the testing problem between two equally likely hypotheses:

$$H_0:\quad q(\cdot\,|\,\cdot), q(\cdot)$$
$$H_1:\quad p(\cdot\,|\,\cdot), p(\cdot)$$

where $p(\cdot\,|\,\cdot)$, $p(\cdot)$ and $q(\cdot\,|\,\cdot)$, $q(\cdot)$ are the transition probabilities and initial distributions of the Markov chains under hypothesis H_1 and H_0, respectively. On the basis of n observed samples $\{\mathbf{x}_1, \mathbf{x}_2, \ldots, \mathbf{x}_n\} \equiv \mathbf{X}_n$, we need to decide whether hypothesis H_0 or H_1 is true. To minimize the probability of error, the likelihood ratio test should again be utilized. The test is

$$t_n(\mathbf{X}_n) \underset{H_0}{\overset{H_1}{\gtrless}} 0$$

where

$$t_n(\mathbf{X}_n) \equiv \log \frac{p(\mathbf{x}_1, \mathbf{x}_2, \ldots, \mathbf{x}_n)}{q(\mathbf{x}_1, \mathbf{x}_2, \ldots, \mathbf{x}_n)} = \log \frac{p(\mathbf{x}_1) \prod_{i=1}^{n-1} p(\mathbf{x}_{i+1}\,|\,\mathbf{x}_i)}{q(\mathbf{x}_1) \prod_{i=1}^{n-1} q(\mathbf{x}_{i+1}\,|\,\mathbf{x}_i)}$$
$$= \log \frac{p(\mathbf{x}_1)}{q(\mathbf{x}_1)} + \sum_{i=1}^{n-1} \log \frac{p(\mathbf{x}_{i+1}\,|\,\mathbf{x}_i)}{q(\mathbf{x}_{i+1}\,|\,\mathbf{x}_i)} \tag{8}$$

When $t_n(\mathbf{X}_n) = 0$, one may announce either H_1 or H_0 and not effect the overall probability of error. As long as $p(\mathbf{x}_1)/q(\mathbf{x}_1)$ is bounded, it is obvious that $t_n(\mathbf{X}_n)$ is asymptotic to

$$\bar{t}_n(\mathbf{X}_n) \equiv \sum_{i=1}^{n-1} \log \frac{p(\mathbf{x}_{i+1}\,|\,\mathbf{x}_i)}{q(\mathbf{x}_{i+1}\,|\,\mathbf{x}_i)} \tag{9}$$

We may now compute the type 1 error rate as

$$\lim_{n\to\infty} \frac{1}{n} \log P(\bar{t}_n(\mathbf{X}_n) < 0\,|\,H_1 \text{ is true})$$
$$= \lim_{n\to\infty} \frac{1}{n} \log P\left(\frac{1}{n-1} \sum_{i=1}^{n-1} \log \frac{p(\mathbf{x}_{i+1}\,|\,\mathbf{x}_i)}{q(\mathbf{x}_{i+1}\,|\,\mathbf{x}_i)} < 0\,|\,H_1 \text{ is true}\right)$$
$$= -I_1(x)|_{x=0} = -I_1(0) \tag{10}$$

where

$$I_1(x) \equiv \sup_{\theta} \{\theta x - \log \lambda_1(\theta)\} \tag{11}$$

[here $I_1(0) = -\inf_\theta \log \lambda_1(\theta)$] and $\lambda_1(\theta)$ is the largest eigenvalue of the operator

$$T_{\theta,1}(g)(y) \equiv \sum_x \exp\left[\theta \log \frac{p(x \mid y)}{q(x \mid y)}\right] g(x) p(x \mid y) = \sum_x q(x \mid y)^{-\theta} p(x \mid y)^{1+\theta} g(x)$$

For the type II error rate, we may calculate in an identical fashion

$$\lim_{n\to\infty} \frac{1}{n} \log P(\bar{t}_n(\mathbf{X}_n) > 0 \mid H_0 \text{ is true}) = \inf_\theta \log \lambda_0(\theta) \tag{12}$$

where $\lambda_0(\theta)$ is the largest eigenvalue of the operator

$$\begin{aligned} T_{\theta,0}(g)(y) &\equiv \sum_x \exp\left[\theta \log \frac{p(x \mid y)}{q(x \mid y)}\right] g(x) q(x \mid y) \\ &= \sum_x p(x \mid y)^\theta q(x \mid y)^{1-\theta} g(x) = T_{\theta-1,1}(g)(y) \end{aligned}$$

Therefore,

$$\inf_\theta \lambda_0(\theta) = \inf_\theta \lambda_1(\theta) \tag{13}$$

or the asymptotic rates of the type I and type II errors are equal—which, intuitively, should be true for a minimum probability of error detector. Hence from (10)–(13), we have

$$\lim_{n\to\infty} \frac{1}{n} \log P(\text{error}) = \inf_\theta \log \lambda_0(\theta) = \inf_\theta \log \lambda_1(\theta) \tag{14}$$

Note that when the Markov chains are degenerate iid sequences, $q(x \mid y) = q(x)$ and $p(x \mid y) = p(x)$. Then

$$T_{\theta,1}(g)(y) = \sum_x q(x)^{-\theta} p(x)^{1+\theta} g(x)$$

is a constant function independent of y. Therefore, the eigenvector can be taken to be the vector of all 1's and

$$\lambda_1(\theta) = \sum_x q(x)^{-\theta} p(x)^{1+\theta}$$

It is simple to show that the infimum of $\lambda_1(\theta)$ is attained in the range

$\theta \in [-1, 0]$ (see exercise 5 of chapter III), so we obtain the finite state space case of Chernoff's theorem as a special case of the above.

C. MEMORYLESS DETECTION THEORY

Consider, as in section B, the hypothesis testing problem between two irreducible Markov chains:

$$H_0: \quad q(\cdot \mid \cdot), q(\cdot)$$
$$H_1: \quad p(\cdot \mid \cdot), p(\cdot)$$

where $q(\cdot \mid \cdot)$, $p(\cdot \mid \cdot)$ are the transition probabilities and $q(\cdot)$, $p(\cdot)$ are the initial distributions for the Markov chains. If the chain is to be stationary, then $q(\cdot)$ and $p(\cdot)$ should be taken to be the stationary distributions of the transition probabilities. We now know that the asymptotically best way to discriminate between these two hypotheses is to form the sum

$$\sum_{i=1}^{n-1} \log \frac{p(\mathbf{x}_{i+1} \mid \mathbf{x}_i)}{q(\mathbf{x}_{i+1} \mid \mathbf{x}_i)}$$

However, to form (as data arrive) the function

$$\log \frac{p(\mathbf{x}_{i+1} \mid \mathbf{x}_i)}{q(\mathbf{x}_{i+1} \mid \mathbf{x}_i)}$$

requires that the previous sample be remembered before the next sample is processed. One may very well want the detector to be memoryless, i.e., be able to process data samples without recourse to storing past values in memory. This requirement leads one to consider statistical tests of the following form:

$$t_n \equiv \sum_{i=1}^{n} f(\mathbf{x}_i) \tag{15}$$

where f is a mapping from the state space S (assumed finite) of the Markov chain into $\mathbb{R}$. How should f be chosen to minimize the asymptotic probability of error? For the degenerate Markov case of iid sequences, the optimum choice is

$$f(x) = \log \frac{p(x)}{q(x)}$$

Is this also the answer in the nondegenerate case? We shall find that it is not.

Suppose that f has been chosen so that at least the test t_n is exponentially consistent, i.e., P(error based on n observations) approaches zero exponentially fast. As long as the stationary distributions of the two chains are different, i.e., $p(x) \neq q(x)$ for all $x \in S$, this could be achieved, for example, by choosing

$$f(x) = \log \frac{p(x)}{q(x)}$$

as above. The type I and type II error rates are (from arguments identical in nature to the derivations of (10)–(12))

$$-I_i(0) = \inf_\theta \log \lambda_i(\theta) \qquad (i = 0, 1) \tag{16}$$

where $\lambda_1(\theta)$ is the largest eigenvalue of the operator

$$T_{\theta,1}(g)(y) \equiv \sum_x \exp[\theta f(x)]g(x)p(x \mid y)$$

and $\lambda_0(\theta)$ is the largest eigenvalue of the operator

$$T_{\theta,0}(g)(y) \equiv \sum_x \exp[\theta f(x)]g(x)q(x \mid y)$$

The overall rate with which the probability of error approaches zero is then

$$-\min\{I_1, I_0\} = \min\left\{\inf_\theta \log \lambda_1(\theta), \inf_\theta \log \lambda_0(\theta)\right\} \tag{17}$$

One can show (and it is intuitive) that the optimal f, call it f_{opt}, exists and furthermore, that $I_1 = I_0$ for this optimal nonlinearity. Therefore (assuming that the infimums are attained), we have

$$\inf_\theta \log \lambda_1(\theta) = \log \lambda_1(\theta_1) = \inf_\theta \log \lambda_0(\theta) = \log \lambda_0(\theta_0) \tag{18}$$

for some pair θ_0, θ_1. It is easy to show, since $\Sigma_x f_{\text{opt}}(x)p(x) > 0$, that $\theta_1 < 0$ (and conversely, $\theta_0 > 0$). It is clear that f_{opt} has the same performance as cf_{opt} for any positive constant c. Hence, by a consideration of how θ and f_{opt} enter into the definitions of $T_{\theta,1}$ and $T_{\theta,0}$, we may scale f_{opt} so that $\theta_0 - \theta_1 = 1$. In fact, it turns out that the important parameter for θ_0 and θ_1 will be their ratio,

$$R \equiv \frac{\theta_0}{\theta_1} \tag{19}$$

Assume for now that we know what R for f_{opt} is. Let us consider perturbing f_{opt} to $f_{\text{opt}} + \varepsilon m$, where m is some function and $\varepsilon > 0$. How does this perturb the principal eigenvalues $\lambda_1(\theta_1)$ and $\lambda_0(\theta_0)$? Consider the perturbed operator

$$\begin{aligned} P^{\varepsilon}_{m,\theta_1,1}(g)(y) &\equiv \sum_x \exp[\theta_1(f_{\text{opt}}(x) + \varepsilon m(x))]g(x)p(x \mid y) \\ &= \sum_x \exp[\theta_1 f_{\text{opt}}(x)][1 + \theta_1 \varepsilon m(x)]g(x)p(x \mid y) + o(\varepsilon) \\ &= T_{\theta_1,1}(g)(y) + \varepsilon \sum_x \theta_1 \exp[\theta_1 f_{\text{opt}}(x)]m(x)g(x)p(x \mid y) + o(\varepsilon) \end{aligned}$$

The first order perturbation of this operator's principal eigenvalue (see appendix C.5) appears as

$$\begin{aligned} \lambda_{P^{\varepsilon}_{m,\theta_1,1}} &= \lambda_1(\theta_1) + \varepsilon \sum_x \sum_y \exp[\theta_1 f_{\text{opt}}(x)] r_{f_{\text{opt}},1}(x) r^*_{f_{\text{opt}},1}(y)\theta_1 m(x)p(x \mid y) + o(\varepsilon) \\ &= \lambda_1(\theta_1) + \varepsilon\theta_1 \left[\lambda_1(\theta_1) \sum_x r_{f_{\text{opt}},1}(x) r^*_{f_{\text{opt}},1}(x) m(x) \right] + o(\varepsilon) \\ &\equiv \lambda_1(\theta_1) + \varepsilon\theta_1 D_{m,1} + o(\varepsilon) \end{aligned} \tag{20}$$

where $r_{f_{\text{opt}},1}(\cdot)$ is the eigenvector associated with the eigenvalue $\lambda_1(\theta_1)$ and $r^*_{f_{\text{opt}},1}(\cdot)$ is the eigenvector associated with the adjoint (recall that for matrix representations this is merely the transpose) of the operator $T_{\theta_1,1}$ (with the same eigenvalue). $r^*_{f_{\text{opt}},1}(\cdot)$ can also be thought of as the left eigenvector of $T_{\theta_1,1}$. Recall that these eigenvectors are always nonnegative (by the Perron–Frobenius theorem); furthermore, we may always normalize them so that $(r_{f_{\text{opt}}}, r^*_{f_{\text{opt}}}) = 1$. With obvious notation, we perform the same operations to the $T_{\theta_0,0}$ operator, to find

$$\begin{aligned} \lambda_{P^{\varepsilon}_{m,\theta_0,0}} &= \lambda_0(\theta_0) + \varepsilon \sum_x \sum_y \exp[\theta_0 f_{\text{opt}}(x)] r_{f_{\text{opt}},0}(x) r^*_{f_{\text{opt}},0}(y)\theta_0 m(x)q(x \mid y) + o(\varepsilon) \\ &= \lambda_0(\theta_0) + \varepsilon\theta_0 \left[\lambda_0(\theta_0) \sum_x r_{f_{\text{opt}},0}(x) r^*_{f_{\text{opt}},0}(x) m(x) \right] + o(\varepsilon) \\ &\equiv \lambda_0(\theta_0) + \varepsilon\theta_0 D_{m,0} + o(\varepsilon) \end{aligned} \tag{21}$$

Now if one principal eigenvalue increases, the other must decrease. If both increased, we could not have an optimum nonlinearity. If both decreased, the negative of that perturbation (replace m by $-m$) would have both increasing, which again cannot occur for the optimum nonlinearity. This implies that $D_{m,1}$ and $D_{m,0}$ must have the same sign (since θ_0 and θ_1 have opposite signs).

Furthermore, consider what happens when m is replaced by $m + c$, where c is a constant. Then (20) gives

$$\lambda_{P^{\varepsilon}_{m+c,\theta_1,1}} = \lambda_1(\theta_1) + \varepsilon\theta_1 \left[\lambda_1(\theta_1) \sum_x r_{f_{\mathrm{opt}},1}(x) r^*_{f_{\mathrm{opt}},1}(x)[m(x) + c] \right] + o(\varepsilon)$$

$$= \lambda_1(\theta_1) + \varepsilon\theta_1 D_{m,1} + \varepsilon\theta_1 c \lambda_1(\theta_1) \sum_y r_{f_{\mathrm{opt}},1}(y) r^*_{f_{\mathrm{opt}},1}(y) + o(\varepsilon)$$

$$= \lambda_1(\theta_1) + \varepsilon\theta_1 D_{m,1} + \varepsilon\theta_1 c \lambda_1(\theta_1) + o(\varepsilon) \tag{22}$$

Similarly, (21) gives

$$\lambda_{P^{\varepsilon}_{m+c,\theta_0,0}} = \lambda_0(\theta_0) + \varepsilon\theta_0 D_{m,0} + \varepsilon\theta_0 c \lambda_0(\theta_0) + o(\varepsilon) \tag{23}$$

Consider then the *change* in the eigenvalues as a function of c; i.e., fix $\varepsilon > 0$ and consider the functions of c:

$$f_1(c) \equiv \lambda_{P^{\varepsilon}_{m+c,\theta_1,1}} - \lambda_1(\theta_1) = \varepsilon\theta_1 D_{m,1} + \varepsilon\theta_1 c \lambda_1(\theta_1) \tag{24}$$

$$f_0(c) \equiv \lambda_{P^{\varepsilon}_{m+c,\theta_0,0}} - \lambda_0(\theta_0) = \varepsilon\theta_0 D_{m,0} + \varepsilon\theta_0 c \lambda_0(\theta_0) \tag{25}$$

Equations (24)–(25), of course, plot out two lines as c is varied (see figure 4). As pointed out previously, whenever $f_1(c)$ is negative, $f_0(c)$ must be positive, and vice versa. We emphasize that (24)–(25) represent the change in the eigenvalues due to a perturbation of the optimum nonlinearity and hence they cannot have the same sign for any value of c. Furthermore, we know that $\lambda_0(\theta_0) = \lambda_1(\theta_1)$ and $D_{m,0}$ and $D_{m,1}$ have the same sign. Therefore, $f_1(c)$ and

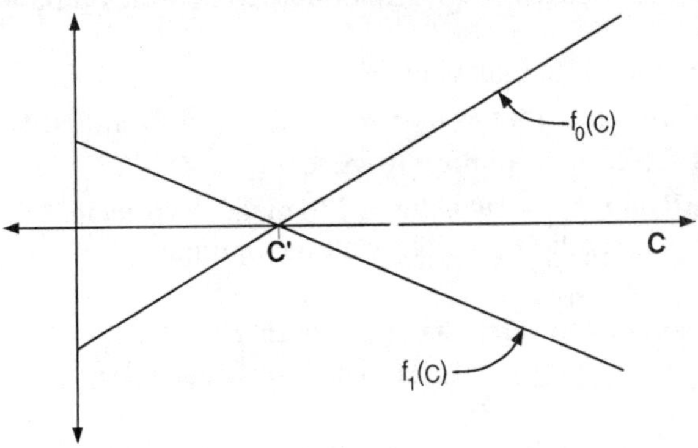

Figure 4

$f_0(c)$ intersect on the c axis at the point c', say. This condition may be written as $f_1(c') = f_0(c') = 0$. $f_0(c') = 0$ implies with (25) that

$$c' = -\frac{\varepsilon\theta_0 D_{m,0}}{\varepsilon\theta_0\lambda_0(\theta_0)} = -\frac{D_{m,0}}{\lambda_0(\theta_0)}$$

Substituting this value into (24) gives

$$f_1(c') = \varepsilon\theta_1 D_{m,1} - \frac{D_{m,0}\varepsilon\theta_1\lambda_1(\theta_1)}{\lambda_0(\theta_0)} = 0$$

or

$$D_{m,1} = D_{m,0} \tag{26}$$

or equivalently, from the defining relations (20)–(21), (26) is rewritten as

$$\lambda_1(\theta_1)\sum_x r_{f_{\text{opt}},1}(x) r^*_{f_{\text{opt}},1}(x) m(x) = \lambda_0(\theta_0)\sum_x r_{f_{\text{opt}},0}(x) r^*_{f_{\text{opt}},0}(x) m(x)$$

and since $\lambda_1(\theta_1) = \lambda_0(\theta_0)$, we have

$$\sum_x r_{f_{\text{opt}},1}(x) r^*_{f_{\text{opt}},1}(x) m(x) = \sum_x r_{f_{\text{opt}},0}(x) r^*_{f_{\text{opt}},0}(x) m(x) \tag{27}$$

This expression can be true for all perturbations $m(\cdot)$ if and only if

$$r_{f_{\text{opt}},1}(x) r^*_{f_{\text{opt}},1}(x) = r_{f_{\text{opt}},0}(x) r^*_{f_{\text{opt}},0}(x) \tag{28}$$

for all x. For a fixed θ_0, θ_1 such that $\theta_0 > 0$, $\theta_1 < 0$, $\theta_0 - \theta_1 = 1$, and $\theta_0/\theta_1 = R$, one can show that there exists a unique solution to the foregoing system. This then suggests a methodology to find the optimum nonlinearity:

0. Choose an initial guess for R.
1. Fix θ_0, θ_1 such that $\theta_0 > 0$, $\theta_1 < 0$, $\theta_0 - \theta_1 = 1$, and $\theta_0/\theta_1 = R$.
2. Find the unique solution to system (28).
3. Vary R and repeat steps 1 and 2 to make the principal eigenvalue of this system as small as possible. Call the optimum R, R_{opt}, and the smallest eigenvalue, λ_{opt}.
4. The solution of (28) that goes with R_{opt} will be the optimum nonlinearity. The error rate will be $-\log \lambda_{\text{opt}}$.

***Example* 1.** Consider the following Markov chain: Stay at state i with probability $1 - \mu$ or jump to a new state in the finite state space S chosen each time from a state independent probability distribution $p(x)$ (hypothesis

H_1) or $q(x)$ (hypothesis H_0) with probability μ. Let us consider the eigenvector problem for the appropriate operators under H_1, say:

$$\sum_x \exp[\theta_1 f(x)] r_f(x) p(x \mid y) = \lambda_1(\theta_1) r_f(y)$$

where

$$p(x \mid y) = \begin{cases} 1 - \mu + \mu p(y) & \text{if } x = y \\ \mu p(x) & \text{if } x \neq y \end{cases}$$

The eigenvector problem then becomes

$$\mu \sum_x \exp[\theta_1 f(x)] r_f(x) p(x) + \exp[\theta_1 f(y)] r_f(y)[1 - \mu] = \lambda_1(\theta_1) r_f(y)$$

Solving this equation for $r_f(y)$, one obtains

$$r_f(y) = \frac{\Sigma_x \exp[\theta_1 f(x)] r_f(x) \mu p(x)}{\lambda_1(\theta_1) - \exp[\theta_1 f(y)][1 - \mu]}$$

We always have freedom to normalize an eigenvector. Let us choose for now the normalization constant to be such that the numerator of the expression above is 1, i.e.,

$$r_f(y) = \frac{1}{\lambda_1(\theta_1) - \exp[\theta_1 f(y)][1 - \mu]}$$

The normalization condition then requires that

$$\sum_x \frac{\exp[\theta_1 f(x)] \mu p(x)}{\lambda_1(\theta_1) - \exp[\theta_1 f(x)][1 - \mu]} = 1$$

$\lambda_1(\theta_1)$ will then be the largest solution to this equation.

The adjoint operator eigenvector problem is

$$\begin{aligned} \sum_x \exp[\theta_1 f(y)] r_f^*(x) p(y \mid x) &= \lambda_1(\theta_1) r_f^*(y) \\ &= \exp[\theta_1 f(y)] \sum_x r_f^*(x) p(y \mid x) \\ &= \exp[\theta_1 f(y)] \left[\sum_x r_f^*(x) \mu p(y) + (1 - \mu) r_f^*(y) \right] \end{aligned}$$

One then obtains

$$r_f^*(y) = \frac{\exp[\theta_1 f(y)]\,\Sigma_x r_f^*(x)\mu p(y)}{\lambda_1(\theta_1) - \exp[\theta_1 f(y)][1-\mu]}$$

$$= \frac{\mu p(y)\exp[\theta_1 f(y)]\,\Sigma_x r_f^*(x)}{\lambda_1(\theta_1) - \exp[\theta_1 f(y)][1-\mu]}$$

Now we may normalize by a constant c_1 so that

$$\sum_x r_f^*(x) = c_1$$

and obtain

$$r_f^*(y) = \frac{\mu p(y)\exp[\theta_1 f(y)]c_1}{\lambda_1(\theta_1) - \exp[\theta_1 f(y)][1-\mu]}$$

The normalization condition implies then that

$$\sum_x \frac{\mu p(x)\exp[\theta_1 f(x)]}{\lambda_1(\theta_1) - \exp[\theta_1 f(x)][1-\mu]} = 1$$

$\lambda_1(\theta_1)$ already has been chosen to solve this equation anyway. This verifies what we already knew—that the principal eigenvalue of an operator is the same as that of the operator's adjoint.

Therefore, our necessary condition (28) becomes

$$\frac{\mu p(y)\exp[\theta_1 f_{\text{opt}}(y)]c_1}{[\lambda_1(\theta_1) - \exp[\theta_1 f_{\text{opt}}(y)][1-\mu]]^2} = \frac{\mu q(y)\exp[\theta_0 f_{\text{opt}}(y)]c_0}{[\lambda_0(\theta_0) - \exp[\theta_0 f_{\text{opt}}(y)][1-\mu]]^2}$$

Using the fact that $\theta_0 - \theta_1 = 1$, we find that f_{opt} must satisfy for some c

$$\exp[f_{\text{opt}}(y)]\frac{[\lambda_1(\theta_1) - \exp[\theta_1 f_{\text{opt}}(y)][1-\mu]]^2}{[\lambda_0(\theta_0) - \exp[\theta_0 f_{\text{opt}}(y)][1-\mu]]^2} = c\frac{p(y)}{q(y)} \tag{29}$$

This, unfortunately, is a transcendental equation. It can be solved iteratively, however. For example, choose initially

$$f_1(y) = \log\frac{p(y)}{q(y)}$$

Then for each i choose $\lambda^{(i)}$ to be the largest solution to the equation

$$\sum_x \frac{\mu p(x)\exp[\theta_1 f_i(x)]}{\lambda^{(i)} - \exp[\theta_1 f_i(x)][1-\mu]} = 1$$

where f_i is chosen iteratively by the equation

$$f_i(y) = \log \frac{p(y)}{q(y)} + \log \frac{[\lambda^{(i-1)} - \exp[\theta_1 f_{i-1}(y)][1-\mu]]^2}{[\lambda^{(i-1)} - \exp[\theta_0 f_{i-1}(y)][1-\mu]]^2}$$

When this system converges, it will be to the unique solution of the system (after constants are adjusted). See exercises 3–6 for further examples and commentary on the theory of this section. It is important to note from (29) that the optimum memoryless nonlinearity is not merely the log of the ratio of the stationary distributions under the two hypotheses. □

D. GAUSSIAN PROCESSES AND QUADRATIC FORM DETECTORS

In this section we perform analyses of two traditionally difficult to analyze nonlinear detectors: the square law (radiometer) detector and a differential phase shift keyed (DPSK) system. Our main analytic tools will be the Gärtner–Ellis theorem and the Toeplitz distribution theorem (see appendix C.3). For simplicity we will assume that we are working with discrete-time systems.

(1) Square Law Detector

Our test statistic is assumed to be of the following form:

$$\frac{1}{n} \sum_{i=1}^{n} \mathbf{y}_i^2 \underset{H_0}{\overset{H_1}{\gtrless}} \gamma$$

where under hypothesis H_0, $\{\mathbf{y}_i\}$ are samples from a stationary Gaussian random process background noise $\{\mathbf{N}_i\}$ with power spectrum $f(\omega)$ and under hypothesis H_1, $\{\mathbf{y}_i\}$ are samples $\{s_i + \mathbf{N}_i\}$, where $\{s_i\}$ is a known deterministic sequence with (deterministic) power spectrum $P(\omega)$. Define

$$\underline{\mathbf{N}} \equiv (\mathbf{N}_1 \quad \mathbf{N}_2 \quad \cdots \quad \mathbf{N}_n)^{\mathrm{T}}$$

and

$$\underline{\mathbf{s}} \equiv (s_1 \quad s_2 \quad \cdots \quad s_n)^{\mathrm{T}}$$

Consider the transformations

$$\underline{\mathbf{N}}_1 \equiv \Lambda^{-1/2} \phi^{\mathrm{T}} \underline{\mathbf{N}}$$

and

$$\underline{\mathbf{s}}_1 \equiv \Lambda^{-1/2} \phi^{\mathrm{T}} \underline{\mathbf{s}}$$

where Λ is the diagonal eigenvalue matrix ($\Lambda \equiv \text{diag}\{\lambda_1, \lambda_2, \ldots, \lambda_n\}$) and ϕ is the unitary eigenvector matrix of the covariance matrix V of the vector $\underline{\mathbf{N}}$. The notation ϕ^{T} means conjugate and transpose when applied to a matrix with complex entries. It is easy to check that $\underline{\mathbf{N}}_1$'s covariance matrix is merely the identity matrix. Hence the transformation $\Lambda^{-1/2}\phi^{\mathrm{T}}$ is sometimes called a *whitening* transformation. Note that $\phi\Lambda^{1/2}$ is the inverse of the whitening transformation ($\phi^{\mathrm{T}} = \phi^{-1}$ since ϕ is unitary).

Now let us consider the probability of error given that H_1 is true. First consider the event

$$\left\{\frac{1}{n}\sum_{i=1}^{n}(s_i + \mathbf{N}_i)^2 \leqslant \gamma\right\} = \left\{\frac{1}{n}(\underline{\mathbf{s}} + \underline{\mathbf{N}})^{\mathrm{T}}(\underline{\mathbf{s}} + \underline{\mathbf{N}}) \leqslant \gamma\right\}$$

$$= \left\{\frac{1}{n}(\underline{\mathbf{s}}_1 + \underline{\mathbf{N}}_1)^{\mathrm{T}}\Lambda^{1/2}\phi^{\mathrm{T}}\phi\Lambda^{1/2}(\underline{\mathbf{s}}_1 + \underline{\mathbf{N}}_1) \leqslant \gamma\right\}$$

$$= \left\{\frac{1}{n}(\underline{\mathbf{s}}_1 + \underline{\mathbf{N}}_1)^{\mathrm{T}}\Lambda(\underline{\mathbf{s}}_1 + \underline{\mathbf{N}}_1) \leqslant \gamma\right\}$$

$$= \left\{\frac{1}{n}\sum_{i=1}^{n}\lambda_i(s_{1i} + \mathbf{N}_{1i})^2 \leqslant \gamma\right\}$$

Let us now place this result in the context of the Gärtner–Ellis theorem. Define

$$\mathbf{Y}_n \equiv \sum_{i=1}^{n}\lambda_i(\mathbf{N}_{1i} - s_{1i})^2$$

[Note we can place a minus sign here because the $\mathbf{N}_{1i}$ are zero mean Gaussian (and hence symmetric).] Thus

$$\varphi_n(\theta) \equiv \frac{1}{n}\log E_1\{\exp[\theta\mathbf{Y}_n]\} = \frac{1}{n}\log E_1\left\{\prod_{i=1}^{n}\exp[\theta\lambda_i(N_{1i} - s_{1i})^2]\right\}$$

$$= \frac{1}{n}\sum_{i=1}^{n}\log E_1\{\exp[\theta\lambda_i(N_{1i} - s_{1i})^2]\}$$

where we have invoked the independence of the $\mathbf{N}_{1i}$, $i = 1, \ldots, n$ random variables. One can compute the expectation by a standard "completing the square" type of argument to obtain that

$$\varphi_n(\theta) = \frac{1}{n}\sum_{i=1}^{n}\log\left\{[1 - 2\theta\lambda_i]^{-1/2}\exp\left[\frac{\theta\lambda_i s_{1i}^2}{1 - 2\theta\lambda_i}\right]\right\}$$

$$= \frac{1}{n}\sum_{i=1}^{n} -\tfrac{1}{2}\log(1 - 2\theta\lambda_i^{(n)}) + \frac{1}{n}\sum_{i=1}^{n}\frac{\theta\lambda_i^{(n)}s_{1i}^2}{1 - 2\theta\lambda_i^{(n)}} \tag{30}$$

where in the final expression the dependence of the eigenvalues on n is made explicit. Now as n approaches infinity (by the Toeplitz distribution theorem; see appendix C.3)

$$\lim_{n\to\infty} \frac{1}{n} \sum_{i=1}^{n} -\tfrac{1}{2}\log(1 - 2\theta\lambda_i^{(n)}) = -\frac{1}{4\pi}\int_0^{2\pi} \log(1 - 2\theta f(\omega))\, d\omega \tag{31}$$

Recall that the $\underline{\mathbf{s}}_1$ vector was obtained by $\Lambda^{-1/2}\phi^{\mathrm{T}}\underline{\mathbf{s}}$. From the discussion in appendix C.3, for large n, ϕ the eigenvector matrix "looks" more and more like the discrete Fourier transform (DFT) matrix. In other words, the columns of ϕ become samples of complex sinusoids. Hence we expect s_{1i}^2 to be asymptotically like $P(\omega_i)/f(\omega_i)$ where ω_i is the frequency corresponding to the ith column of the DFT matrix ($\omega_i = 2\pi i/n$). By this argument and the Toeplitz distribution theorem we expect that

$$\lim_{n\to\infty} \frac{1}{n} \sum_{i=1}^{n} \frac{\theta\lambda_i^{(n)} s_{1i}^2}{1 - 2\theta\lambda_i^{(n)}} = \frac{\theta}{2\pi}\int_0^{2\pi} \frac{P(\omega)}{1 - 2\theta f(\omega)}\, d\omega \tag{32}$$

and hence, from (30)–(32),

$$\varphi(\theta) \equiv \lim_{n\to\infty} \varphi_n(\theta) = -\frac{1}{4\pi}\int_0^{2\pi} \log(1 - 2\theta f(\omega))\, d\omega + \frac{\theta}{2\pi}\int_0^{2\pi} \frac{P(\omega)}{1 - 2\theta f(\omega)}\, d\omega \tag{33}$$

We may now invoke the Gärtner–Ellis theorem to have from (33) that

$$\begin{aligned} \lim_{n\to\infty} \frac{1}{n} \log P\left(\frac{1}{n}\sum_{i=1}^{n} (s_i + N_i)^2 \leqslant \gamma\right) &= -\sup_{\theta} \{\theta\gamma - \varphi(\theta)\} \\ &= -\sup_{\theta} \left\{\theta\gamma + \frac{1}{4\pi}\int_0^{2\pi} \log(1 - 2\theta f(\omega))\, d\omega - \frac{\theta}{2\pi}\int_0^{2\pi} \frac{P(\omega)}{1 - 2\theta f(\omega)}\, d\omega\right\} \end{aligned}$$

To find the optimum θ, θ_γ, we differentiate with respect to θ and set the result to zero to find that θ_γ solves

$$\gamma = \frac{1}{2\pi}\int_0^{2\pi} \frac{P(\omega) + f(\omega)(1 - 2\theta_\gamma f(\omega))}{(1 - 2\theta_\gamma f(\omega))^2}\, d\omega$$

In general, this is a transcendental equation and cannot be solved in closed form. In exercise 7 you are asked to solve this system for the case of white noise [$f(\omega) = 1$].

Note that we have been concentrating on the probability of error given that signal is present. To obtain the rate of the error when signal is absent, just replace $P(\omega)$ in the foregoing expression by zero.

(2) Differential Phase Shift Keying

In differential phase shift keying (DPSK), a change in sign of the signal from the preceding signaling interval signifies a logic 0 (hypothesis H_0). If the same sign is maintained, this signifies a logic 1 (hypothesis H_1). This is a popular modulation scheme when the communication channel's phase characteristics change with time but remain fairly stable over a single bit period (see Stremler [1990]). In continuous time the test statistic appears as

$$\frac{1}{T}\int_0^T [s(t)+\mathbf{N}(t)][s(t+T)+\mathbf{N}(t+T)]\,dt \underset{H_0}{\overset{H_1}{\gtrless}} \gamma$$

where usually $s(t) = \pm\cos(w_0 t)$ but we will merely assume that it is periodic with fundamental period w_0. (γ is usually zero.)

Let us discretize this problem and consider the test statistic

$$\frac{2}{n}\sum_{i=0}^{n-1}\sum_{j=0}^{n-1} a_{ij}\mathbf{y}_i\mathbf{y}_j \underset{H_0}{\overset{H_1}{\gtrless}} \gamma$$

where the $\{\mathbf{y}_i\}$ should be considered to be samples of our received sequence over an interval of length $2T$, n is an even multiple of the period of the signal (i.e., $n = 2\pi r/w_0$, where r is a positive even integer), and the matrix $A \equiv (a_{ij})$ $(i, j \in \{0, 1, \ldots, n-1\})$ is given by

$$A \equiv \frac{1}{2}\left[\begin{array}{c|c} 0 & \begin{matrix} 1 & & & \\ & 1 & & \\ & & \ddots & \\ & & & 1 \end{matrix} \\ \hline \begin{matrix} 1 & & & \\ & 1 & & \\ & & \ddots & \\ & & & 1 \end{matrix} & 0 \end{array}\right]$$

This matrix is circulant and its eigenvalues (as always) are given by the DFT of the first row (which in our setting has index zero to match up with standard DFT notation). Thus

$$\sigma_k \equiv \sum_{j=0}^{n-1} a_{0j}\exp\left[-\frac{i2\pi kj}{n}\right] = \tfrac{1}{2}(-1)^k \qquad k = 0, 1, \ldots, n-1$$

are calculated to be the eigenvalues of A. Let Σ denote the diagonal eigenvalue matrix of A so that

$$A = \Psi\Sigma\Psi^{\mathrm{T}} \tag{34}$$

where Ψ is the DFT matrix and is the eigenvector matrix common to all circulant matrices. Let V be the $n \times n$ covariance matrix of the noise samples and $\underline{\mathbf{y}}$ our vector of observations. Let ϕ and Λ be the eigenvector and eigenvalue matrices of V (as in the preceding subsection) and we immediately have the diagonalization

$$V = \phi \Lambda \phi^{\mathrm{T}} \tag{35}$$

Consider again our "whitening" transformation

$$\underline{\mathbf{y}}_1 \equiv \Lambda^{-1/2} \phi^{\mathrm{T}} \underline{\mathbf{y}} \tag{36}$$

(and $\mathbf{N}_1, s_1$ are defined exactly as in the preceding subsection). Then from (36) we have

$$\left\{\frac{2}{n} \sum_{i=1}^{n} \sum_{j=1}^{n} a_{ij} \mathbf{y}_i \mathbf{y}_j \leqslant \gamma\right\} = \left\{\frac{2}{n} \underline{\mathbf{y}}^{\mathrm{T}} A \underline{\mathbf{y}} \leqslant \gamma\right\} = \left\{\frac{2}{n} \underline{\mathbf{y}}_1^{\mathrm{T}} \Lambda^{1/2} \phi^{\mathrm{T}} A \phi \Lambda^{1/2} \underline{\mathbf{y}}_1 \leqslant \gamma\right\} \tag{37}$$

Now as n approaches infinity ϕ should be approaching the DFT matrix Ψ. In that case $\phi^{\mathrm{T}} A \phi$ should be approaching $\Psi^{\mathrm{T}} A \Psi$, which equals Σ, the diagonal eigenvalue matrix of A. Hence (37) becomes (for large n)

$$\left\{\frac{2}{n} \underline{\mathbf{y}}_1^{\mathrm{T}} \Lambda^{1/2} \phi^{\mathrm{T}} A \phi \Lambda^{1/2} \underline{\mathbf{y}}_1 \leqslant \gamma\right\} \approx \left\{\frac{2}{n} \underline{\mathbf{y}}_1^{\mathrm{T}} \Lambda^{1/2} \Sigma \Lambda^{1/2} \underline{\mathbf{y}}_1 \leqslant \gamma\right\}$$

$$= \left\{\frac{2}{n} \sum_{k=0}^{n-1} \sigma_k \lambda_k |\mathbf{y}_{1k}|^2 \leqslant \gamma\right\} \tag{38}$$

Assume that there is no sign shift; i.e., logic 1 has been sent. Due to symmetry the asymptotic probabilities of error will be the same under either hypothesis. In this case it is clear that the signal component in $\mathbf{y}_{1k}$ will be zero (as n approaches infinity) at every odd value of k. This is due to the fact that n has been chosen to be an even multiple of the fundamental period of the signal. Hence

$$\begin{aligned}
\varphi_n(\theta) &= \frac{1}{n} \log E\left\{\exp\left\{2\theta \sum_{i=0}^{n-1} \sum_{j=0}^{n-1} a_{ij} \mathbf{y}_i \mathbf{y}_j\right\}\right\} \\
&= \frac{1}{n} \log\left\{\prod_{\substack{k=0 \\ k \text{ odd}}}^{n-1} E\{\exp[-\theta \lambda_k \mathbf{N}_{1k}^2]\} \prod_{\substack{k=0 \\ k \text{ even}}}^{n-1} E\{\exp[\theta \lambda_k [\mathbf{N}_{1k} + s_{1k}]^2\}\right\} \\
&= \frac{1}{n} \sum_{k \text{ odd}} \log[1 + 2\theta \lambda_k]^{-1/2} + \frac{1}{n} \sum_{k \text{ even}} \log\left[[1 - 2\theta \lambda_k]^{-1/2} \exp\left[\frac{\lambda_k \theta s_{1k}^2}{1 - 2\theta \lambda_k}\right]\right]
\end{aligned} \tag{39}$$

It should be clear that summation over half the values of k merely reduces the asymptotic limits by half. Thus, from (39),

$$\begin{aligned}\varphi(\theta) &= \lim_{n\to\infty} \varphi_n(\theta)\\ &= \frac{1}{4\pi}\int_0^{2\pi} -\tfrac{1}{2}\log[1 + 2\theta f(\omega)]\, d\omega\\ &\quad + \frac{1}{4\pi}\int_0^{2\pi} -\frac{1}{2}\log[1 - 2\theta f(\omega)]\, d\omega + \frac{\theta}{2\pi}\int_0^{2\pi} \frac{P(\omega)}{1 - 2\theta f(\omega)}\, d\omega \end{aligned} \tag{40}$$

Thus from the Gärtner–Ellis theorem and (40), we obtain

$$\begin{aligned}&\lim_{n\to\infty} \frac{2}{n}\log P\left(\frac{2}{n}\underline{\mathbf{y}}^{\mathrm{T}} A \underline{\mathbf{y}} \leqslant \gamma\right)\\ &\quad= -2\sup_{\theta}\{\theta\gamma - \varphi(\theta)\}\\ &\quad= -\sup_{\theta'}\{\theta'\gamma - 2\varphi(\tfrac{1}{2}\theta')\}\\ &\quad= -\sup_{\theta'}\left\{\theta'\gamma + \frac{1}{4\pi}\int_0^{2\pi}\log[1 - (\theta' f(\omega))^2]\, d\omega - \frac{\theta'}{2\pi}\int_0^{2\pi}\frac{P(\omega)}{1 - \theta' f(\omega)}\, d\omega\right\}\end{aligned} \tag{41}$$

In the case of a sinusoidal signal of peak amplitude B and frequency ω_0, (41) becomes

$$-\sup_{\theta}\left\{\theta\gamma + \frac{1}{4\pi}\int_0^{2\pi}\log[1 - (\theta f(\omega))^2]\, d\omega - \frac{\theta B^2}{2[1 - \theta f(\omega_0)]}\right\} \tag{42}$$

By taking the derivative, we can find the maximizing θ_γ as

$$\gamma = \frac{B^2}{2(1 - \theta_\gamma f(\omega_0))^2} + \frac{1}{2\pi}\int_0^{2\pi}\frac{\theta_\gamma f^2(\omega)}{1 - \theta_\gamma^2 f^2(\omega)}\, d\omega$$

***Example* 2.** The threshold $\gamma = 0$ is of greatest interest. Suppose that the noise is white $[f(\omega) = 1]$. One can then verify that

$$\theta_0 = \frac{1}{2}\left(1 + \frac{B^2}{2}\right) - \frac{1}{2}\left(\left(1 + \frac{B^2}{2}\right)^2 + 2B^2\right)^{1/2}$$

The asymptotic rate of the probability of error to 0 is then

$$\frac{1}{2}\log[1-\theta_0^2]-\frac{B^2\theta_0}{1-\theta_0}$$

For example, $B=10$ gives -23.2 and $B=1$ gives -0.0686. There is a decided threshold effect for this type of nonlinear receiver. □

E. THE CHANGEPOINT PROBLEM

In recent years a body of statistical work known as "changepoint problems" has emerged as an important application area in statistical signal processing. It deals with the general situation of the occurrence of a change in the statistics of a random time series. The key question to be answered is: Has a change occurred? This question (and its answer) lie at the heart of many engineering and scientific endeavors.

Changepoint problems in statistics arose historically through various sorts of quality control/production line problems. For example, one observes some parameter on a manufactured device and determines whether the device should be accepted or rejected. When the assembly line is performing well, there will still be a certain percentage of devices that will be rejected. What is of concern is if some component in the assembly line itself fails or changes, causing the failure rate to change. One wants to detect this changepoint as quickly as possible in order to shut production down so that the assembly line may be fixed. However, one does not want to shut down unless it is absolutely necessary to do so.

Aside from these quality control procedures, changepoint problems have arisen in segmentation of speech [Milosavljevic, 1988], edge detection in images [Basseville et al., 1981; Basseville, 1981], geophysical signal processing [Basseville and Benveniste, 1983], electrocardiogram analysis [Gustafson et al., 1978a, 1978b], link failure in communication networks [Papantoni-Kazakos, 1979], to name just a sampling of the possible application areas (see also the survey articles by Shaban [1980] and Willsky [1976]).

Suppose that we observe n independent samples. The hypothesis testing problem is

H_0: $\{\mathbf{x}_i\}_{i=1}^{n}$ is distributed with density $q(x)$

H_1: $\{\mathbf{x}_i\}_{i=1}^{n\lambda}$ is distributed with density $q(x)$ and $\{x_i\}_{i=n\lambda+1}^{n}$ is distributed with density $p(x)$

where $\lambda \in \{1/n, 2/n, \ldots, (n-1)/n\}$. Hypothesis H_0 essentially is the hypothesis that $\lambda \equiv 1$. This problem is not a simple binary hypothesis testing problem. There are in reality $n - 1$ hypotheses "hidden" in H_1 depending on the value of λ. If the value of λ is known, the optimal procedure is to use the likelihood ratio test

$$\frac{1}{n}\sum_{i=1}^{[n\lambda]} \log \frac{q(\mathbf{x}_i)}{q(\mathbf{x}_i)} + \frac{1}{n}\sum_{i=[n\lambda+1]}^{n} \log \frac{p(\mathbf{x}_i)}{q(\mathbf{x}_i)} = \frac{1}{n}\sum_{i=[n\lambda+1]}^{n} \log \frac{p(\mathbf{x}_i)}{q(\mathbf{x}_i)} \underset{H_0}{\overset{H_1}{\gtrless}} g(\lambda)$$

where $g(\lambda)$ is chosen to fix a certain false-alarm rate. The structure of this test suggests the form of a test that we might use when we do not know a priori what λ is. We will consider the test

$$\max_{\lambda \in [0,1)} \left[\frac{1}{n} \sum_{i=[n\lambda+1]}^{n} \log \frac{p(\mathbf{x}_i)}{q(\mathbf{x}_i)} - g(\lambda) \right] \underset{H_0}{\overset{H_1}{\gtrless}} 0 \tag{43}$$

where $g(\lambda)$ is to be chosen.

To choose $g(\cdot)$, we will wish to formulate our problem in a Neyman–Pearson context. Unfortunately, the Neyman–Pearson lemma is no longer applicable since (as alluded to above) we are no longer considering a simple binary hypothesis problem. In general, there will not be a single test that will have false alarm rate α and perform *uniformly* good (in λ) for the detection probability. For example, suppose that the observations are Gaussian, zero mean, unit variance up to the changepoint and then Gaussian, mean s (>0), unit variance, after the change. If an optimal test existed, it should have the same performance as the Neyman–Pearson test for a known λ. This test for a known λ would be

$$\frac{1}{(1-\lambda)n} \sum_{i=[n\lambda]}^{n} \mathbf{x}_i \underset{H_0}{\overset{H_1}{\gtrless}} T(\alpha)$$

where $T(\alpha)$ depends only on α (since the left-hand side is Gaussian, zero mean, variance $1/[(1-\lambda)n]$ under H_0) and is set to give the desired false-alarm rate. Since the very form of the test (i.e., which random variables even come into consideration) is heavily dependent on λ, one may see that no uniformly optimal test can exist for this problem.

Regardless of this lack of optimality, we may still use the Neyman–Pearson formulation to optimize the form of the suboptimal test that we have chosen. Hence it will be required that the type I error probability (error probability given that H_0 is true) will go to zero as n approaches infinity at rate α. It will be desired that the type II error rate (error rate given that H_1 is true) be made as small as possible subject to the first condition.

For simplicity, let us reindex our observations by time reversing (i.e., define the "new" $\mathbf{x}_i$ as equal to the "old" $\mathbf{x}_{n-i}$). Furthermore, replace λ by $(1-\lambda)$ and consider a statistic of the form

$$\max_{\lambda \in [0,1]} \left[\frac{1}{n} \sum_{i=1}^{[n\lambda]} \log \frac{p(\mathbf{x}_i)}{q(\mathbf{x}_i)} - g(\lambda) \right] \underset{H_0}{\overset{H_1}{\gtrless}} 0 \tag{44}$$

To find the large deviations of this statistic (under H_0), we need only recall the slow Markov walk model. Recall that model was

$$\mathbf{x}_{k+1}^n = \mathbf{x}_k^n + \frac{1}{n} V(\mathbf{x}_k^n, \xi_k)$$

where $\{\xi_k\}$ are iid random variables. In our setting

$$\mathbf{x}_{k+1}^n \text{ corresponds to } \frac{1}{n} \sum_{i=1}^{[n\lambda+1]} \log \frac{p(\mathbf{x}_i)}{q(\mathbf{x}_i)}$$

$$\mathbf{x}_k^n \text{ corresponds to } \frac{1}{n} \sum_{i=1}^{[n\lambda]} \log \frac{p(\mathbf{x}_i)}{q(\mathbf{x}_i)}$$

and

$$V(x, \xi) \text{ corresponds to } \log \frac{p(\xi)}{q(\xi)}$$

We now need to calculate

$$\begin{aligned} M_x(\theta) &\equiv E\{\exp[\theta V(x, \xi_1)]\} \\ &= \int q(\xi) \exp\left[\theta \log \frac{p(\xi)}{q(\xi)}\right] d\xi \\ &= \int p(\xi)^\theta q(\xi)^{1-\theta}\, d\xi \end{aligned}$$

Furthermore,

$$I_x(u) \equiv \sup_\theta \{\theta u - \log M_x(\theta)\} = \sup_\theta \left\{\theta u - \log \int p(\xi)^\theta q(\xi)^{1-\theta}\, d\xi\right\}$$

Let u_m denote the value of u at which $I_x(u) = 0$. u_m will be the mean value of the random variable $\log[p(\xi_1)/q(\xi_1)]$. The rate function for the trajectories of the slow Markov walk was then defined as

$$I(f) \equiv \int_0^1 I_{f(t)}(f'(t))\,dt = \int_0^1 \sup_\theta \left\{ \theta f'(t) - \log \int p(\xi)^\theta q(\xi)^{1-\theta}\,d\xi \right\} dt \quad (45)$$

Hence let us define $A \subset C_{1,0}[0,1]$ (recall that $C_{1,x}[a,b]$ is the set of all piecewise differentiable functions on the interval $[a,b]$ with bounded derivative and whose value at a is x) as

$$A \equiv \{ f \in C_{1,0} : f(\lambda) > g(\lambda) \text{ for some } \lambda \in [0,1] \}$$

Then we have (assuming limits exist)

$$\lim_{n\to\infty} \frac{1}{n} \log P \left\{ \max_{\lambda \in [0,1)} \left[\frac{1}{n} \sum_{i=1}^{n\lambda} \log \frac{p(\mathbf{x}_i)}{q(\mathbf{x}_i)} - g(\lambda) \right] > 0 \,|\, H_0 \text{ is true} \right\}$$

$$= -\inf_{f \in A} \int_0^1 \sup_\theta \left\{ \theta f'(t) - \log \int p(\xi)^\theta q(\xi)^{1-\theta}\,d\xi \right\} dt \quad (46)$$

Suppose that the optimum trajectory $f_{\text{opt}}(\cdot)$ is such that at $\lambda_0 \in [0,1]$,

$$f_{\text{opt}}(\lambda_0) = g(\lambda_0)$$

After that point it must be true that

$$f'_{\text{opt}}(t) = u_m \qquad \text{for } t \geqslant \lambda_0$$

since that will add zero additional contribution to the integral of (46) because

$$\int_{\lambda_0}^1 I_{f_{\text{opt}}(t)}(f'_{\text{opt}}(t))\,dt = \int_{\lambda_0}^1 I_x(u_m)\,dt = 0$$

Hence

$$I(f_{\text{opt}}) = \int_0^{\lambda_0} \sup_\theta \left\{ \theta f'(t) - \log \int p(\xi)^\theta q(\xi)^{1-\theta}\,d\xi \right\} dt \quad (47)$$

By a simple calculus of variations exercise (see exercise 9), it is easy to show that the optimal trajectory is a straight line with initial value of zero and slope

$g(\lambda_0)/\lambda_0$. Thus

$$I(f_{\text{opt}}) = \lambda_0 \sup_\theta \left\{ \theta \frac{g(\lambda_0)}{\lambda_0} - \log \int p(\xi)^\theta q(\xi)^{1-\theta} \, d\xi \right\} \tag{48}$$

which of course should equal α.

A moment's thought should indicate that the optimal way to choose $g(\cdot)$ is so that a trajectory has an equal asymptotic chance of crossing its graph for every value of $\lambda \in [0, 1]$. The large deviation probabilities are controlled by the worst-case trajectory, which implies that one should make the worst case uniform over λ. This gives the minimal $g(\cdot)$ consistent will false-alarm rate α and hence must give the best type II error rate performance.

Hence our design criteria is: Choose $g(\cdot)$ so that

$$\lambda \sup_\theta \left\{ \theta \frac{g(\lambda)}{\lambda} - \log \int p(\xi)^\theta q(\xi)^{1-\theta} \, d\xi \right\} \doteq \alpha \text{ for all } \lambda \in [0, 1] \tag{49}$$

Example **3.** Suppose that $q(x)$ and $p(x)$ are Gaussian, unit variance densities with means of zero and $s > 0$, respectively. Then

$$\log \int p(\xi)^\theta q(\xi)^{1-\theta} \, d\xi = s^2 \frac{\theta - \theta^2}{2}$$

and

$$\lambda \sup_\theta \left\{ \theta \frac{g(\lambda)}{\lambda} - \log \int p(\xi)^\theta q(\xi)^{1-\theta} \, d\xi \right\}$$

$$= \lambda \left[\frac{g(\lambda)}{2\lambda} \left[1 - \frac{g(\lambda)}{ts^2} \right] - \frac{s^2}{4} \left[1 - \frac{g(\lambda)^2}{t^2 s^4} \right] \right] = \alpha$$

This is a quadratic form for λ that may be solved to give

$$g(\lambda) = -\frac{s^2}{2}\lambda + \sqrt{2s^2\alpha\lambda}$$

It is interesting to note that although an optimum test does not exist in the classical Neyman–Pearson sense, in certain cases one does exist in this exponential Neyman–Pearson formulation that we are using as our fidelity criterion (see exercise 10). We also leave to exercise 10 the problem of calculating the type II error performance of the above test. □

Notes and Comments

Bahadur first proposed the asymptotic rate of the error probability to zero as a measure of the goodness of a test. Hence this measure is frequently called the *Bahadur efficiency*.

Stein's lemma evidently first appeared in an unpublished manuscript. Bahadur [1971] gives a more complete proof for the case when densities do not exist and randomized tests have to be considered in the Neyman–Pearson formulation. The derivation in the text using the Gärtner–Ellis theorem is new.

Several authors have generalized Chernoff's theorem to the Markov case. Most of the work has been done using the entropic rate formulations of section B of chapter V, although sometimes not explicitly. Koopmans [1960], Boza [1971], Perez [1971, 1974], and more recently Natarajan [1985] and Anantharam [1987] consider the problem in this light.

Memoryless detection theory originated in a paper by Miller and Thomas [1972]. Since then many authors have worked on the problem for various kinds of random processes [Poor and Thomas, 1979; Halverson and Wise, 1984]. The fidelity criterion in these works is almost always the asymptotic relative efficiency (ARE) of statistical decision theory instead of the asymptotic rate of the probability of error or Bahadur efficiency, which we have used. The ARE formulation requires that the two hypotheses converge to one another in some sense. The relative efficiency is then related to the number of samples required to maintain a certain type I and II error. More recently, large deviation analysis (but not design) techniques have appeared in Sadowsky and Bucklew [1986] and Sadowsky [1986]. The design techniques presented here are a reworking of Bucklew and Ney [1988].

The analysis of quadratic form detectors in Gaussian noise is due to Benitz and Bucklew [1987]. This paper gives rigorous alternatives to the heuristics presented in the text. Evaluating the limits involving deterministic signals possessing a power spectrum necessitated a slight extension of the Toeplitz distribution theorem, which is also presented in that paper.

The work presented on the changepoint problem and the associated exercises is a reworking of some pioneering results of Deshayes and Picard [1982, 1986].

EXERCISES

1. Generalize Stein's lemma to the Markov hypothesis testing case.

2. Consider the two transition matrices $\begin{bmatrix} .7 & .3 \\ .3 & .7 \end{bmatrix}$ and $\begin{bmatrix} .3 & .7 \\ .7 & .3 \end{bmatrix}$ in which the

ith row and jth column represents $p(x_{m+1} = j \mid x_m = i)$. What is the form of the asymptotically optimum detector? What is the asymptotic rate to zero of the probability of error?

3. Consider a fixed ratio $\theta_0/\theta_1 \equiv R$ such that $\theta_0 > 0$, $\theta_1 < 0$, $\theta_0 - \theta_1 = 1$. Show that if a solution exists to the system $r_{f_{\text{opt}},1}(x) r^*_{f_{\text{opt}},1}(x) = r_{f_{\text{opt}},0}(x) r^*_{f_{\text{opt}},0}(x)$, it is unique.

4. Consider a fixed ratio $\theta_0/\theta_1 \equiv R$ such that $\theta_0 > 0$, $\theta_1 < 0$, $\theta_0 - \theta_1 = 1$. For any function $f(\cdot)$ on the state space, associate with it a real number λ_f given by the following procedure. Compute the eigenstructure of the operators associated with this function (i.e., $\Sigma_x \exp[\theta_i f(x)] p_i(x \mid y) r_i(x) \equiv \lambda_{f,i} r_i(x) i = \{0, 1\}$). Define the constant c_f to be that constant which when added to $f(\cdot)$ makes $\lambda_{f+c,1} = \lambda_{f+c,0} \equiv \lambda_f$. If the state space of the Markov chain has d values, one may consider $f(\cdot)$ to be a vector in $\mathbb{R}^d$. The mapping $f \to \lambda_f$ (which is a mapping from $\mathbb{R}^d$ to $\mathbb{R}$) is differentiable. Find an expression for the derivative and from that find an expression for the (unique) critical point. What is the steepest descent direction to the critical point starting from point f? Suppose that the critical value f is obtained for some θ_0 and θ_1. In what direction should θ_1 (and θ_0) be perturbed in order to bring about a lowering of the eigenvalue? What is a condition for a critical point for the θ perturbations?

5. Generalize the theory of section C to the following cases

(a) Consider the Neyman–Pearson problem. Fix the type I error to be α, $0 < \alpha < 1$. Find the optimum nonlinearity to maximize the rate of the type II error to zero.

(b) Consider continuous-time finite state space Markov processes.

(c) Consider the case of discrete-time m-Markov chains, i.e., $p(x_n \mid x_{n-1}, x_{n-2}, \ldots) = p(x_n \mid x_{n-1}, x_{n-2}, \ldots, x_{n-m})$.

6. Consider the Markov chain transition matrix:

$$\begin{bmatrix} 0 & 1 & 0 & 0 & 0 \\ 1-\mu_i & 0 & \mu_i & 0 & 0 \\ 0 & 1-\mu_i & 0 & \mu_i & 0 \\ 0 & 0 & 1-\mu_i & 0 & \mu_i \\ 0 & 0 & 0 & 1-\mu_i & \mu_i \end{bmatrix}$$

Let $\mu_0 = .3$ and $\mu_1 = .7$. On a computer, find the optimum memoryless nonlinearity for the hypothesis testing problem between μ_0 and μ_1. What is the asymptotic rate to zero of the probability of error? How does this rate compare to the optimal rate?

7. Compute the (closed form) asymptotics of the probability of error for the radiometer in the case of additive white Gaussian noise.

8. [Dacunha-Castelle, 1979] Compute the asymptotics of the probability of error for the problem of discriminating between two zero mean stationary Gaussian random processes with an optimal detection scheme. (*Hint:* Use a Gärtner–Ellis theorem type of formulation on the likelihood ratio test.)

9. Show that $\inf_{f \in C_\Gamma} \int_0^{\lambda_0} \sup_\theta \{\theta f'(t) - \log \int p(\xi)^\theta q(\xi)^{1-\theta}\, d\xi\}$ is a straight line with initial value of zero and slope $g(\lambda_0)/\lambda_0$.

10. Compute the type II error performance for the detector in example 3. Show that this detector is (asymptotically) uniformly most powerful for this problem.

CHAPTER VII

Asymptotic Expansions

A. IID REAL VALUED CASE

Often one needs to know more than just the asymptotic rate with which a probability approaches zero. For example, one may actually have to give a number for the value of the probability of error in a digital communications link. The tools and methodologies of large deviation theory that we have investigated so far have not touched on problems of this type. In this chapter we indicate some of the results that are available for finding the constants in front of the exponential rates.

Let $\mathbf{x}_1, \mathbf{x}_2, \ldots$ be an independent, identically distributed sequence of random variables with distribution μ. Define

$$M(\theta) \equiv E\{\exp[\theta \mathbf{x}_1]\} = \int \exp[\theta x]\mu(dx)$$

Suppose that θ_a exists so that

$$I(a) \equiv \sup_{\theta} \{\theta a - \log M(\theta)\} = \theta_a a - \log M(\theta_a)$$

where θ_a would be given by the solution to the equation

$$a = \frac{E\{\mathbf{x}_1 \exp[\theta_a \mathbf{x}_1]\}}{M(\theta_a)} = \frac{M'(\theta_a)}{M(\theta_a)} \tag{1}$$

We are interested in the asymptotics of

$$P(\mathbf{S}_n \geqslant a) \equiv P\left(\frac{1}{n} \sum_{i=1}^{n} \mathbf{x}_i \geqslant a\right) \tag{2}$$

Define the twisted probability distribution

$$\mu_\gamma(dx) \equiv \frac{\exp[\gamma x]\mu(dx)}{\int \exp[\gamma x]\mu(dx)} = \frac{\exp[\gamma x]\mu(dx)}{M(\gamma)} \tag{3}$$

Considering the moment generating function of this distribution:

$$\int \exp[\theta x]\mu_\gamma(dx) = \frac{\int \exp[\theta + \gamma]\mu(dx)}{M(\gamma)} = \frac{M(\theta + \gamma)}{M(\gamma)} \tag{4}$$

From (4), we can compute the mean value of the twisted distribution as

$$\frac{M'(\gamma)}{M(\gamma)} \tag{5}$$

and the variance as

$$\frac{M''(\gamma)}{M(\gamma)} - \left(\frac{M'(\gamma)}{M(\gamma)}\right)^2 \tag{6}$$

It also follows that the moment generating function of the convolution of n copies of $\mu_\gamma(\cdot)$ is given by

$$\left[\frac{M(\theta + \gamma)}{M(\gamma)}\right]^n \tag{7}$$

The inverse transform of (7) gives

$$\mu_\gamma^{*n}(dx) \equiv \text{convolution of } n \text{ copies of } \mu_\gamma(dx) = \frac{\exp[\gamma x]\mu^{*n}(dx)}{M(\gamma)^n} \tag{8}$$

Therefore, from (2) and (8), we have

$$P(\mathbf{S}_n \geqslant a) = \int_{na}^{\infty} \mu^{*n}(dx) = M(\gamma)^n \int_{na}^{\infty} \exp[-\gamma x]\mu_\gamma^{*n}(dx) \tag{9}$$

Now perform a change of variable and replace x by $x + na$ to obtain that (9) is

$$\begin{aligned} P(\mathbf{S}_n \geqslant a) &= M(\gamma)^n \int_0^{\infty} \exp[-\gamma(x + na)]\mu_\gamma^{*n}(dx + na) \\ &= M(\gamma)^n \exp[-\gamma na] \int_0^{\infty} \exp[-\gamma x]\mu_\gamma^{*n}(dx + na) \end{aligned} \tag{10}$$

Now choose $\gamma = \theta_a$ to obtain from (10) that

$$P(\mathbf{S}_n \geqslant a) = [M(\theta_a)\exp[-\theta_a a]]^n \int_0^\infty \exp[-\theta_a x]\mu_{\theta_a}^{*n}(dx + na)$$

$$= \exp[-nI(a)] \int_0^\infty \exp[-\theta_a x]\mu_{\theta_a}^{*n}(dx + na) \tag{11}$$

Now from (1) and (5) the mean value of $\mu_{\theta_a}(\cdot)$ is just

$$\frac{M'(\theta_a)}{M(\theta_a)} = a \tag{12}$$

The convolution $\mu_{\theta_a}^{*n}(\cdot)$ then has mean value na or the mean value of $\mu_{\theta_a}^{*n}(\cdot + na)$ is zero. The variance of $\mu_{\theta_a}(\cdot)$ is, from (6) and (12),

$$\sigma^2 \equiv \frac{M''(\theta_a)}{M(\theta_a)} - \left[\frac{M'(\theta_a)}{M(\theta_a)}\right]^2 = \frac{M''(\theta_a)}{M(\theta_a)} - a^2 \tag{13}$$

which implies that the variance of $\mu_{\theta_a}^{*n}(\cdot)$ [which is the same as that of $\mu_{\theta_a}^{*n}(\cdot + na)$] is $n\sigma^2$.

Let $1_A(\cdot)$ be the indicator function of the set A, i.e., $1_A(x) \equiv 1$ if $x \in A$ and $1_A(x) \equiv 0$ otherwise. Therefore, the integral in (11),

$$\int_0^\infty \exp[-\theta_a x]\mu_{\theta_a}^{*n}(dx + na) \tag{14}$$

is just the expectation of $\exp[-\theta_a \mathbf{x}]1_{[0,\infty)}(\mathbf{x})$ under the distribution $\mu_{\theta_a}^{*n}(dx + na)$. This, of course, is the distribution of the sum of n independent, zero mean, variance σ^2, random variables. Let us normalize this sum by dividing by $\sqrt{n}\sigma$. The resulting normalized sum has mean zero and variance 1. Let us denote its distribution by $H_n(dx)$. Hence

$$\int_0^\infty \exp[-\theta_a x]\mu_{\theta_a}^{*n}(dx + na) = \int_0^\infty \exp[-\theta_a x\sqrt{n}\sigma]H_n(dx) \tag{15}$$

(Note that if we upper bound this quantity by the constant 1, we immediately obtain the celebrated *Chernoff bound* first mentioned in section A of chapter II.) Now $H_n(dx)$ is converging (by the Central Limit Theorem) in distribution to a mean zero, variance 1, normal distribution. We now need to distinguish between two separate situations in order to determine the asymptotic behavior of (15). The separate cases to be considered are:

1. The $\{\mathbf{x}_n\}$ are lattice random variables, i.e., $\mathbf{x}_i$ only takes on values in the set $\{c + kd: k = 0, \pm 1, \pm 2, \ldots\}$, where c is a fixed constant and d is the lattice spacing. Without loss of generality (for our asymptotics) we will assume that $c \equiv 0$.
2. The $\{\mathbf{x}_n\}$ are not lattice-type random variables.

***Case* 1.** Let us return to (15). It is simple to show by an integration by parts that

$$\int_0^\infty \exp[-\theta_a x\sqrt{n\sigma}]H_n(dx)$$

$$= \theta_a\sqrt{n\sigma}\int_0^\infty \exp[-\theta_a x\sqrt{n\sigma}][H_n(x) - H_n(0)]\,dx$$

$$= \theta_a\sqrt{n\sigma}\sum_{k=0}^{\infty}\int_{kd/\sqrt{n\sigma}}^{(k+1)d/\sqrt{n\sigma}} \exp[-\theta_a x\sqrt{n\sigma}]\,dx\left[H_n\left(\frac{(k+1)d}{\sqrt{n\sigma}}\right) - H(0)\right]$$

$$= \sum_{k=0}^{\infty}\left[H_n\left(\frac{(k+1)d}{\sqrt{n\sigma}}\right) - H(0)\right][\exp[-\theta_a kd] - \exp[-\theta_a(k+1)d]] \qquad (16)$$

Now for a fixed k (by the Central Limit Theorem),

$$\left[H_n\left(\frac{(k+1)d}{\sqrt{n\sigma}}\right) - H(0)\right] \approx \left[\mathcal{N}\left(\frac{(k+1)d}{\sqrt{n\sigma}}\right) - \mathcal{N}(0)\right]$$

$$\approx \int_0^{(k+1)d/\sqrt{n\sigma}} \frac{\exp[-x^2/2]}{\sqrt{2\pi}}\,dx \approx \frac{(k+1)d}{\sqrt{2\pi n\sigma}} \qquad (17)$$

where $\mathcal{N}(\cdot)$ denotes the Gaussian mean zero, variance 1 probability distribution. Formally substituting (17) into (16) yields

$$\int_0^\infty \exp[-\theta_a x\sqrt{n\sigma}]H_n(dx)$$

$$= \sum_{k=0}^{\infty}\frac{(k+1)d}{\sqrt{2\pi n\sigma}}[\exp[-\theta_a kd] - \exp[-\theta_a(k+1)d]] \qquad (18)$$

Note that by summing the geometric series, we have

$$\sum_{k=0}^{\infty}(k+1)\exp[-\theta_a(k+1)d] = \frac{\exp[-\theta_a d]}{(1 - \exp[-\theta_a d])^2} \qquad (19)$$

and

$$\sum_{k=0}^{\infty} (k+1)\exp[-\theta_a kd] = \frac{1}{(1-\exp[-\theta_a d])^2} \tag{20}$$

Hence, substituting expressions (19)–(20) into (18), we have

$$\int_0^{\infty} \exp[-\theta_a x\sqrt{n}\sigma] H_n(dx) \approx \frac{d}{\sqrt{2\pi n}\,\sigma(1-\exp[-\theta_a d])} \tag{21}$$

Therefore from (11), (15), and (21), we obtain

$$\lim_{n\to\infty} P(\mathbf{S}_n \geqslant a)\exp[nI(a)]\sqrt{n} = \frac{d}{\sqrt{2\pi}\,\sigma(1-\exp[-\theta_a d])} \tag{22}$$

***Case* 2.** It is intuitive that we can approximate any sequence $\{\mathbf{x}_i\}$ of iid random variables by a sequence of lattice-type random variables $\{\mathbf{x}_i'\}$ with lattice spacing d. [The obvious approximation would be $\mathbf{x}_i' = kd$ if and only if $(k-1)d < \mathbf{x}_i \leqslant kd$.] The approximation will be more precise and should become exact as d is allowed to approach zero. Based on this heuristic argument, one would expect that the nonlattice constant should be attained by taking the limit as d approaches zero of (22). Performing this formal operation yields

$$\lim_{n\to\infty} P(\mathbf{S}_n \geqslant a)\exp[nI(a)]\sqrt{n} = \frac{1}{\sqrt{2\pi}\,\sigma\theta_a} \tag{23}$$

It turns out that this is indeed the correct limit, with the rigorous statement of the theorem as follows:

Theorem (Bahadur–Rao)
Let $\mathbf{x}_1, \mathbf{x}_2, \ldots$ be an iid sequence of random variables such that $E\{\mathbf{x}_1\} = 0$ and $M(\theta) < \infty$ for all $\theta \in \mathbb{R}$. Let $a > 0$ and assume that $P(\mathbf{x}_1 > a) > 0$. Then, if $\mathbf{x}_1$ is not of lattice type,

$$\lim_{n\to\infty} P(\mathbf{S}_n \geqslant a)\exp[nI(a)]\sqrt{n} = (\sqrt{2\pi}\,\sigma\theta_a)^{-1} \tag{24}$$

If $\mathbf{x}_1$ is of lattice type and $P(\mathbf{x}_1 = a) > 0$, then

$$\lim_{n\to\infty} P(\mathbf{S}_n \geqslant a)\exp[nI(a)]\sqrt{n} = \frac{d}{\sigma(1-\exp[-\theta_a d])\sqrt{2\pi}} \tag{25}$$

A rigorous proof of this theorem requires delicate error estimates in the Central Limit Theorem (known as Berry–Esseen inequalities).

B. FUNCTIONAL OF A MARKOV CHAIN CASE

Deriving the asymptotic expansion for the Markov chain case is more technically difficult than in the iid case. The same ideas, however, are used in the two derivations. We first "tilt" or "twist" the original probability distribution so that the resulting "twisted" mean value is at the dominating point of the set of interest. One then invokes a Central Limit Theorem to obtain the full asymptotic expansion.

Suppose that we are interested in the asymptotic expansion of

$$P(\mathbf{S}_n = a) \equiv P\left(\frac{1}{n}\sum_{i=1}^{n} f(\mathbf{x}_i) = a\right) \tag{26}$$

where $\{\mathbf{x}_i\}$ is a finite state space S irreducible Markov chain with initial distribution $p(x_1)$ and transition probabilities

$$p(x \mid y) \equiv P\{\mathbf{x}_{k+1} = x \mid \mathbf{x}_k = y\} \tag{27}$$

and $f(\cdot)$ is an arbitrary function mapping from S into Z (the integers). Hence $f(\mathbf{x}_i)$ is a lattice-type random variable (with lattice spacing $d = 1$) by the definition of the preceding section. We will assume that $P(\mathbf{S}_n = a) > 0$ for all n. We know that a large deviation principle exists for $\mathbf{S}_n$ with rate function given by

$$I(a) = \sup_{\theta} \{\theta a - \log \lambda(\theta)\} \tag{28}$$

where $\lambda(\theta)$ is the largest eigenvalue of the operator

$$T_\theta(g)(y) \equiv \sum_x \exp[\theta f(x)] g(x) p(x \mid y) \tag{29}$$

or, equivalently, the adjoint operator

$$T_\theta^*(g)(y) \equiv \exp[\theta f(y)] \sum_x g(x) p(y \mid x)$$

We will denote the eigenvectors associated with this eigenvalue for the operators T_θ and T_θ^* as $r_\theta(\cdot)$ and $r_\theta^*(\cdot)$, respectively. In this relatively nice setting we can certainly guarantee that

$$I(a) = \theta_a a - \log \lambda(\theta_a) \tag{30}$$

where θ_a is the solution to the equation

$$a = \frac{\lambda'(\theta_a)}{\lambda(\theta_a)} \tag{31}$$

Define a new transition probability (the *twisted transition probability*) as

$$t(x \mid y) \equiv \exp[\theta_a f(x)] r_{\theta_a}(x) \frac{p(x \mid y)}{\lambda(\theta_a) r_{\theta_a}(y)} \tag{32}$$

It is easy to verify that $t(x \mid y)$ is the transition probability of a different Markov chain (merely note that for every fixed y, $t(x \mid y)$ is a valid probability distribution on x since by definition

$$T_\theta(r_{\theta_a})(y) \equiv \sum_x \exp[\theta_a f(x)] r_{\theta_a}(x) p(x \mid y) = \lambda(\theta_a) r_{\theta_a}(y)) \tag{33}$$

Now let us consider the problem of finding the stationary distribution of the twisted chain. If T is the transition matrix for the twisted chain, then the stationary distribution, π, satisfies the matrix eigenvector equation $\pi T = \pi$. $t(x \mid y)$ is the entry in the T matrix for the xth column and yth row. The matrix equation may equivalently be written

$$\sum_y \pi(y) t(x \mid y) = \pi(x) \tag{34}$$

From (32) the left-hand side of (34) is

$$\sum_y \pi(y) \exp[\theta_a f(x)] r_{\theta_a}(x) \frac{p(x \mid y)}{\lambda(\theta_a) r_{\theta_a}(y)} \tag{35}$$

Let us try the solution

$$\pi(y) = r_{\theta_a}(y) r^*_{\theta_a}(y) \tag{36}$$

Substituting (36) into (35), we obtain

$$\lambda(\theta_a)^{-1} r_{\theta_a}(x) \exp[\theta_a f(x)] \sum_y r^*_{\theta_a}(y) p(x \mid y) \tag{37}$$

The term in the sum times the exponential is just the adjoint operator operating on its eigenvector $r^*_{\theta_a}$. Thus (37) becomes

$$r_{\theta_a}(x)r^*_{\theta_a}(x)\frac{\lambda(\theta_a)}{\lambda(\theta_a)} = r_{\theta_a}(x)r^*_{\theta_a}(x) = \pi(y)$$

Hence we have verified that $r_{\theta_a}(x)r^*_{\theta_a}(x)$ is the stationary distribution of the twisted chain.

Consider the probability

$$\begin{aligned} P(\mathbf{S}_n = a \mid \mathbf{x}_1 = x, \mathbf{x}_n = y) &= P\left(\sum_{i=1}^{n} f(\mathbf{x}_i) = na \mid \mathbf{x}_1 = x,\, \mathbf{x}_n = y\right) \\ &= \sum\sum\cdots\sum_{x_1,\ldots,x_n:S_n=a} p(x_1, x_2, \ldots, x_n \mid \mathbf{x}_1 = x, \mathbf{x}_n = y) \quad (38) \end{aligned}$$

where $p(x_1, x_2, \ldots, x_n \mid \mathbf{x}_1 = x, \mathbf{x}_n = y) \equiv P\{\mathbf{x}_1 = x_1, \mathbf{x}_2 = x_2, \ldots, \mathbf{x}_n = x_n \mid \mathbf{x}_1 = x, \mathbf{x}_n = y\}$. However, we have for conditional probabilities

$$\begin{aligned} p(x_1, \ldots, x_n \mid \mathbf{x}_1 = x, \mathbf{x}_n = y) &= \frac{p(x, x_2, \ldots, x_{n-1}, y)}{p(x)p^{(n)}(y \mid x)} \\ &= p(x)\frac{\prod_{i=1}^{n-1} p(x_{i+1} \mid x_i)}{p(x)p^{(n)}(y \mid x)} \\ &= \frac{\prod_{i=1}^{n-1} p(x_{i+1} \mid x_i)}{p^{(n)}(y \mid x)} \quad (39) \end{aligned}$$

where $p(x)$ is the probability distribution of $\mathbf{x}_1$ and $p^{(n)}(y \mid x)$ is the probability of the chain hitting y after n steps given that it starts off at x. Hence substituting (39) into (38) and using (32), we obtain

$$\begin{aligned} P(\mathbf{S}_n = a \mid \mathbf{x}_1 = x, \mathbf{x}_n = y) &= \sum\sum\cdots\sum_{x_1,\ldots,x_n:S_n=a} \frac{\prod_{i=1}^{n-1} p(x_{i+1} \mid x_i)}{p^{(n)}(y \mid x)} \\ &= \frac{\lambda(\theta_a)^{n-1}}{p^{(n)}(y \mid x)} \sum\sum\cdots\sum_{x_1,\ldots,x_n:S_n=a} \exp\left[-\theta_a \sum_{i=2}^{n} f(x_i)\right]\left[\prod_{i=1}^{n-1} \frac{r_{\theta_a}(x_i)}{r_{\theta_a}(x_{i+1})}\right]\left[\prod_{i=1}^{n-1} t(x_{i+1} \mid x_i)\right] \\ &= \left[\frac{\lambda(\theta_a)^{n-1} r_{\theta_a}(x)}{p^{(n)}(y \mid x) r_{\theta_a}(y)}\right] \sum\sum\cdots\sum_{x_1,\ldots,x_n S_n=a} \exp[-\theta_a(na - f(x))] \prod_{i=1}^{n-1} t(x_{i+1} \mid x_i) \\ &= \left[\frac{\lambda(\theta_a)^{n-1} r_{\theta_a}(x)}{p^{(n)}(y \mid x) r_{\theta_a}(y)}\right] \sum\sum\cdots\sum_{x_1,\ldots,x_n:S_n=a} \\ &\quad \cdot \exp[-\theta_a(na - f(x))] t^{(n)}(y \mid x) t(x_1, x_2, \ldots, x_n \mid \mathbf{x}_1 = x, \mathbf{x}_n = y) \quad (40) \end{aligned}$$

Now $t^{(n)}(y \mid x)$ should converge as $n \to \infty$ to $r_{\theta_a}(y) r^*_{\theta_a}(y)$ the stationary distribution of the twisted chain. Therefore, for large n, we may replace $t^{(n)}(y \mid x)$ by $r_{\theta_a}(y) r^*_{\theta_a}(y)$ to obtain

$$P(S_n = a \mid \mathbf{x}_1 = x, \mathbf{x}_n = y) \approx \left[\lambda(\theta_a)^{n-1} r_{\theta_a}(x) \frac{r^*_{\theta_a}(y)}{p^{(n)}(y \mid x)} \right] \sum \sum_{x_1, \ldots, x_n : S_n = a} \cdots \sum$$
$$\cdot \exp[-\theta_a(na - f(x))] t(x_1, x_2, \ldots, x_n \mid \mathbf{x}_1 = x, \mathbf{x}_n = y) \tag{41}$$

Note also, by a similar argument, that we have for any constant $s \in Z$ (including $s = 0$)

$$P\left(S_n = a + \frac{s}{n} \mid \mathbf{x}_1 = x, \mathbf{x}_n = y\right) \approx \left[\lambda(\theta_a)^{n-1} r_{\theta_a}(x) \frac{r^*_{\theta_a}(y)}{p^{(n)}(y \mid x)} \right] \exp[-\theta_a(s - f(x))]$$
$$\cdot \sum \sum_{x_1, \ldots, x_n : S_n = a + s/n} \cdots \sum \exp[-\theta_a na] t(x_1, x_2, \ldots, x_n \mid \mathbf{x}_1 = x, \mathbf{x}_n = y)$$
$$= \exp[-n(\theta_a a - \log \lambda(\theta_a))] \left[r_{\theta_a}(x) \frac{r^*_{\theta_a}(y)}{p^{(n)}(y \mid x)} \right]$$
$$\cdot \exp[-\theta_a(s - f(x)] \lambda(\theta_a)^{-1} \sum \sum_{x_1, \ldots, x_n : S_n = a + s/n} \cdots \sum t(x_1, x_2, \ldots, x_n \mid \mathbf{x}_1 = x, \mathbf{x}_n = y)$$
$$= \exp[-nI(a)] \left[r_{\theta_a}(x) \frac{r^*_{\theta_a}(y)}{p^{(n)}(y \mid x)} \right]$$
$$\cdot \exp[-\theta_a(s - f(x))] \lambda(\theta_a)^{-1} \sum \sum_{x_1, \ldots, x_n : S_n = a + s/n} \cdots \sum t(x_1, x_2, \ldots, x_n \mid \mathbf{x}_1 = x, \mathbf{x}_n = y) \tag{42}$$

Since the invariant distribution of the twisted chain is $r_{\theta_a}(\cdot) r^*_{\theta_a}(\cdot)$, we must have that the asymptotic mean value of $\mathbf{S}_n$ (under the twisted distribution) is

$$\lim_{n \to \infty} E_t\{\mathbf{S}_n\} = \sum_x f(x) r_{\theta_a}(x) r^*_{\theta_a}(x) \tag{43}$$

From perturbation theory arguments, we have that the first order approximation to

$$\lambda(\theta + \Delta\theta) = \lambda(\theta) + \Delta\theta \sum_x \sum_y \exp[\theta f(x)] f(x) r_\theta(x) r^*_\theta(y) p(x \mid y) + o(\Delta\theta)$$
$$= \lambda(\theta) + \lambda(\theta) \Delta\theta \sum_x f(x) r_\theta(x) r^*_\theta(x) + o(\Delta\theta) \tag{44}$$

Hence

$$\lambda'(\theta) = \lim_{\Delta\theta \to 0} \frac{\lambda(\theta + \Delta\theta) - \lambda(\theta)}{\Delta\theta} = \sum_x \lambda(\theta) f(x) r_\theta(x) r^*_\theta(x) \tag{45}$$

Therefore,

$$\frac{\lambda'(\theta)}{\lambda(\theta)} = \sum_x f(x) r_{\theta_a}(x) r^*_{\theta_a}(x) \tag{46}$$

or, from (31),

$$\lim_{n\to\infty} E_t\{\mathbf{S}_n\} = \frac{\lambda'(\theta_a)}{\lambda(\theta_a)} = a \tag{47}$$

by the definition of θ_a. By using properties of the moment generating functions, one cannot only verify the above but can also check that (see exercise 1)

$$\sigma^2 \equiv \lim_{n\to\infty} \mathrm{var}_t\{\mathbf{S}_n\sqrt{n}\} = \frac{\lambda''(\theta_a)}{\lambda(\theta_a)} - a^2 \tag{48}$$

One of our main results now follows by an application of the Central Limit Theorem to the summation portion of (42):

$$\sum\sum\cdots\sum_{x_1,\ldots,x_n:S_n=a+s/n} t(x_1, x_2, \ldots, x_n \,|\, \mathbf{x}_1 = x, \mathbf{x}_n = y) \tag{49}$$

First perform a change of variable to shift the mean value of $\mathbf{S}_n$ to the left by $[a+(f(x)+f(y)+s)/n]$ and scale the result by $\sqrt{n}/\sigma$. $(\mathbf{S}_n - a - (f(x)+f(y)+s)/n)\sqrt{n}/\sigma$ should be converging in distribution to a normal law with zero mean and variance 1 for every fixed $s \in Z$ and x, y in the state space of the chain. Hence

$$\sum\sum\cdots\sum_{x_1,\ldots,x_n:S_n=a+s/n} t(x_1, x_2, \ldots, x_n \,|\, \mathbf{x}_1 = x, \mathbf{x}_n = y)$$

$$\approx \mathcal{N}\left(\frac{1}{2\sqrt{n}\sigma}\right) - \mathcal{N}\left(-\frac{1}{2\sqrt{n}\sigma}\right) \approx \frac{1}{\sqrt{2\pi n}\,\sigma} \tag{50}$$

Note that this result will depend on the lattice spacing d in general (see exercise 4). Note also that the result is independent of the initial and final conditions. The intuitive reason is that these initial and final conditions will have very little probabilistic effect on the large number of indices in the middle. As $n \to \infty$ these middle indices effectively account for the whole sum. The asymptotic probability distribution of $\mathbf{S}_n$ will essentially be as if the underlying distribution were $t(x_1, x_2, \ldots, x_n)$ instead of $t(x_1, x_2, \ldots, x_n \,|\, \mathbf{x}_1 = x, \mathbf{x}_n = y)$. The Central Limit approximation is very good near the mean value of a sum, which is where we are operating.

Equation (42) and (50) then give the result:

Theorem (Miller)
Let $\{\mathbf{x}_i\}$, S_n, $I(\cdot)$, $f(\cdot)$, be as above. Denote the principal eigenvalue of the complex operator $T_{i\gamma}(\cdot)$ ($i = \sqrt{-1}$ and $\gamma \in \mathbb{R}$) as $\mu(\gamma)$. We assume that $|\mu(\gamma)| < 1$ for $-\pi < \gamma < \pi$. Then

$$\lim_{n\to\infty} p^{(n)}(y \mid x) P\left(\mathbf{S}_n = a + \frac{s}{n} \middle| \mathbf{x}_1 = x, \mathbf{x}_n = y\right)[\exp[nI(a)]\sqrt{n}]$$

$$= r_{\theta_a}(x) r^*_{\theta_a}(y) \frac{\exp[-\theta_a(s - f(x))]}{\lambda(\theta_a)\sqrt{2\pi}\,\sigma}$$

Note that the result *does* depend (somewhat unintuitively) on the initial and final conditions. The residual effects of the initial and final conditions are not negligible because we are far away from the mean value of $\mathbf{S}_n$ (in fact, a large deviation away!). Central Limit approximations are not valid here at all; in fact, a Central Limit approximation to the above probability would not have any x, y dependence. The extra condition involving $\mu(\cdot)$ is very mild and is to prevent pathologies involving periodicities in the sums under consideration.

For our next limit result, let us return to (42) and sum from s equal 0 to ∞. We obtain

$$P(\mathbf{S}_n \geqslant a \mid \mathbf{x}_1 = x, \mathbf{x}_n = y) \approx \exp[-nI(a)]\left[r_{\theta_a}(x)\frac{r^*_{\theta_a}(y)}{p^{(n)}(y \mid x)}\right]\lambda(\theta_a)^{-1}\exp[\theta_a f(x)]$$

$$\cdot \sum_{s=0}^{\infty} \exp[-\theta_a s] \sum\sum\cdots\sum_{x_1,\ldots,x_n : S_n = a + s/n} t(x_1, x_2, \ldots, x_n \mid \mathbf{x}_1 = x, \mathbf{x}_n = y) \tag{51}$$

Now let us denote the probability distribution of $[S_n - a]\sqrt{n}/\sigma$ under the distribution $t(x_1, x_2, \ldots, x_n \mid \mathbf{x}_1 = x, \mathbf{x}_n = y)$ as $H_n(\cdot)_{xy}$. We then have with a change of variable (and integration by parts) that the summation portion of (51) becomes

$$\sum_{s=0}^{\infty} \exp[-\theta_a s] \sum\sum\cdots\sum_{x_1,\ldots,x_n : S_n = a + s/n} t(x_1, x_2, \ldots, x_n \mid \mathbf{x}_1 = x, \mathbf{x}_n = y)$$

$$= \int_0^{\infty} \exp[-\theta_a z\sqrt{n}\sigma] H_n(dz)_{xy}$$

$$= \theta_a\sqrt{n}\sigma \int_0^{\infty} \exp[-\theta_a z\sqrt{n}\sigma][H_n(z)_{xy} - H_n(0)_{xy}]\,dz \tag{52}$$

We may compute the asymptotics of this integral as we did in the lattice case of section A [(16) and (21) with lattice spacing $d = 1$]. The final result will be independent of x and y as $H_n(\cdot)_{xy}$ is converging to a normal, mean zero, unit variance random variable which of course has no x and y dependence. Hence (52) is approximately (for large n)

$$\sum_{s=0}^{\infty} \exp[-\theta_a s] \sum_{x_1,\ldots,x_n:S_n=a+s/n}\sum\cdots\sum t(x_1, x_2, \ldots, x_n \mid \mathbf{x}_1 = x, \mathbf{x}_n = y)$$
$$\approx \frac{1}{\sqrt{2\pi n}\,\sigma(1 - \exp[-\theta_a])} \tag{53}$$

Under the hypotheses of the preceding theorem, combining (51) and (53) gives the following corollary.

Corollary (Miller)

$$\lim_{n\to\infty} p^{(n)}(y \mid x) P(\mathbf{S}_n \geqslant a \mid \mathbf{x}_1 = x, \mathbf{x}_n = y)[\exp[nI(a)]\sqrt{n}]$$
$$= r_{\theta_a}(x) r_{\theta_a}(y) \frac{\exp[\theta_a f(x)]}{\lambda(\theta_a)(1 - \exp[-\theta_a])\sqrt{2\pi}\,\sigma}$$

One may sum out the remaining x and y dependency to write the previous corollary alternatively as

Corollary (Miller)

$$P(\mathbf{S}_n \geqslant a) = \sum_x \sum_y p^{(n)}(y \mid x) P(\mathbf{S}_n \geqslant a \mid \mathbf{x}_1 = x, \mathbf{x}_n = y) p(x)$$
$$\approx \exp[-nI(a)]\left(\sum_x p(x)\exp[\theta_a f(x)] r_{\theta_a}(x)\right) \frac{\Sigma_y r^*_{\theta_a}(y)}{\lambda(\theta_a)(1 - \exp[-\theta_a])\sqrt{2\pi n}\,\sigma} \tag{54}$$

Notes and Comments

Cramér [1938], in his original paper, gave full asymptotic expansions for sums of iid nonlattice-type real valued random variables. Hoeffding [1965] and Bahadur and Rao [1960] extended this work to lattice-type random variables (among other results). Work has also been done in the iid vector valued ($\mathbb{R}^d$) random variable setting to try to obtain the asymptotic constants. Ney [1983, 1984] has considered the asymptotics of $P(\mathbf{S}_n \in B)$, where B is a convex set in $\mathbb{R}^d$ and the $\{\mathbf{x}_i\}$ are d-dimensional random vectors. The asymptotic expansion of this probability depends very much on the *curvature* of the set near the dominating point of the set. The "constant" is of the form $\exp[-nI(B)]n^{-\gamma}c$, where $d/2 \leq \gamma \leq \frac{1}{2}$. The actual value of γ is the curvature dependent part. c appears to be computable as in the one-dimensional case but is not computed explicitly in these references.

The material in section B on Markov chains follows the seminal work of Miller [1961]. Miller restricted himself to finite state space Markov chains and lattice-type functionals with lattice spacing 1. The twisted Markov chain is first introduced in this paper. Ney and Nummelin [1987a, 1987b] extend via regeneration theory techniques much of this work to very general Markov processes in discrete and continuous time but only for $\mathbb{R}^d$ valued functionals of the process.

EXERCISES

1. Verify for the Markov chain case that under the twisted distribution $\lim_{n\to\infty} \text{var}_t\{\mathbf{S}_n\sqrt{n}\} = \lambda''(\theta_a)/\lambda(\theta_a) - a^2 \equiv \sigma^2$.

2. Let $\{\mathbf{x}_i\}$ be an iid sequence of zero mean variance 1 Gaussian random variables. Compute the full asymptotics of $P\{(1/n)\Sigma_{i=1}^n \mathbf{x}_i \geq a\}$. Does this match up with exercise 8 of chapter II?

3. Let $\{\mathbf{x}_i\}$ be a sequence of iid zero mean, variance σ^2, lattice-type random variables. Define $\mathbf{S}_n \equiv (1/n)\Sigma_{i=1}^n \mathbf{x}_i$. Suppose that $P(\mathbf{S}_n = 0) > 0$ for all n. Evaluate the limit

$$\lim_{n\to\infty} P(\mathbf{S}_n = 0)\sqrt{n}$$

4. Suppose that the Markov chain functionals considered in section B are not on the integers but on some other lattice with lattice spacing d. How should the formulas for the asymptotics be modified in this case? What about in the nonlattice case?

5. Consider exercise 6 in chapter VI. Compute the full asymptotic expansion of the type I error for the optimal memoryless detector.

CHAPTER VIII

Quick Simulation

In chapter VII we saw how to derive some of the fine structure of the asymptotic probability of a large deviation. The formulas we derived (with quite a bit of effort) are valid only as the number of samples approaches infinity. One may hope that for finite n, the equations give reasonably good approximations, but sometimes that still is not adequate. Since the test statistic's probability distribution is usually very difficult or impossible to compute in closed form, one is faced with the problem of computer simulation in order to find the probability of interest.

How does one go about simulating a rare event in order to find its probability? This is a more difficult problem than one may expect. Consider a sequence $\{\mathbf{x}_i\}$ of iid Bernoulli random variables with

$$P\{\mathbf{x}_1 = 1\} = \rho = 1 - P\{\mathbf{x}_1 = 0\}$$

Suppose that we wish to estimate from the observed sequence the parameter ρ. We wish to have at most a 20% error on ρ with 95% confidence. This means that we must have

$$P\{|\rho - \hat{\boldsymbol{\rho}}| \leqslant .2\rho\} = .95$$

where $\hat{\boldsymbol{\rho}}$ is the estimate of ρ. For example, the maximum likelihood estimate would be

$$\hat{\boldsymbol{\rho}} \equiv \frac{1}{k} \sum_{i=1}^{k} \mathbf{x}_i$$

The variance of a Bernoulli ρ random variable is $\rho(1 - \rho)$. If ρ is very small (i.e., $\mathbf{x}_1 = 1$ is a rare event), the variance of $\mathbf{x}_1$ is approximately ρ. Hence the variance of $\hat{\mathbf{p}}$ is

$$\frac{\rho(1-\rho)}{k} \approx \frac{\rho}{k}$$

[the mean value of $\hat{\mathbf{p}}$ is ρ which (by definition) means that the estimate is *unbiased*]. Hence using a Central Limit Theorem approximation, we have that

$$P\{|\rho - \hat{\mathbf{p}}| \leqslant .2\rho\} = P\left\{\frac{1}{k}\sum_{i=1}^{k} \frac{\mathbf{x}_i - \rho}{\sqrt{\rho}} \leqslant .2\sqrt{\rho}\right\}$$

$$\approx P\{|\mathbf{Z}| \leqslant .2\sqrt{\rho k}\}$$

where $\mathbf{Z}$ is a Gaussian, mean zero, variance 1, random variable. Now for a standard Gaussian we have that $P\{|\mathbf{Z}| \leqslant z\} = .95$ implies (from tables) that $z \approx 2$, i.e., two standard deviations about the mean captures 95% of the probability of a Gaussian distribution. Hence $.2\sqrt{\rho k} = 2$ implies that $k = 100/\rho$. Therefore, if ρ is somewhere on the order of 10^{-6}, we would need $10^8 \approx 2^{27}$ number of samples to estimate it to the desired level of precision. Typical random number generators have a period of anywhere from 2^{15} to 2^{32}, the latter number considered a long period. The number of "good" random numbers that one generates is typically much less than the period of the generator. On the other hand, error probabilities on the order of 10^{-6} are typical in digital communication system performance. Hence it is clear that a direct simulation of such a system would require an unrealistically large amount of computation and impose severe, if not impossible demands on the random number generator.

One way out of this quandary is to utilize a notion from simulation theory known as *importance sampling*. Suppose that we wish to estimate

$$\rho = E\{\eta(\mathbf{X})\}$$

where $\mathbf{X}$ is a random variable describing some observation on a random system. η is typically the indicator function of some set; hence ρ would be the probability of that set. Suppose that the observation random variable $\mathbf{X}$ is controlled by a probability density function $p(x)$. The direct simulation method would be to generate a sequence of iid random numbers $\mathbf{x}_1, \mathbf{x}_2, \ldots, \mathbf{x}_k$ from the density $p(x)$ and form the estimate

$$\hat{\rho}_p \equiv \frac{1}{k}\sum_{i=1}^{k} \eta(\mathbf{x}_i) \tag{1}$$

Alternatively, we could generate a sequence of iid random numbers $\mathbf{y}_1$,

$\mathbf{y}_2, \ldots, \mathbf{y}_k$ distributed with density $q(x)$. We could then form the estimate

$$\hat{\boldsymbol{\rho}}_q \equiv \frac{1}{k} \sum_{i=1}^{k} \frac{p(\mathbf{y}_i)}{q(\mathbf{y}_i)} \eta(\mathbf{y}_i) \tag{2}$$

The expected value of $\hat{\boldsymbol{\rho}}_q$ under the density $q(x)$ is just

$$\frac{1}{k} \sum_{i=1}^{k} \int \frac{p(y_i)}{q(y_i)} \eta(y_i) q(y_i)\, dy_i = \int p(y)\eta(y)\, dy = E\{\eta(\mathbf{X})\} = \rho \tag{3}$$

Therefore, the estimate $\hat{\boldsymbol{\rho}}_q$ is unbiased and, as $k \to \infty$, we expect it to be converging by the law of large numbers to its mean value ρ. The question is: Are there better choices for $q(\cdot)$ than just $p(\cdot)$?

Let us consider the variance of $\hat{\boldsymbol{\rho}}_q$:

$$\begin{aligned}
\operatorname{var}(\hat{\boldsymbol{\rho}}_q) &= \frac{1}{k} \int \left[\left(p(x)\frac{\eta(x)}{q(x)}\right) - \rho\right]^2 q(x)\, dx \\
&= \frac{1}{k} \int \left[\left[\left(\frac{(p(x)\eta(x))^2}{q(x)}\right)\right] - 2\rho[p(x)\eta(x)]\right] dx + \frac{\rho^2}{k} \\
&= \frac{1}{k} \int \left[\left(\frac{(p(x)\eta(x))^2}{q(x)}\right)\right] dx - \frac{\rho^2}{k} \qquad \text{(4a)} \\
&\equiv \frac{1}{k}[ft(q) - \rho^2] \qquad \text{(4b)}
\end{aligned}$$

Where we have emphasized that $k \operatorname{var}(\hat{\boldsymbol{\rho}}_q)$ may be expressed as a difference of a first term, $ft(q)$, and a second term ρ^2. Using Lagrange multiplier techniques, it is easy to show that the optimal choice is for $q(x)$ to be proportional to $|p(x)\eta(x)|$ (see exercise 1). Unfortunately, this is not a very practical result for a couple of reasons. $p(\cdot)$ in our problems is not usually specified in closed form. It is typically the distribution of a large sum of iid or Markov distributed random variables. We may generate samples from it easily enough but explicit expressions for it are generally not available. Second (and more important), even if $p(\cdot)$ were known, the constant of proportionality is exactly ρ^{-1}, precisely the parameter that we are trying to estimate! Hence in computing the weighting factor for the importance sampling estimator of ρ, we first must know what ρ is. Clearly, we must search for other methodologies. We will find that by making a good choice for $q(\cdot)$, significant (orders of magnitude even) reduction of estimator variance over that of direct simulation is not only possible but to be expected. Utilizing large deviation theory to make a good choice for $q(\cdot)$ is the subject of the following sections.

A. MARKOV CHAIN METHOD (TWISTED SIMULATION METHOD)

Let $\{\mathbf{x}_i\}$ be the samples from a finite state space irreducible Markov chain with transition probabilities $p(x \mid y)$ and steady state (stationary) distribution $\pi(x)$. Define the set A as

$$A \equiv \left\{x_1, x_2, \ldots, x_n : \sum_{i=1}^{n} f(x_i) \geqslant 0\right\} \tag{5}$$

Suppose that we are interested in

$$\rho \equiv P(A) = P\left\{\frac{1}{n}\sum_{i=1}^{n} f(x_i) \geqslant 0\right\} \tag{6}$$

where

$$\sum_{x \in S} f(x)\pi(x) < 0 \tag{7}$$

and $f(\cdot)$ is a lattice-type function from the state space of the chain into the set $\{\ldots, -2d, -d, 0, d, 2d, \ldots\}$. Let us suppose that we wish to estimate this probability via simulation.

Now we know that

$$\lim_{n \to \infty} \frac{1}{n} \log P\left\{\frac{1}{n}\sum_{i=1}^{n} f(\mathbf{x}_i) \geqslant 0\right\} = \log \lambda(\theta_0) \tag{8}$$

where $\lambda(\theta)$ is the largest eigenvalue of the operator

$$T_\theta(g)(y) \equiv \sum_x \exp[\theta f(x)] g(x) p(x \mid y) \tag{9}$$

[Again, as a technical condition to prevent pathologies, assume that the principal eigenvalue $\mu(\gamma)$ of the complex operator $T_{i\gamma}(\cdot)$ ($i = \sqrt{-1}$, $\gamma \in \mathbb{R}$) satisfies $|\mu(\gamma)| < 1$ for $-\pi/d < \gamma < \pi/d$.] Furthermore,

$$\frac{\lambda'(\theta_0)}{\lambda(\theta_0)} = 0$$

or

$$\lambda'(\theta_0) = 0 \tag{10}$$

is the defining equation for θ_0. Let us consider a simulation carried out under the "twisted" distribution

$$t(x \mid y) \equiv p(x \mid y) \exp[\theta_0 f(x)] \frac{r(x)}{r(y)\lambda(\theta_0)} \tag{11}$$

where $r(\cdot)$ is the eigenvector associated with $\lambda(\theta_0)$. Suppose that the initial value distribution under the original or twisted distribution is $\mathbf{x}_1 = x$ with probability 1 [with immediate consequence $p(x_1)/t(x_1) = 1$]. Consider the ratio of the probability distributions

$$\begin{aligned}
\frac{p(x_1, x_2, \ldots, x_n)}{t(x_1, x_2, \ldots, x_n)} &= \frac{p(x_1) \prod_{i=1}^{n-1} p(x_{i+1} \mid x_i)}{t(x_1) \prod_{i=1}^{n-1} t(x_{i+1} \mid x_i)} \\
&= \exp\left[-\theta_0 \sum_{i=1}^{n-1} f(x_{i+1}) \right] r(x_1) \frac{\lambda(\theta_0)^{n-1}}{r(x_n)} \\
&= \exp\left[-\theta_0 \sum_{i=1}^{n-1} f(x_{i+1}) \right] r(x) \frac{\lambda(\theta_0)^{n-1}}{r(x_n)} \qquad (12)
\end{aligned}$$

Therefore, to obtain k "samples" we need to generate k independent realizations of the first n steps of the chain, denoted as $\mathbf{x}_1^{(j)}, \mathbf{x}_2^{(j)}, \ldots, \mathbf{x}_n^{(j)}$, $j = 1, \ldots, k$. Denote

$$\mathbf{X}^j \equiv (\mathbf{x}_1^{(j)}, \mathbf{x}_2^{(j)}, \ldots, \mathbf{x}_n^{(j)}) \tag{13}$$

We then form the estimate of ρ (or $P(A)$) as

$$\begin{aligned}
\hat{\boldsymbol{\rho}}_t &\equiv \frac{1}{k} \sum_{j=1}^{k} 1_A(\mathbf{X}^j) \frac{p(\mathbf{x}_1^{(j)}, \mathbf{x}_2^{(j)}, \ldots, \mathbf{x}_n^{(j)})}{t(\mathbf{x}_1^{(j)}, \mathbf{x}_2^{(j)}, \ldots, \mathbf{x}_n^{(j)})} \\
&= \frac{1}{k} \sum_{j=1}^{k} 1_A(\mathbf{X}^j) \exp\left[-\theta_0 \sum_{i=1}^{n-1} f(\mathbf{x}_{i+1}^{(j)}) \right] r(x) \frac{\lambda(\theta_0)^{n-1}}{r(\mathbf{x}_n^{(j)})} \qquad (14)
\end{aligned}$$

where $1_A(\cdot)$ is the indicator function of the set A. We will call this estimate the *twisted estimate* or the *twisted simulation estimate.*

It is easy to check that this estimator is unbiased, but what is its variance? Using the techniques of chapter VII, we may calculate at least the asymptotics of the variance for large n. *For simplicity in the calculations suppose that* $f(x) = 0$, *where* x *is the initial value of the chain* (*see exercise 5 for general case*). Now k times the variance of our estimate, $k \operatorname{var}(\hat{\boldsymbol{\rho}}_t)$, will be given from (4) by

$$k \operatorname{var}(\hat{\boldsymbol{\rho}}_t)$$

$$= \sum_{\Sigma f(x_i) \geq 0} \left[\exp\left[-\theta_0 \sum_{i=1}^{n-1} f(x_{i+1}) \right] r(x) \frac{\lambda(\theta_0)^{n-1}}{r(x_n)} \right]^2 t(x_1, x_2, \ldots, x_n \mid \mathbf{x}_1 = x) - P(A)^2 \tag{15}$$

Let us consider the first term, denoted $ft(t)$:

$$ft(t) \equiv \sum_{\Sigma f(x_i) \geq 0} \left[\exp\left[-\theta_0 \sum_{i=1}^{n-1} f(x_{i+1}) \right] r(x) \frac{\lambda(\theta_0)^{n-1}}{r(x_n)} \right]^2 t(x_1, x_2, \ldots, x_n \mid \mathbf{x}_1 = x)$$

$$= \sum_{s=0}^{\infty} \sum_{\Sigma f(x_i) = sd} \left[\exp\left[-\theta_0 \sum_{i=1}^{n-1} f(x_{i+1}) \right] r(x) \frac{\lambda(\theta_0)^{n-1}}{r(x_n)} \right]^2 t(x_1, x_2, \ldots, x_n \mid \mathbf{x}_1 = x) \tag{16}$$

Now conditioning on the event that $\mathbf{x}_n = y$, we may expand (16) as

$$ft(t) = \sum_{s=0}^{\infty} \sum_{y} t^{(n)}(y \mid x) \exp[-2\theta_0 sd] \frac{r(x)^2}{r(y)^2} \lambda(\theta_0)^{2(n-1)} \cdot \left[\sum_{\Sigma f(x_i) = sd} t(x_1, \ldots, x_n \mid \mathbf{x}_1 = x, \mathbf{x}_n = y) \right] \tag{17}$$

Now as n gets large $t^{(n)}(y \mid x)$ should be converging (in some sense) to the stationary distribution, $r(y)r^*(y)$, of the twisted chain. Substituting and changing the order of summations gives that (17) is asymptotically equal to

$$ft(t) \approx r(x)^2 \lambda(\theta_0)^{2(n-1)} \sum_{y} \frac{r^*(y)}{r(y)} \sum_{s=0}^{\infty} \exp[-2\theta_0 sd] \cdot \left[\sum_{\Sigma f(x_i) = sd} t(x_1, \ldots, x_n \mid \mathbf{x}_1 = x, \mathbf{x}_n = y) \right] \tag{18}$$

Now denote the distribution of $\Sigma f(\mathbf{x}_i)$ under the distribution $t(x_1, \ldots, x_n \mid \mathbf{x}_1 = x, \mathbf{x}_n = y)$ as $\tilde{H}_n(\cdot)_{xy}$. Equation (18) may be rewritten as

$$ft(t) \approx r(x)^2 \lambda(\theta_0)^{2(n-1)} \sum_{y} \frac{r^*(y)}{r(y)} \int_0^{\infty} \exp[-2\theta_0 z] \tilde{H}_n(dz)_{xy} \tag{19}$$

Performing a change of variable (let $v = z/\sqrt{n}\sigma$) obtains

$$ft(t) \approx r(x)^2 \lambda(\theta_0)^{2(n-1)} \sum_{y} \frac{r^*(y)}{r(y)} \int_0^{\infty} \exp[-2\theta_0 \sqrt{n}\sigma v] H_n(dv)_{xy} \tag{20}$$

where $H_n(\cdot)_{xy}$ is the distribution of $\Sigma f(\mathbf{x}_i)/\sqrt{n}\sigma$ under the distribution $t(x_1, \ldots, x_n \mid \mathbf{x}_1 = x, \mathbf{x}_n = y)$. We have worked out the asymptotics of integrals like the above in the text and exercises of chapter VII. [See derivations of (21) and (50) and solution to exercise 4 of chapter VII.] The result to be recalled is

$$\int_0^\infty \exp[-2\theta_0\sqrt{n}\sigma v]H_n(dv)_{xy} \approx \frac{d}{\sqrt{2\pi n}\,\sigma[1-\exp[-2\theta_0 d]]} \tag{21}$$

Hence, from (20)–(21), we obtain that the first term of (15) asymptotics are

$$\begin{aligned} ft(t) &= \sum_{\Sigma f(x_i)\geqslant 0}\left[\exp\left[-\theta_0\sum_{i=1}^{n-1} f(x_{i+1})\right] r(x)\frac{\lambda(\theta_0)^{n-1}}{r(x_n)}\right]^2 t(x_1, x_2, \ldots, x_n \mid \mathbf{x}_1 = x) \\ &\approx r(x)^2\lambda(\theta_0)^{2(n-1)}\frac{d}{\sqrt{2\pi n}\,\sigma}\frac{\Sigma_y r^*(y)/r(y)}{1-\exp[-2\theta_0 d]} \end{aligned} \tag{22}$$

Let us consider the asymptotics of the second term of (15), $P(A)^2$. Now

$$P(A) = \sum_{\Sigma f(x_i)\geqslant 0} p(x_1, \ldots, x_n) \tag{23}$$

In chapter VII [see (54) and exercise 4 of chapter VII], we developed an approximation to the quantity above as

$$\begin{aligned} P(S_n \geqslant a) &\approx \exp[-nI(a)]\left(\sum_x p(x)\exp[\theta_a f(x)]r_{\theta_a}(x)\right)\left(\sum_y r^*_{\theta_a}(y)\right) \\ &\quad\times \frac{d}{\lambda(\theta_a)(1-\exp[-\theta_a d])\sqrt{2\pi n}\,\sigma} \end{aligned} \tag{24}$$

In our present setting, $r_{\theta_a}(\cdot) = r(\cdot)$, $r^*_{\theta_a}(\cdot) = r^*(\cdot)$, $a = 0$, $p(\cdot) = \delta(\cdot - x)$, $f(x) = 0$, $I(a) = -\log\lambda(\theta_0)$, which gives

$$P(A) \approx \lambda(\theta_0)^{n-1} r(x) d \frac{\Sigma_y r^*(y)}{(1-\exp[-\theta_0 d])\sqrt{2\pi n}\,\sigma} \tag{25}$$

It is easy to see that $P(A)^2$ is negligible for large n compared to the first term of (15). Hence the variance asymptotics are controlled entirely by the first term.

Now consider the variance asymptotics for the direct simulator. In this case we have

$$k\,\mathrm{var}(\hat{\boldsymbol{\rho}}_p) = P(A)(1 - P(A)) = P(A) - P(A)^2$$

Hence $ft(p) = P(A)$. For large n, $P(A)$ is very small. This implies for the direct simulation estimator that $k\,\mathrm{var}(\hat{\boldsymbol{\rho}}_p) \approx P(A)$. We have thus sketched the proof of the following theorem.

Theorem

Let $p(x \mid y)$ be the transition probabilities of a finite state space irreducible (nondegenerate) Markov chain with initial condition $P\{x_1 = x\} = 1$ and let $|\mu(\gamma)| < 1$ for $-\pi/d < \gamma < \pi/d$. Let k denote the number of simulation runs for a twisted simulation estimate of $P\{S_n \geqslant 0\} = P\{\Sigma_{i=1}^{n} f(x_i) \geqslant 0\}$, where $f(\cdot)$ is a function of lattice type with $f(x) = 0$. Denote the variances of the twisted and direct estimators by $\text{var}(\hat{\boldsymbol{\rho}}_t)$ and $\text{var}(\hat{\boldsymbol{\rho}}_p)$, respectively. Then if $P\{S_n = 0) > 0$ for all n, we have

$$\lim_{n\to\infty} k\,\text{var}(\hat{\boldsymbol{\rho}}_t)\sqrt{n}\lambda(\theta_0)^{-2(n-1)} = r(x)^2 \sum_y \frac{r^*(y)}{r(y)} d[[1 - \exp[-2\theta_0 d]]\sqrt{2\pi}\,\sigma]^{-1}$$

and

$$\lim_{n\to\infty} k\,\text{var}(\hat{\boldsymbol{\rho}}_p)\sqrt{n}\,\lambda(\theta_0)^{n-1} = r(x)\sum_y r^*(y) d[[1 - \exp[-\theta_0 d]]\sqrt{2\pi}\,\sigma]^{-1}$$

We see from the theorem above that the direct simulator's variance goes to zero like $\lambda(\theta_0)^{n-1}/\sqrt{n}$. The twisted distribution simulation estimator has variance that goes to zero like $\lambda(\theta_0)^{2(n-1)}/\sqrt{n}$. Since $\lambda(\theta_0) < 1$, for large enough n, the twisted estimator can give an orders of magnitude better performance. Hence the smaller $\lambda(\theta_0)$ is, the better the twisted estimator is over a direct Monte Carlo simulation.

We note from (22) and (25) that the logarithmic rate (in n) of the first term, $ft(t)$, of (15) is the same as that of the second term, $P(A)^2$. Since a variance must always be nonnegative, we immediately have that the logarithmic rate of the twisted simulators ft term is the best possible. By our analysis of the constants in front of the exponential rate terms, we have concluded that $P(A)^2$ is negligible compared to $ft(t)$ and $ft(p)$. However, it is conceivable that another importance sampling methodology's $ft(q)$ term would have the same logarithmic rate *and* asymptotically equivalent front end constants as $P(A)^2$. This would allow the possibility of a superexponential convergence to zero of the overall variance. In fact, this is precisely what happens when the "optimal" (but impractical) importance sampling distribution (see exercise 1) is utilized [in that case $ft = P(A)^2$ for all $n \geqslant 1$].

We now define a quantity, a figure of merit, called the *speed factor*, SF.

$$\text{SF}(\hat{\boldsymbol{\rho}}_q) \equiv \lim_{n\to\infty} \frac{1}{n} \log[k\,\text{var}(\hat{\boldsymbol{\rho}}_q)] \tag{26}$$

An immediate corollary of the above is the following:

Corollary
Under the same hypotheses as the theorem above, we have

$$\mathrm{SF}(\hat{\boldsymbol{\rho}}_t) = 2\log[\lambda(\theta_0)]$$
$$\mathrm{SF}(\hat{\boldsymbol{\rho}}_p) = \log[\lambda(\theta_0)]$$

In other words, there is a factor of 2 difference in the variance rates between the two estimators.

***Example* 1 (*General iid Case*).** In this example we consider the general iid case explicitly. We assume probability densities for convenience, but this is not essential and the precise statements with Lebesgue–Stieltjes integrations with respect to the distributions may be substituted when needed. The theory given above does not cover this case since we no longer have a finite state space. However, an extension via the Bahadur–Rao theorem of chapter VII is straightforward (see Notes and Comments). We now merely indicate the results to be expected. Since there is no dependency, the left and right eigenvectors are constants, which leads to the twisted distribution

$$t(x) \equiv p(x)\frac{\exp[\theta_0 f(x)]}{\lambda(\theta_0)} \tag{27}$$

where

$$\lambda(\theta_0) \equiv \int p(x)\exp[\theta_0 f(x)]\,dx \tag{28}$$

and θ_0 is chosen so that $\int f(x)t(x)\,dx = 0$ or equivalently, $\lambda'(\theta_0) = 0$. Our estimate is then to generate "blocks" of iid data from $t(\cdot)$ and create the new iid sequence $\mathbf{J}^{(1)}, \mathbf{J}^{(2)}, \ldots, \mathbf{J}^{(k)}$, where

$$\mathbf{J}^{(m)} \equiv \frac{1}{n}\sum_{i=1}^{n} f(\mathbf{y}_i^{(m)}) \tag{29}$$

We then form our estimate of

$$\rho \equiv P(A) = P\left\{\frac{1}{n}\sum_{i=1}^{n} f(\mathbf{x}_i) \geqslant 0\right\} \tag{30}$$

as

$$\hat{\mathbf{p}}_t \equiv \frac{1}{k}\sum_{m=1}^{k} 1_{[0,\infty)}(\mathbf{J}^{(m)})\lambda(\theta_0)^n \exp\left[-\theta_0 \sum_{i=1}^{n} f(\mathbf{y}_i^{(m)})\right]$$

$$= \frac{1}{k}\sum_{m=1}^{k} 1_{[0,\infty)}(\mathbf{J}^{(m)})\lambda(\theta_0)^n \exp[-\theta_0 n \mathbf{J}^{(m)}] \tag{31}$$

The asymptotic variance of this twisted estimate depends on the lattice spacing (if any) of the random variable $f(\mathbf{x}_1)$. If the spacing is d, we have that k times the asymptotic variance is given by

$$k \operatorname{var}(\hat{\mathbf{p}}_t) \approx \lambda(\theta_0)^{2n} \frac{d}{(1 - \exp[-2\theta_0 d])\sqrt{2\pi n}\,\sigma} \tag{32}$$

where

$$\sigma^2 = \frac{\lambda''(\theta_0)}{\lambda(\theta_0)} \tag{33}$$

When $f(\mathbf{x}_1)$ is nonlattice, we can find the asymptotic variance formally by taking the limit as $d \to 0$ to find k times the asymptotic variance is given by

$$k \operatorname{var}(\hat{\mathbf{p}}_t) \approx \frac{\lambda(\theta_0)^{2n}}{2\theta_0\sqrt{2\pi n}\,\sigma} \tag{34}$$

(a) *Gaussian case*, $f(x) = x$. Suppose $\{\mathbf{x}_i\}$ is a sequence of iid mean $-m$ $(m > 0)$, variance s, Gaussian random variables.

$$t(x) = p(x)\frac{\exp[\theta_0 x]}{\lambda(\theta_0)} = \frac{\exp[-(x+m)^2/2s + \theta_0 x]}{\lambda(\theta_0)\sqrt{2\pi s}}$$

where

$$\lambda(\theta_0) = \exp\left[\frac{\theta_0^2 s - 2\theta_0 m}{2}\right]$$

and

$$\lambda'(\theta_0) = 0$$

or

$$\theta_0 = \frac{m}{s}$$

in this case we have

$$\lambda(\theta_0) = \exp\left[-.5\frac{m^2}{s}\right]$$

and

$$\sigma^2 = \frac{\lambda''(\theta_0)}{\lambda(\theta_0)} = s$$

Hence the twisted distribution corresponds to a simple shift to the origin of the original distribution. $\mathrm{SF}(\hat{\boldsymbol{\rho}}_t) = -m^2/s$.

(b) *Laplacian case*, $f(x) = x$. Suppose that $\{\mathbf{x}_i\}$ is a sequence of iid Laplacian mean $-m$ ($m > 0$), variance $2a^2$ random variables.

$$t(x) = p(x)\frac{\exp[\theta_0 x]}{\lambda(\theta_0)} = \frac{\exp[-|x + m|/a + \theta_0 x]}{\lambda(\theta_0)}$$

where

$$\theta_0 = -\frac{1}{m} + \sqrt{\frac{1}{m^2} + \frac{1}{a^2}}$$

and

$$\lambda(\theta_0) = m^2\frac{\exp[-C]}{2}a^2 C$$

with

$$C = \sqrt{1 + \frac{m^2}{a^2}} - 1$$

If the squared mean to variance ratio $m^2/2a^2 = 1$, then $\lambda(\theta_0) = \exp[1 - \sqrt{3}]/[1 - \sqrt{3}] \approx .657$ and $\mathrm{SF}(\hat{\boldsymbol{\rho}}_t) = 2\log(\lambda(\theta_0)) \cong -.84$. By comparison, for the Gaussian case, $\mathrm{SF}(\hat{\boldsymbol{\rho}}_t) = -1$, which implies that more speedup is possible with Gaussian densities.

(c) *Bernoulli case*, $f(x) = x - a$. Let $\{\mathbf{x}_i\}$ be iid Bernoulli random variables with $P\{\mathbf{x}_1 = 1\} = p(1) \equiv p = 1 - P\{\mathbf{x}_1 = 0\} \equiv 1 - p(0)$ and where $0 < p < a < 1$. Then

$$E\{\exp[\theta \mathbf{x}_1]\} = 1 - p + p\exp[\theta]$$

and

$$\lambda(\theta) = \exp[-a\theta][1 - p + p\exp[\theta]]$$

θ_0 is the solution to the equation $0 = \lambda'(\theta_0)/\lambda(\theta_0)$, which yields

$$\theta_0 = \log\frac{a(1-p)}{(1-a)p}$$

Substituting this value back into the expression for $\lambda(\theta)$ gives

$$\lambda(\theta_0) = \frac{1-p}{1-a}\bigg/\left[\frac{a(1-p)}{(1-a)p}\right]^a$$

Furthermore, after performing the necessary differentiations we find that

$$\begin{aligned}\sigma^2 &= \frac{\lambda''(\theta_0)}{\lambda(\theta_0)} - 0^2 \\ &= a^2 + a(1-2a)\frac{(1-p)}{(1-p)(1-a) + a(1-p)} \\ &= a(1-a)\end{aligned}$$

The twisted distribution is

$$t(x) = p(x)\frac{\exp[\theta_0(x-a)]}{\lambda(\theta_0)}$$

which upon substitution gives $t(0) = 1 - a$ and $t(1) = a$. Therefore, the twisted distribution is Bernoulli distributed with success probability $a > p$. If $a = .999$ and $p = .5$, we find that $\lambda(\theta_0) \cong \frac{1}{2}$ and $\mathrm{SF}(\hat{\mathbf{p}}_t) = 2\log(\lambda(\theta_0)) \cong -1.37$. □

***Example* 2 (*Two-state* {0, 1} *Markov chain*, $f(x) = 1 - x - a$, $1 > a > .5$).** Let the transition probability matrix be

$$\begin{bmatrix} p & q \\ q & p \end{bmatrix}$$

i.e., the chain stays where it is with probability $p < .5$ and flips state with probability $q = 1 - p$. The stationary distribution is obviously $(\frac{1}{2} \quad \frac{1}{2})^{\mathrm{T}}$. To find $\lambda(\theta)$, we must find the largest eigenvalue of the matrix

$$\begin{bmatrix} \exp[(1-a)\theta]p & \exp[(1-a)\theta]q \\ \exp[-a\theta]q & \exp[-a\theta]p \end{bmatrix}$$

Setting the determinant equal to zero and solving, we obtain

$$\lambda(\theta) = \left[p(\exp[\theta] + 1) + \frac{[p^2(\exp[\theta] + 1)^2 + 4\exp[\theta](1 - 2p)]^{.5}}{2} \right] \exp[-a\theta]$$

This expression can be differentiated and set equal to zero to solve for θ_0 and hence $\lambda(\theta_0)$ and $\mathrm{SF}(\hat{\boldsymbol{\rho}}_t)$ for various values of p and a (see exercise 4). □

***Example* 3 (*Three-state* $\{1, 2, 3\}$ *Markov chain*).** Suppose the transition probability matrix of a three-state Markov chain is

$$P \equiv \begin{bmatrix} 0 & .2 & .8 \\ 0 & .2 & .8 \\ .1 & .1 & .8 \end{bmatrix}$$

The stationary distribution is found by solving for $\underline{x}^{\mathrm{T}}P = \underline{x}$ or $\underline{x}^{\mathrm{T}} = (2 \quad 3 \quad 20)/25$. Suppose that $f(1) = 2.6$, $f(2) = .79$, and $f(3) = -1.29$. Note that $\underline{x}^{\mathrm{T}}f < 0$. Then by computer search we found that θ_0 is approximately -1.085 and $\lambda(\theta_0) = \exp[-.189]$ and hence $\mathrm{SF}(\hat{\boldsymbol{\rho}}_t) \cong -.378$. □

B. VARIANCE RATE AND OPTIMALITY RESULTS

Let $p(x \mid y)$ be, as in section A, the transition probabilities of a finite state space irreducible Markov chain. Let $q(x \mid y)$ denote an importance sampling transition probability distribution. We will not necessarily suppose that $q(\cdot \mid \cdot)$ is the twisted distribution associated with the twisted simulation method. The questions that we focus on in this section are: What is the speed factor for such a choice of $q(\cdot \mid \cdot)$? What is the best choice for $q(\cdot \mid \cdot)$?

Now as stated in the introductory section (see also exercise 1), the optimal (unconstrained) importance sampling distribution is known. One merely chooses $q(x_1, x_2, \ldots, x_n)$ to be $p(x_1, x_2, \ldots, x_n) \cdot 1_{\{A\}}/P(A)$. This optimal simulation distribution not only requires knowledge of $P(A)$ (the parameter we are trying to estimate) but is also not Markov, iid, or stationary even when $p(x_1, x_2, \ldots, x_n)$ itself is. Hence our focus in this section is, if one constrains $q(x_1, x_2, \ldots, x_n)$ to be stationary Markov (or iid), to find the best choice for $q(\cdot|\cdot)$ and evaluate its asymptotic variance performance, $\text{SF}(\hat{\boldsymbol{\rho}}_q)$.

Consider again [from (4)] the variance expression for the estimator $\hat{\boldsymbol{\rho}}_q$ after k simulation runs,

$$\text{var}(\hat{\boldsymbol{\rho}}_q) \equiv \frac{1}{k} \sum\sum \cdots \sum_{x_1, x_2, \ldots, x_n} \frac{p(x_1, x_2, \ldots, x_n)^2}{q(x_1, x_2, \ldots, x_n)} 1_{\{A\}} - \frac{P(A)^2}{k} \tag{35}$$

where again

$$A \equiv \left\{ x_1, x_2, \ldots, x_n : \sum_{i=1}^{n} f(x_i) \geqslant 0 \right\}$$

We know the large n asymptotics of $P(A)$, so let us concentrate our attention on the large n asymptotics of k times the first term, denoted again as $ft(q)$.

$$\begin{aligned} ft(q) &\equiv \sum\sum \cdots \sum_{x_1, x_2, \ldots, x_n} \frac{p(x_1, x_2, \ldots, x_n)^2}{q(x_1, x_2, \ldots, x_n)} 1_{\{A\}} \\ &= \sum\sum \cdots \sum_{x_1, x_2, \ldots, x_n} \prod_{i=1}^{n-1} \frac{p(x_{i+1} | x_i)^2 p(x_1)^2}{q(x_{i+1} | x_i) q(x_1)} 1_{\{A\}} \end{aligned} \tag{36}$$

Note that since the variance must always be positive, we automatically have, from (35),

$$\lim_{n \to \infty} \frac{1}{n} \log[ft(q)] \geqslant \lim_{n \to \infty} \frac{1}{n} \log[P(A)^2] = 2 \log[\lambda(\theta_0)] \tag{37}$$

If $\lim_{n \to \infty} 1/n \log[ft(q)] = 2 \log[\lambda(\theta_0)]$ (as with the twisted simulator), one needs to evaluate the first order expansions of the relevant quantities (as in section A) to be able to determine the rate constants for the ultimate quantity of interest $\lim_{n \to \infty} (1/n) \log[k \, \text{var}(\hat{\boldsymbol{\rho}}_q)]$. In the case of the twisted simulator we had $\text{SF}(\hat{\boldsymbol{\rho}}_t) \equiv \lim_{n \to \infty} (1/n) \log[k \, \text{var}(\hat{\boldsymbol{\rho}}_t)] = \lim_{n \to \infty} (1/n) \log[ft(t)] = 2 \log[\lambda(\theta_0)]$. If, for a particular $q(\cdot|\cdot)$, we have $\lim_{n \to \infty} (1/n) \log[ft(q)] > 2 \log[\lambda(\theta_0)]$, one has immediately that $\text{SF}(\hat{\boldsymbol{\rho}}_q) \equiv \lim_{n \to \infty} (1/n) \log[k \, \text{var}(\hat{\boldsymbol{\rho}}_q)] = \lim_{n \to \infty} (1/n) \log[ft(q)]$ (which was the case with the direct simulator).

Define the operator

$$K_\gamma(g)(y) \equiv \sum_x \exp[\gamma f(x)]g(x)\frac{p(x \mid y)^2}{q(x \mid y)} \tag{38}$$

This is a positive operator and hence by the Perron–Frobenius theorem will have a largest positive eigenvalue $\zeta(\gamma)$ and associated right (left) eigenvector $e_\gamma(\cdot)$ $(e_\gamma^*(\cdot))$. We now have the following theorem.

Theorem

$$\lim_{n\to\infty}\frac{1}{n}\log[ft(q)] = \inf_\gamma \log[\zeta(\gamma)] \equiv \log[\zeta(\gamma_0)] \tag{39}$$

Proof: From similar arguments made in previous chapters it is easy to verify that γ_0 is nonnegative and may be characterized as that value of γ for which equivalently $\zeta'(\gamma_0)=0$ or $\Sigma_x f(x)e_{\gamma_0}(x)e_{\gamma_0}^*(x)=0$ (always normalize so that $\langle e_\gamma, e_\gamma^*\rangle = 1$). Also, let us define a twisted probability distribution

$$k_\gamma(x \mid y) \equiv \frac{p(x \mid y)^2 \exp[\gamma f(x)]e(x)}{q(x \mid y)\zeta(\gamma)e(y)}$$

As usual, $e_\gamma(x)e_\gamma^*(x)$ is the stationary distribution of this chain.

Upper bound. Now for $\gamma > 0$, we have

$$\begin{aligned} ft(q) &= \sum\sum_{x_1,x_2,\ldots,x_n}\cdots\sum \prod_{i=1}^{n-1}\frac{p(x_{i+1} \mid x_i)^2}{q(x_{i+1} \mid x_i)}\frac{p(x_1)^2}{q(x_1)}1_{\{A\}} \\ &\leqslant \sum\sum_{x_1,x_2,\ldots,x_n}\cdots\sum\left[\prod_{i=1}^{n-1}\frac{p(x_{i+1} \mid x_i)^2}{q(x_{i+1} \mid x_i)}\frac{p(x_1)^2}{q(x_1)}\right]\exp\left[\gamma\sum_{i=1}^{n} f(x_i)\right] \\ &= \sum_{x_1} K_\gamma(1)^{n-1}(x_1)\frac{p(x_1)^2}{q(x_1)} \end{aligned}$$

Therefore,

$$\overline{\lim_{n\to\infty}}\,\frac{1}{n}\log[ft(q)] \leqslant \log[\zeta(\gamma)]$$

Minimizing the upper bound over all γ gives the desired result,

$$\overline{\lim_{n\to\infty}}\,\frac{1}{n}\log[ft(q)] \leqslant \inf_\gamma \log[\zeta(\gamma)] = \log[\zeta(\gamma_0)]$$

Lower bound. Now we have that (for $\varepsilon > 0$ and $\gamma > 0$)

$$ft(q) \equiv \sum_{x_1, x_2, \ldots, x_n}\sum\cdots\sum \prod_{i=1}^{n-1} \frac{p(x_{i+1} \mid x_i)^2}{q(x_{i+1} \mid x_i)} \frac{p(x_1)^2}{q(x_1)} \; 1\left\{\frac{1}{n}\sum f(x_i) \geqslant 0\right\}$$

$$\geqslant \sum_{x_1, x_2, \ldots, x_n}\sum\cdots\sum \prod_{i=1}^{n-1} \frac{p(x_{i+1} \mid x_i)^2}{q(x_{i+1} \mid x_i)} \frac{p(x_1)^2}{q(x_1)} \; 1\left\{\varepsilon > \frac{1}{n}\sum f(x_i) \geqslant 0\right\}$$

$$\geqslant \sum_{x_1, x_2, \ldots, x_n}\sum\cdots\sum \prod_{i=1}^{n-1} \frac{p(x_{i+1} \mid x_i)^2}{q(x_{i+1} \mid x_i)} \frac{p(x_1)^2}{q(x_1)} \; 1\left\{\varepsilon > \frac{1}{n}\sum f(x_i) \geqslant 0\right\} \exp\left[\gamma\left(\sum_{i=1}^{n} f(x_i) - n\varepsilon\right)\right]$$

$$= \sum_{x_1, x_2, \ldots, x_n}\sum\cdots\sum \prod_{i=1}^{n-1} \left[k_\gamma(x_{i+1} \mid x_i)\frac{e_\gamma(x_i)}{e_\gamma(x_{i+1})}\zeta(\gamma)\right] \frac{p(x_1)^2}{q(x_1)} \exp[-\gamma n\varepsilon] \; 1\left\{\varepsilon > \frac{1}{n}\sum f(x_i) \geqslant 0\right\}$$

$$= \zeta(\gamma)^{n-1} \exp[-\gamma n\varepsilon] \sum_{x_1, x_2, \ldots, x_n}\sum\cdots\sum \left[\left[\prod_{i=1}^{n-1} k_\gamma(x_{i+1} \mid x_i)\right]\right] \frac{e(x_1)}{e(x_n)} \frac{p(x_1)^2}{q(x_1)} \; 1\left\{\varepsilon > \frac{1}{n}\sum f(x_i) \geqslant 0\right\}$$

Since $e(\cdot)$ is bounded above and below (strictly away from zero), we may define the constants c_u, c_l by $0 < c_l \leqslant e(\cdot) \leqslant c_u < \infty$. Also let $L \equiv \Sigma_x p(x)^2/q(x)$. We may then continue the inequality string above with

$$\geqslant \frac{c_u}{c_l} \zeta(\gamma)^n \exp[-\gamma n\varepsilon] L \sum_{x_1, x_2, \ldots x_n}\sum\cdots\sum \left[\prod_{i=1}^{n-1} [k_\gamma(x_{i+1} \mid x_i)]\right] \frac{p(x_1)^2}{q(x_1)L} \cdot 1\left\{\varepsilon > \frac{1}{n}\sum f(x_i) \geqslant 0\right\}$$

Now consider the term in the sum. $p(x_1)^2/q(x_1)L$ is acting as an initial distribution to the twisted chain specified by the transition probabilities $k_\gamma(\cdot \mid \cdot)$. Under this distribution $(1/n)\Sigma_{i=1}^{n} f(x_i)$ is converging to $\Sigma_x f(x)e_\gamma(x)e_\gamma^*(x)$ the mean value under the stationary distribution of the twisted chain. Hence choose $\gamma = \gamma_\varepsilon$ so that $\varepsilon > \Sigma_x f(x)e_{\gamma_\varepsilon}(x)e_{\gamma_\varepsilon}^*(x) > 0$. In that case the term in the sum is converging to 1 as n approaches infinity. Hence

$$\varliminf_{n\to\infty} \frac{1}{n} \log[ft(q)] \geqslant -\gamma_\varepsilon \varepsilon + \log[\zeta(\gamma_\varepsilon)]$$

Now as ε approaches zero, γ_ε approaches γ_0 and we have

$$\varliminf_{n\to\infty} \frac{1}{n} \log[ft(q)] \geqslant \log[\zeta(\gamma_0)]$$

which finishes the proof of the lower bound and hence the theorem. □

We wish to choose the importance sampling distribution that makes $\log[\zeta(\gamma_0)]$ as small as possible. We know that choosing the twisted distribution will minimize, but is that choice unique? The following theorem shows that the answer is yes and that the twisted distribution is uniquely optimal.

Theorem

$$\lim_{n\to\infty} \frac{1}{n} \log[k \operatorname{var}(\hat{\boldsymbol{\rho}}_q)] = \log[\zeta(\gamma_0)] \geqslant 2 \log[\lambda(\theta_0)]$$

with equality if and only if $q(\cdot\,|\,\cdot)$ is chosen as

$$t(x\,|\,y) = \frac{r_{\theta_0}(x) \exp[\theta_0 f(x)] p(x\,|\,y)}{r_{\theta_0}(y)\lambda(\theta_0)}$$

the twisted simulator importance sampling distribution.

Proof: We wish to view $\zeta(\gamma_0)$ as a functional of $q(\cdot\,|\,\cdot)$. It is desired to find the best choice for $q(\cdot\,|\,\cdot)$, i.e., makes the value of $\zeta(\gamma_0)$ as small as possible. Consider the perturbed operator

$$K_\gamma^\varepsilon(g)(y) \equiv \sum_x \exp[\gamma f(x)] \frac{p(x\,|\,y)^2}{q(x\,|\,y) + \varepsilon m(x, y)} g(x) \tag{40}$$

where $\Sigma_x m(x, y) = 0$ for every y. We have, after expanding the denominator of (40) in a Taylor series in ε,

$$K_\gamma^\varepsilon(g)(y) = K_\gamma(g)(y) - \varepsilon \sum_x \exp[\gamma f(x)] g(x) \frac{p(x\,|\,y)^2}{q(x\,|\,y)^2} m(x, y) + O(\varepsilon^2) \tag{41}$$

Hence we may write down the first order perturbation expansion (see appendix C.5) of the relevant eigenvalue as

$$\zeta_\varepsilon(\gamma) = \zeta(\gamma) + \varepsilon \sum_x \sum_y m(x, y) \exp[\gamma f(x)] e_\gamma(x) e_\gamma^*(y) \frac{p(x\,|\,y)^2}{q(x\,|\,y)^2} + O(\varepsilon^2) \tag{42}$$

For a minimizing $q(\cdot\,|\,\cdot)$, we should have the term multiplying ε be zero for all allowed perturbations $m(\cdot, \cdot)$. □

Claim 1. For a minimizing $q(\cdot\,|\,\cdot)$,

$$d(x, y) \equiv \exp[\gamma f(x)] e_\gamma(x) e_\gamma^*(y) \frac{p(x\,|\,y)^2}{q(x\,|\,y)^2} = c(y) \tag{43}$$

i.e., $d(x, y)$ depends only on y.

Proof of Claim 1: Suppose that the claim is false. Then $d(x', y_0) \neq d(x'', y_0)$ for some y_0. Let us choose $m(x', y_0) = 1$, $m(x'', y_0) = -1$, and $m(x, y) = 0$ for all other values. One then has immediately that the perturbation

$$\varepsilon \sum_x \sum_y m(x, y) d(x, y) = \varepsilon[d(x', y_0) - d(x'', y_0)] \neq 0$$

However, for an optimum $q(\cdot \mid \cdot)$, this term must be zero. Hence contradiction, and the original supposition is false. □

Claim 2. For a minimizing $q(\cdot \mid \cdot)$,

$$c(y) = \zeta(\gamma) e_\gamma(y) e_\gamma^*(y) \tag{44}$$

Proof of Claim 2: From (43) we have that

$$q(x \mid y) = p(x \mid y) \exp\left[f(x) \frac{\gamma}{2} \right] \sqrt{e_\gamma(x) \frac{e_\gamma^*(y)}{c(y)}} \tag{45}$$

Also,

$$\sum_x q(x \mid y) = 1$$

implies from (45) that

$$\sum_x p(x \mid y) \exp\left[f(x) \frac{\gamma}{2} \right] \sqrt{e_\gamma(x)} = \sqrt{\frac{c(y)}{e_\gamma^*(y)}} \tag{46}$$

Furthermore, substituting (45) into (38) yields

$$\begin{aligned} K_\gamma(e_\gamma)(y) &= \sum_x \exp[\gamma f(x)] p(x \mid y)^2 e_\gamma(x) \frac{\sqrt{c(y)/e_\gamma(x) e_\gamma^*(y)}}{p(x \mid y) \exp[\gamma f(x)/2]} \\ &= \sum_x \exp\left[\gamma \frac{f(x)}{2} \right] p(x \mid y) e_\gamma(x) \sqrt{\frac{c(y)}{e_\gamma(x) e_\gamma^*(y)}} \\ &= \sqrt{\left[\frac{c(y)}{e_\gamma^*(y)} \right]} \sum_x \exp\left[\gamma \frac{f(x)}{2} \right] p(x \mid y) \sqrt{e_\gamma(x)} \end{aligned} \tag{47}$$

Also, we must have $K_\gamma(e_\gamma)(y) = \zeta(\gamma) e_\gamma(y)$, or

$$\sum_x \exp\left[f(x) \frac{\gamma}{2} \right] p(x \mid y) \sqrt{e_\gamma(x)} = \zeta(\gamma) e_\gamma(y) \sqrt{\frac{e_\gamma^*(y)}{c(y)}} \tag{48}$$

Substituting (46) for the left side of (48) obtains

$$\sqrt{\frac{c(y)}{e_\gamma^*(y)}} = \zeta(\gamma)e_\gamma(y)\sqrt{\frac{e_\gamma^*(y)}{c(y)}}$$

which immediately gives the claim. □

Therefore, from (44)–(45), we obtain

$$q(x \mid y) = p(x \mid y)\exp\left[f(x)\frac{\gamma}{2}\right]\frac{\sqrt{e_\gamma(x)}}{\sqrt{\zeta(\gamma)e_\gamma(y)}} \tag{49}$$

Furthermore, from (47), we have

$$K_\gamma(e_\gamma(y)) = \sum_x \exp\left[f(x)\frac{\gamma}{2}\right]p(x \mid y)\sqrt{\zeta(\gamma)e_\gamma(x)e_\gamma(y)} = \zeta(\gamma)e_\gamma(y) \tag{50}$$

or

$$\sum_x \exp\left[f(x)\frac{\gamma}{2}\right]p(x \mid y)\sqrt{e_\gamma(x)} = \sqrt{\zeta(\gamma)e_\gamma(y)}$$

or, from (9),

$$T_{\gamma/2}(\sqrt{e_\gamma})(y) = \sqrt{\zeta(\gamma)e_\gamma(y)} \tag{51}$$

which implies that

$$\sqrt{e_\gamma(x)} = r_{\gamma/2}(x) \tag{52}$$

and

$$\zeta(\gamma) = \lambda^2\left(\frac{\gamma}{2}\right) \tag{53}$$

Hence, from (49) and (52)–(53), we have

$$q(x \mid y) = p(x \mid y)\exp\left[f(x)\frac{\gamma}{2}\right]\frac{r_{\gamma/2}(x)}{\lambda(\gamma/2)r_{\gamma/2}(y)} \tag{54}$$

Therefore, an optimizing $q(\cdot \mid \cdot)$ is of the form of an exponential twist. We now need to search for the optimum among all exponential twists. Consider a general exponential twist

$$q(x \mid y) = p(x \mid y) \exp[\theta f(x)] \frac{r_\theta(x)}{\lambda(\theta) r_\theta(y)} \tag{55}$$

Hence we wish to consider the largest eigenvalue of the operator associated with this simulation distribution. Therefore, from (38), we need to consider

$$K_\gamma(g)(y) = \sum_x \exp[\gamma f(x)] p(x \mid y) \exp[-\theta f(x)] g(x) \lambda(\theta) \frac{r_\theta(y)}{r_\theta(x)} \tag{56}$$

It is straightforward to verify that the largest positive eigenvalue of this operator is $\lambda(\theta)\lambda(\gamma - \theta)$ with associated eigenvector $r_\theta(x) r_{\gamma-\theta}(x)$. Now it is true that $\inf_\theta \lambda(\theta) = \lambda(\theta_0)$. Hence a lower bound to $\lambda(\theta)\lambda(\gamma - \theta)$ is $\lambda(\theta_0)^2$. Now from the arguments of section A, we know that this eigenvalue is attained when $\theta = \theta_0$ (and consequently, $\gamma_0 = 2\theta_0$). Furthermore, this solution is unique due to convexity of $\lambda(\theta)$ as a function of θ. □

***Example* 4.** Suppose that $f(x) = x$, where $\{\mathbf{x}_i\}$ are iid Laplacian mean $-m$ $(m > 0)$ random variables, i.e., $p(x \mid y) = p(x) = \exp[-|x + m|]/2$. Then suppose that $q(x) = \exp[-|x|]/2$, a simple mean shift. Therefore,

$$\begin{aligned}\zeta(\gamma) &= \int_{-\infty}^{\infty} \exp[\gamma x] \exp[-|x + m| 2] \frac{\exp[|x|]}{2} dx \\ &= \frac{\exp[-2m]}{\gamma^2 - 4\gamma + 3} - \frac{2\exp[m(1-\gamma)]}{\gamma^2 - 2\gamma - 3}\end{aligned}$$

Let $m = \sqrt{2}$ (which implies a squared mean to variance ratio of 1). By computer search we found $\gamma_0 \approx .826$ and $\zeta(\gamma_0) \approx .8006$ and $\mathrm{SF}(\hat{p}_q) = \log(\zeta(\gamma_0)) \cong -.2224$. However, from example 1b we had found that $2\log(\lambda(\theta_0)) \approx (.657)^2 = -.84$, which indicates a substantial speedup for the twisted simulator in comparison to a simple mean shift. □

C. SLOW MARKOV WALK METHOD (QUICK SIMULATION METHOD)

Sometimes the Markov process of interest does not have a finite state space. In such a case, it is much more difficult to solve for the twisted distribution and find the relevant eigenvalues and eigenvectors. In some important cases, however, one can adequately model the relevant dynamics as a slow Markov walk (see section C of chapter IV).

The ideas that one desires to utilize for simulation are the same. From the relevant large deviation principles we find the dominating point of the set in question. We then "twist" the original probability measure so that it has mean value at this dominating point. We then use the twisted probability measure to carry out the simulation.

Recall the slow Markov walk model:

$$\mathbf{x}_{k+1}^n = \mathbf{x}_k^n + \frac{1}{n} V(\mathbf{x}_k^n, \xi_k)$$

for $k \geqslant 0$ and $\mathbf{x}_0^n \equiv x$. $\mathbf{x}^n(\cdot)$ is the continuous-time piecewise linear interpolation random process ($\mathbf{x}^n(k/n) \equiv \mathbf{x}_k^n$). $V(\cdot,\cdot)$ maps $\mathbb{R}^2$ into $\mathbb{R}$ and the $\{\xi_k\}$ are iid. Let $F_x(\cdot)$ denote the distribution function of $V(x, \xi_1)$ and

$$I_x(y) \equiv \sup_\theta \{\theta y - \log M_x(\theta)\}$$

is the associated large deviation rate function. In section C of chapter IV we argued (subject to some technical conditions) that $\mathbf{x}^n(\cdot)$ obeys a large deviation principal with rate function

$$I(f) \equiv \int_0^T I_{f(t)}(f'(t))\, dt$$

for $f \in C_{1,x}[0, T]$ [the set of all continuous piecewise differentiable functions with bounded derivatives such that $f(0) = x$]. Let $A \subset C_{1,x}[0, \infty)$ be a set of trajectories that have some property of interest. We are interested in simulating to obtain a good estimate of $P(A)$. The large deviation asymptotics are (assuming limits exist)

$$\lim_{n\to\infty} \frac{1}{n} \log P(x^n(\cdot) \in A) = -\inf_{f \in A} I(f)$$

Suppose for simplicity that

$$-\inf_{f \in A} I(f) = -I(f_{\text{opt}})$$

f_{opt}, of course, is the dominating point of the set A in which we are interested. Note f_{opt}, in this setting, is a function, not just a point in $\mathbb{R}^d$. Our problem then is how to twist the original probability measure so that it has f_{opt} as its mean trajectory.

Define

$$dF_x^*(y) \equiv \exp[\theta_x y] \frac{dF_x(y)}{M_x(\theta_x)}$$

Let $m^*(x)$ denote the mean value of $F_x^*(\cdot)$. Note that

$$m^*(x) = \frac{M_x'(\theta_x)}{M_x(\theta_x)}$$

Now we must have

$$n[\mathbf{x}_{k+1}^n - \mathbf{x}_k^n] = n\left[\mathbf{x}^n\left(\frac{k+1}{n}\right) - \mathbf{x}^n\left(\frac{k}{n}\right)\right] = V\left(\mathbf{x}\left(\frac{k}{n}\right), \xi_k\right) = \mathbf{x}^{n\prime}\left(\frac{k}{n}\right)$$

Therefore, $F_x(\cdot)$ is controlling the distribution of the *derivative* of the trajectory that the random process $\mathbf{x}^n(\cdot)$ exhibits. Thus we wish to select θ_x along f_{opt} in such a way that $f_{\text{opt}}'(t) = m^*(f_{\text{opt}}(t))$. In other words, in following the optimal trajectory at time t we wish to twist the jump distribution so that on the average we jump in the optimal direction.

***Example* 5.** Consider a continuation of example 2 of chapter IV.

$$\mathbf{x}_{k+1} = x_k + V(\mathbf{x}_k, \xi_k)$$

$\mathbf{x}_0 = x$ and for $x > 0$,

$$P\{V(x, \xi_k) = 1\} = \lambda = 1 - P\{V(x, \xi_k) = -1\} < .5$$

and

$$\bar{\lambda} \equiv 1 - \lambda$$

Furthermore, $P\{V(0, \xi_k) = 1\} = 1$, which means automatically that $\mathbf{x}_1 = 1$. The set A will be the set of trajectories of this chain that starting from zero reach N before hitting zero again. We rescale to

$$\mathbf{x}_{k+1}^N = \mathbf{x}_k^N + \frac{1}{N} V(\mathbf{x}_k^N, \xi_k)$$

$$x_0^N = 0$$

The A set now corresponds to the set of trajectories that reach 1 before hitting zero again. This chain is useful as a model for a simple M/M/1 queue. $P(A)$ is then the probability that the waiting line starts off at zero and exceeds N in number before emptying again.

$$M_x(\theta) = \lambda \exp[\theta] + \bar{\lambda} \exp[-\theta]$$

Previously, we found that

$$f_{\text{opt}}(t) = (1 - 2\lambda)t \qquad 0 \leqslant t \leqslant (1 - 2\lambda)^{-1}$$

and

$$m^*(x) = \frac{M'_x(\theta_x)}{M_x(\theta_x)} = \frac{\lambda \exp[\theta_x] - \bar{\lambda} \exp[-\theta_x]}{\lambda \exp[\theta_x] + \bar{\lambda} \exp[-\theta_x]}$$

Now since

$$m^*(f_{\text{opt}}(t)) = f'_{\text{opt}}(t) = 1 - 2\lambda$$

we have

$$m^*(x) = 1 - 2\lambda$$

Solving this relationship for θ_x yields

$$\theta_x = \log \frac{\bar{\lambda}}{\lambda}$$

Hence

$$dF^*_x(y) = \left(\frac{\bar{\lambda}}{\lambda}\right)^y \frac{dF_x(y)}{\lambda \exp[\log(\bar{\lambda}/\lambda)] + \bar{\lambda} \exp[-\log(\bar{\lambda}/\lambda)]} = \left(\frac{\bar{\lambda}}{\lambda}\right)^y dF_x(y)$$

for $y \in \{1, -1\}$. This corresponds to simulating with $\{\tilde{V}(x, \xi_k)\}$ the iid twisted random variables with $P\{\tilde{V}(x, \xi_k) = 1\} = \bar{\lambda} = 1 - P\{\tilde{V}(x, \xi_k) = -1\}$. In a queuing context, this implies that we should simulate a stable queue with arrival rate $\lambda < .5$ and service rate $1 - \lambda$ with an unstable queue with arrival rate $\bar{\lambda} > .5$ and service rate λ. The unstable queue's backlog usually builds up very quickly to exceed N (before hitting zero again). For large N this is a much more efficient way to simulate than is a direct method. □

Notes and Comments

In this chapter we introduce what is still very much an active research area. Cottrell et al. [1983] first proposed the slow Markov walk large deviation method as a way to speed up simulation of various stochastic algorithms. They explicitly simulate an Aloha communications network via their methodology. Parekh and Walrand [1986] use some of these ideas to simulate backlogs in Jackson queuing networks. Our example 6 is but one of many such examples worked out by these authors. They point out some unsolved problems with their methodology and suggest some heuristics to overcome these problems. The methodology of section A is a synthesis of the ideas found in Cottrell et al. [1983] with the twisted distributions developed for Markov chains by Miller [1961]. The asymptotic variance analysis appears to be new, although some first steps toward such an analysis were undertaken by Cottrell et al. This material (including a consideration of the iid case) is presented in Hunkel and Bucklew [1988]. The section B material of rate results and optimality is a simplified special case of Bucklew et al. [1988].

Importance sampling has had a long history in the statistical sampling literature. For an excellent overview of simulation techniques (including random number generation), see the text by Ripley [1987]. For an overview of importance sampling and communication systems analysis, see Hahn and Jeruchim [1987]. These last references make one keep in mind that there are a number of other methods besides large deviation theory techniques that one may use for choosing $q(\cdot)$ to yield gains in estimator performance.

The main competitor to the exponential shifts for importance sampling applications appears to be various translation and stretching methodologies. These methods merely translate or stretch the generated random variables by a constant in order to make rare events more likely. There are several disadvantages to these formulations. In the finite state space case, a simple translation is not possible without enlarging the number of states. Variance analysis as in section A of this chapter is not nearly as simple, due to the complexities of computing the weighting factors. Finally, dependency, while easily handled in the Markov setting with exponential shifts, presents tremendous theoretical and practical difficulties for the translation methodologies. However, there are situations where translation methods outperform twisted simulators. One must keep in mind that not every rare event is a large deviation rare event. One cannot expect the twisted simulator to be optimal or near optimal if such is the case. Each simulation problem is unique and must be understood before a good decision of which methodology to employ can be made. Both methods (and others to be discovered) will have their uses. However, whenever the rare event is being controlled by a large deviation principle, it is now clear that exponential twists are the superior method.

EXERCISES

1. Show that the optimum sampling density for importance sampling is proportional to $|p(x)\eta(x)|$.

2. Consider exercise 6 in chapter VI. Simulate the type I error probability for the optimum memoryless detector. Compare to exercise 5 in chapter VII.

3. Consider $P\{(1/n)\,\Sigma_{i=1}^{n}\,\mathbf{x}_i \geqslant 10\}$, where the $\{\mathbf{x}_i\}$ are iid zero mean, variance 1, normal random variables. Simulate this probability directly and via quick simulation for values of $n = 2$, 10. How close are the simulated values compared to the true values (from asymptotic expressions)?

4. Consider example 3. Compute for various values of p and a some representative values $\lambda(\theta_0)$.

5. In the text it is assumed that $f(x) = 0$, where x is the initial starting point of the chain. How do the asymptotic variance expressions change if $f(x) \neq 0$?

6. Suppose that it is desired to estimate $P((1/n)\,\Sigma_{i=1}^{n}\,f(\mathbf{x}_i) \geqslant 0)$, where the $\{\mathbf{x}_i\}$ are samples from an irreducible Markov chain with transition probabilities $p(x\,|\,y)$. It is desired that the importance sampling distribution $q(\cdot)$ be product type (i.e., iid samples are generated). Characterize the optimal (smallest speed factor) choice for $q(\cdot)$.

CHAPTER IX

Applications to Parameter Estimation

A. LOWER PERFORMANCE BOUNDS FOR PARAMETER ESTIMATES

One usually desires that a parameter estimator have the property of being *consistent*, i.e., as the number of observations approaches infinity, the estimated value of the parameter approaches the true parameter. In many important situations this convergence will be exponentially fast. In such cases, our criterion for estimator design will be to try to make this rate of convergence as fast as possible. In this chapter we present a result that sets a lower bound on the rate performance of any consistent parameter estimator. It may be thought of as a large deviation Cramér–Rao bound (see appendix F). However, we first need to prove a very general large deviation lower bound theorem which will provide the crucial intermediate step in route to our main result.

Let $\{p^n(x)\}_{n=1}^{\infty}$ and $\{q^n(x)\}_{n=1}^{\infty}$ be two sequences of probability densities where $p^k(\cdot)$, $q^k(\cdot)$ are densities on $\mathbb{R}^k$. For any t, $-\infty \leqslant t \leqslant \infty$, let

$$C_n(t) \equiv \left\{x \in \mathbb{R}^n : \frac{1}{n} \log \frac{q^n(x)}{p^n(x)} \leqslant t\right\} \tag{1}$$

Further define

$$Q_n(A) \equiv \int_A q^n(x)\, dx \tag{2}$$

$$P_n(A) \equiv \int_A p^n(x)\, dx \tag{3}$$

and

$$K(q, p) \equiv \inf\left\{t : \lim_{n \to \infty} Q_n(C_n(t)) = 1\right\} \tag{4}$$

Note that

$$Q_n(C_n(t)) = \int_{\{x:q^n(x)/p^n(x)\leqslant \exp[nt]\}} q^n(x)\,dx \leqslant \int_{C_n(t)} p^n(x)\exp[nt]\,dx \leqslant \exp[nt] \tag{5}$$

Therefore, for $t < 0$, $\lim_{n\to\infty} Q_n(C_n(t)) = 0$. Hence $0 \leqslant K(q, p) \leqslant \infty$.

Theorem (Bahadur–Zabell–Gupta)
Let $\{A_n\}$ be a sequence of sets such that $A_k \subset \mathbb{R}^k$ for $k = 1, 2, \ldots$ and such that

$$\varliminf_{n\to\infty} \int_{A_n} q^n(x)\,dx > 0 \tag{6}$$

Then

$$\varliminf_{n\to\infty} \frac{1}{n}\log \int_{A_n} p^n(x)\,dx \geqslant -K(q, p) \tag{7}$$

Proof: Suppose that $0 \leqslant K(q,p) < \infty$. If $K(q,p) = \infty$, the theorem is trivial. Let $t > K(q,p)$ and define $B_n(t) \equiv C_n(t)^c$, i.e., the set complementation of $C_n(t)$. Then

$$\lim_{n\to\infty} Q_n(B_n(t)) = \lim_{n\to\infty} [1 - Q_n(C_n(t))] = 1 - \lim_{n\to\infty} Q_n(C_n(t))$$

[and since $t > K(q, p)$]

$$= 1 - 1 = 0$$

Now

$$\begin{aligned} P_n(A_n) \geqslant P_n(A_n \cap C_n(t)) &= \int_{A_n\cap C_n(t)} p^n(x)\,dx \\ &\geqslant \int_{A_n\cap C_n(t)} \exp[-nt]\frac{q^n(x)}{p^n(x)}\,p^n(x)\,dx \\ &= \exp[-nt]Q_n(A_n \cap C_n(t)) \\ &\geqslant \exp[-nt][Q_n(A_n) - Q_n(C_n(t)^c)] \\ &= \exp[-nt][Q_n(A_n) - Q_n(B_n(t))] \end{aligned} \tag{8}$$

Now take $(1/n)\log$ and limit as n approaches infinity of both sides of (8). Note that since $\underline{\lim}_{n\to\infty} Q_n(A_n) > 0$ and $Q_n(B_n(t)) \xrightarrow[n\to\infty]{} 0$, we have $\lim_{n\to\infty} (1/n)\log[Q_n(A_n) - Q_n(B_n(t))] = 0$. Therefore,

$$\underline{\lim}_{n\to\infty} \frac{1}{n} \log P_n(A_n) \geqslant -t$$

where t is arbitrary as long as it is greater than $K(q, p)$. Choosing the infimum of all such t gives the theorem statement. □

***Example* 1.** Suppose that $\{\mathbf{x}_i\}$ is an iid sequence. Let the density function of n random variables be denoted as

$$p^n(x) = \prod_{j=1}^{n} p^1(x_j)$$

where $x = (x_1 \quad x_2 \quad \cdots \quad x_n)$. Let

$$q^n(x) = \prod_{j=1}^{n} q^1(x_j)$$

Then

$$C_n(t) = \left\{x \in \mathbb{R}^n : \frac{1}{n} \sum_{j=1}^{n} \log \frac{q^1(x_j)}{p^1(x_j)} \leqslant t\right\}$$

By the law of large numbers (if $\mathbf{x}$ is distributed as $q^n(x)$),

$$\frac{1}{n} \sum_{j=1}^{n} \log \frac{q^1(\mathbf{x}_j)}{p^1(\mathbf{x}_j)} \to \int_{\mathbb{R}^1} q^1(y) \log \frac{q^1(y)}{p^1(y)}\, dy$$

Therefore,

$$\lim_{n\to\infty} Q_n(C_n(t)) = 1$$

if

$$t \geqslant \int_{\mathbb{R}^1} q^1(y) \log \frac{q^1(y)}{p^1(y)}\, dy$$

and equals zero otherwise. Hence

$$K(q, p) = \int_{\mathbb{R}^1} q^1(y) \log \frac{q^1(y)}{p^1(y)} \, dy$$

which is the Sanov's theorem rate function (also known as the *Kullback–Liebler number*, *generalized Shannon entropy*, *divergence*, etc.). □

Suppose that we have a collection of probability densities $\{p^n(x, \gamma), x \in \mathbb{R}^n\}_{n=1}^{\infty}$ and $\gamma \in \otimes$, where $\otimes$ is some index set. Let $g(\cdot)$ be a mapping from $\otimes$ into some other set Γ (usually, $\Gamma = \otimes = \mathbb{R}$). Our problem then is to estimate $g(\gamma)$ given that we observe the random vector $\mathbf{x} \in \mathbb{R}^n$. Denote our estimate as $T_n(\mathbf{x})$. The sequence $\{T_n\}_{n=1}^{\infty}$ is said to be a *consistent* estimate of g if for each $\gamma \in \otimes$,

$$\lim_{n\to\infty} \int_{\{x \in \mathbb{R}^n : T_n(x) \in A\}} p^n(x, \gamma) \, dx = 1 \tag{9}$$

for every open set $A \subset \Gamma$ with $g(\gamma) \in A$.

Let $d(\cdot, \cdot)$ be a continuous distance measure on the set Γ. For example, if $\Gamma = \mathbb{R}$, then $d(x, y) = |x - y|$ would be a common choice. A measure of how close (or good) our estimate is to the true value would then be $d(T_n(\mathbf{x}), g(\gamma))$. Define

$$a_n(\varepsilon, \gamma) \equiv \int_{\{x \in \mathbb{R}^n : d(T_n(x), g(\gamma)) \geqslant \varepsilon\}} p^n(x, \gamma) \, dx \tag{10}$$

and

$$\Delta(\varepsilon, \gamma) \equiv \{\delta : \delta \in \otimes,\ d(g(\delta), g(\gamma)) > \varepsilon\} \tag{11}$$

Then, if $\Delta(\varepsilon, \gamma)$ is not empty, define

$$b(\varepsilon, \gamma) \equiv \inf\{K(p(\cdot, \delta), p(\cdot, \gamma)) : \delta \in \Delta(\varepsilon, \gamma)\} \tag{12}$$

If $\Delta(\varepsilon, \gamma)$ is empty, define $b(\varepsilon, \gamma) = \infty$. The following remarkable theorem is then true.

Theorem (Bahadur–Zabell–Gupta)
Suppose that $\{T_n\}$ is a consistent estimate of g. Then

$$\underline{\lim}_{n\to\infty} \frac{1}{n} \log a_n(\varepsilon, \gamma) \geqslant -b(\varepsilon, \gamma) \tag{13}$$

Proof: Choose and fix γ and ε. If $b(\varepsilon, \gamma)$ is infinity, the theorem is trivial so suppose that $b(\varepsilon, \gamma) < \infty$. In that case $\Delta(\varepsilon, \gamma)$ is nonempty, so we may choose a δ from it. Suppose that the observations are coming from the $\{p^n(x, \delta)\}$ distributions. Then $d(T_n(\mathbf{x}), g(\gamma)) \xrightarrow[n\to\infty]{} d(g(\delta), g(\gamma)) > \varepsilon$ since T_n is a consistent estimate [and $d(\cdot, \cdot)$ is continuous]. The convergence above is in probability. Let $A_n = \{x : d(T_n(x), g(\gamma)) \geqslant \varepsilon\}$. Then

$$\lim_{n\to\infty} \int_{A_n} p^n(x, \delta)\, dx = 1 \tag{14}$$

Now the previous theorem expressions (6)–(7) can be invoked with $\{p^n(x, \delta)\}$ and $\{p^n(x, \gamma)\}$ taking the place of $\{q^n(x)\}$ and $\{p^n(x)\}$, respectively. Hence

$$\varliminf_{n\to\infty} \frac{1}{n} \log \int_{A_n} p^n(x, \gamma)\, dx \geqslant -K(p(\cdot, \delta), p(\cdot, \gamma)) \tag{15}$$

δ is arbitrary as long as $\delta \in \Delta(\varepsilon, \gamma)$. Choosing the value of δ to give the largest lower bound then gives the theorem statement. □

***Example* 2.** Suppose that $\otimes = (0, \infty)$ and $\{\mathbf{x}_i\}$ is an iid sequence of uniform $[0, \gamma]$ random variables. We suppose that $g(\gamma) \equiv \gamma$ and $d(x, y) \equiv |x - y|$.

$$K(p(\cdot, \delta), p(\cdot, \gamma)) = \int p(x, \delta) \log \frac{p(x, \delta)}{p(x, \gamma)}\, dx = \log \frac{\gamma}{\delta}$$

for $\delta \leqslant \gamma$ and equals infinity otherwise.

$$\Delta(\varepsilon, \gamma) = \{\delta : \delta \in (0, \infty) \text{ and } |\delta - \gamma| > \varepsilon\}$$

$$b(\varepsilon, \gamma) = \inf\left\{\log\frac{\gamma}{\delta} : \delta \in \Delta(\varepsilon, \gamma) \text{ and } \delta \leqslant \gamma\right\} = \log \frac{\gamma}{\gamma - \varepsilon}$$

for $0 < \varepsilon < \gamma$ (and equals infinity for $\varepsilon > \gamma$). Now also

$$a_n(\varepsilon, \gamma) = \int_{\{x : |T_n(x) - \gamma| \geqslant \varepsilon\}} p^n(x, \gamma)\, dx$$

Suppose that $T_n(\mathbf{x}) = \max\{\mathbf{x}_1, \mathbf{x}_2, \ldots, \mathbf{x}_n\}$. It is clear that $\lim_{n\to\infty} T_n(\mathbf{x}) = \gamma$.

The distribution function of $T_n(\mathbf{x})$ is

$$\begin{aligned} P\{T_n(\mathbf{x}) \leqslant \alpha\} &= P\{\max\{\mathbf{x}_1, \ldots, \mathbf{x}_n\} \leqslant \alpha\} \\ &= P\{\text{all } n \text{ iid samples} \leqslant \alpha\} \\ &= P\{\text{one sample} \leqslant \alpha\}^n = \left(\frac{\alpha}{\gamma}\right)^n \end{aligned}$$

for $0 \leqslant \alpha \leqslant \gamma$. Therefore, the probability density for $T_n(\mathbf{x})$ is

$$p_{T_n}(\alpha) = n\left(\frac{\alpha}{\gamma}\right)^{n-1}\frac{1}{\gamma}$$

Suppose that $\varepsilon < \gamma$. Then

$$a_n(\varepsilon, \gamma) = \int_0^{\gamma-\varepsilon} p_{T_n}(\alpha)\, d\alpha = \frac{n}{\gamma^n}\int_0^{\gamma-\varepsilon} \alpha^{n-1}\, d\alpha = \left(\frac{\gamma-\varepsilon}{\gamma}\right)^n$$

Therefore,

$$\frac{1}{n}\log a_n(\varepsilon, \gamma) = \log\frac{\gamma-\varepsilon}{\gamma} = -b(\varepsilon, \gamma)$$

For $\varepsilon \geqslant \gamma$, we again have the same equality. Therefore, in this example, the lower bound is actually attained for all values of n, ε, and γ. It is a simple exercise to show that T_n is the maximum likelihood estimate. □

B. MAXIMUM LIKELIHOOD ESTIMATORS ("SMALL ε" ASYMPTOTICS)

Suppose that $\{\mathbf{x}_i\}$ is an iid sequence of random observations and $g(\gamma) \equiv \gamma$ is the parameter of interest. Let T_n be a consistent estimate of γ. Then

$$a_n(\varepsilon, \gamma) \equiv P_\gamma(|T_n(\mathbf{x}_1) - \gamma| \geqslant \varepsilon) \tag{16}$$

where $P_\gamma(\cdot)$ is the distribution of the $\{\mathbf{x}_i\}$ when γ is the true underlying parameter. From example 1 and the definitions, we have that

$$b(\varepsilon, \gamma) = \inf_{\delta:|\delta-\gamma|>\varepsilon} K(p(\cdot, \delta), p(\cdot, \gamma)) = K(p(\cdot, \gamma+\varepsilon), p(\cdot, \gamma)) \tag{17}$$

[*Note:* $K(\cdot, \cdot)$ is convex and for ε small enough the minimization should be on an interval boundary.] Now for small ε, the right side of (17) is approximately $k(\gamma)\varepsilon^2/2$ (see appendix F), where

$$k(\gamma) \equiv E_\gamma \left\{ \left(\frac{\partial}{\partial \gamma} \log p(\mathbf{x}_1, \gamma) \right)^2 \right\} = -E_\gamma \left\{ \frac{\partial^2}{\partial \gamma^2} \log p(\mathbf{x}_1, \gamma) \right\} \tag{18}$$

is the Fisher information at the point γ (see exercise 4). Hence, by combining (13) with the heuristics above, we have indicated the proof to the following theorem.

Theorem (Bahadur)
Let $\gamma \in \otimes$, where $\otimes$ is an open set in $\mathbb{R}$. Let T_n be a consistent estimate of γ (from the iid data). Then subject to the regularity condition that

$$K(p(\cdot, \gamma + \varepsilon), p(\cdot, \gamma)) = \frac{\varepsilon^2}{2} k(\gamma) + o(\varepsilon^2) \tag{19}$$

we have

$$\lim_{\varepsilon \to 0} \lim_{n \to \infty} \frac{1}{\varepsilon^2 n} \log P_\gamma(|T_n(\mathbf{x}_1) - \gamma| > \varepsilon) \geqslant \frac{-k(\gamma)}{2} \tag{20}$$

Our task is now to show that a class of estimates, the maximum likelihood estimators, can actually attain this bound.

Let $\hat{\gamma}_{\mathrm{ml},n}$ be the maximum likelihood estimate based on n iid samples. The likelihood function (see appendix F) appears as

$$\prod_{i=1}^{n} p(\mathbf{x}_i, \gamma) \tag{21}$$

Then by definition of the maximum likelihood estimate,

$$\hat{\gamma}_{\mathrm{ml},n} \equiv \text{value of } \gamma \text{ that maximizes } \prod_{i=1}^{n} p(\mathbf{x}_i, \gamma) \tag{22}$$

The hope is that there exists a unique value of the parameter that maximizes the likelihood function. By monotonicity of the logarithm, we equivalently have

$$\hat{\gamma}_{\mathrm{ml},,n} \equiv \text{value of } \gamma \text{ that maximizes } \sum_{i=1}^{n} \log p(\mathbf{x}_i, \gamma) \tag{23}$$

If quantities are differentiable, one performs the maximization by taking the derivative and setting equal to zero or

$$\sum_{i=1}^{n} \frac{\partial}{\partial \gamma} \log p(\mathbf{x}_i, \gamma)|_{\gamma=\hat{\gamma}_{\mathrm{ml},n}} = 0 \tag{24}$$

We assume as a technical condition needed in the sequel that

$$E_\gamma \left\{ \frac{\partial}{\partial \gamma} \log p(\mathbf{x}_1, \gamma) \right\} = 0 \tag{25}$$

(along with all of the differentiability conditions needed so far for finding the maximum likelihood estimate via differentiation).

Consider the event $\{\hat{\gamma}_{\mathrm{ml},,n} \geqslant \gamma + \varepsilon\}$. The only way for this event to occur is if in the differentiation above, the critical point occurs at a value greater than or equal to $\gamma + \varepsilon$. For small ε this should mean that the derivative evaluated at the point $\gamma + \varepsilon$ should be positive (or at least nonnegative). Hence we have

$$P_\gamma(\hat{\gamma}_{\mathrm{ml},n} \geqslant \gamma + \varepsilon) \approx P\left(\frac{1}{n} \sum_{i=1}^{n} \frac{\partial}{\partial \gamma} \log p(\mathbf{x}_i, \gamma + \varepsilon) \geqslant 0 \right) \tag{26}$$

We may now use Cramér's theorem to find the large deviation asymptotics of this last probability, or

$$\begin{aligned} \lim_{n\to\infty} \frac{1}{n} \log P_\gamma(\hat{\gamma}_{\mathrm{ml},n} \geqslant \gamma + \varepsilon) &= -\sup_t \left\{ -\log E_\gamma \left\{ \exp\left[t \frac{\partial}{\partial \gamma} \log p(\mathbf{x}_1, \gamma + \varepsilon) \right] \right\} \right\} \\ &= \inf_t \log E_\gamma \left\{ \exp\left[t \frac{\partial}{\partial \gamma} \log p(\mathbf{x}_1, \gamma + \varepsilon) \right] \right\} \\ &= \log\left[\inf_t E_\gamma \left\{ \exp\left[t \frac{\partial}{\partial \gamma} \log p(\mathbf{x}_1, \gamma + \varepsilon) \right] \right\} \right] \end{aligned} \tag{27}$$

To find thc optimal t, wc takc thc dcrivative with respect to t and set to zero to find that

$$0 = E_\gamma \left\{ \left(\frac{\partial}{\partial \gamma} \log p(\mathbf{x}_1, \gamma + \varepsilon) \right) \exp\left[t \frac{\partial}{\partial \gamma} \log p(\mathbf{x}_1, \gamma + \varepsilon) \right] \right\} \tag{28}$$

Now as $\varepsilon \to 0$, the optimal t will also get small. Hence we can expand the exponential into its Taylor series to find that

$$0 = E_\gamma\left\{\left(\frac{\partial}{\partial\gamma}\log p(\mathbf{x}_1, \gamma + \varepsilon)\right)\left[1 + t\frac{\partial}{\partial\gamma}\log p(\mathbf{x}_1, \gamma + \varepsilon) + O(t^2)\right]\right\} \tag{29}$$

Hence

$$t_{\text{opt}} \approx -E_\gamma\left\{\frac{\partial}{\partial\gamma}\log p(\mathbf{x}_1, \gamma + \varepsilon)\right\}\Big/ E_\gamma\left\{\left(\frac{\partial}{\partial\gamma}\log p(\mathbf{x}_1, \gamma + \varepsilon)\right)^2\right\}$$

$$\approx -E_\gamma\left\{\frac{\partial}{\partial\gamma}\log p(\mathbf{x}_1, \gamma + \varepsilon)\right\}\Big/ E_\gamma\left\{\left(\frac{\partial}{\partial\gamma}\log p(\mathbf{x}_1, \gamma)\right)^2\right\}$$

(perturbations in a denominator are negligible as long as everything is nonzero)

$$= -E_\gamma\left\{\frac{\partial}{\partial\gamma}\log p(\mathbf{x}_1, \gamma + \varepsilon)\right\}\Big/ k(\gamma)$$

$$= -E_\gamma\left\{\frac{\partial}{\partial\gamma}\log p(\mathbf{x}_1, \gamma) + \varepsilon\frac{\partial^2}{\partial\gamma^2}\log p(\mathbf{x}_1, \gamma) + O(\varepsilon^2)\right\}\Big/ k(\gamma)$$

$$= -\left[\varepsilon E_\gamma\left\{\frac{\partial^2}{\partial\gamma^2}\log p(\mathbf{x}_1, \gamma)\right\} + O(\varepsilon^2)\right]\Big/ k(\gamma)$$

[the first term of the numerator is zero by (25)]

$$= \varepsilon\frac{k(\gamma)}{k(\gamma)} + O(\varepsilon^2)$$

$$= \varepsilon + O(\varepsilon^2)$$

or

$$t_{\text{opt}} \approx \varepsilon \tag{30}$$

Therefore, substituting (30) back into (27), we obtain

$$\lim_{n\to\infty}\frac{1}{n}\log P_\gamma(\hat{\gamma}_{\text{ml},n} \geqslant \gamma + \varepsilon) = \log\inf_t E_\gamma\left\{\exp\left[t\frac{\partial}{\partial\gamma}\log p(\mathbf{x}_1, \gamma + \varepsilon)\right]\right\}$$

$$\approx \log E_\gamma\left\{\exp\left[\varepsilon\frac{\partial}{\partial\gamma}\log p(\mathbf{x}_1, \gamma + \varepsilon)\right]\right\}$$

$$= \log E_\gamma\left\{1 + \varepsilon\frac{\partial}{\partial\gamma}\log p(\mathbf{x}_1, \gamma + \varepsilon)\right.$$

$$+ \frac{\varepsilon^2}{2}\left(\frac{\partial}{\partial\gamma}\log p(\mathbf{x}_1, \gamma + \varepsilon)\right)^2 + O(\varepsilon^3)\Bigg\}$$

$$\approx \log\Bigg[1 + \varepsilon E_\gamma\left\{\frac{\partial}{\partial\gamma}\log p(\mathbf{x}_1, \gamma + \varepsilon)\right\}$$

$$+ \frac{\varepsilon^2}{2} E_\gamma\left\{\left(\frac{\partial}{\partial\gamma}\log p(\mathbf{x}_1, \gamma + \varepsilon)\right)^2\right\}\Bigg]$$

$$\approx \log\left[1 + \varepsilon[-\varepsilon k(\gamma)] + \frac{\varepsilon^2}{2}[k(\gamma)]\right]$$

$$\approx \log\left[1 - \frac{\varepsilon^2}{2}k(\gamma)\right] \approx \frac{\varepsilon^2}{2}k(\gamma) \tag{31}$$

for small ε. Equation (31) then implies the following theorem.

Theorem (Bahadur)
Let $\gamma \in \otimes$ be an open set in $\mathbb{R}$. Let $\hat{\gamma}_{\mathrm{ml},n}$ be the maximum likelihood estimator of γ from n iid observations. Then subject to various technical conditions mentioned above,

$$\lim_{\varepsilon\to 0}\lim_{n\to\infty}\frac{1}{\varepsilon^2 n}\log P_\gamma(|\hat{\gamma}_{\mathrm{ml},n} - \gamma| \geqslant \varepsilon) = \frac{-k(\gamma)}{2} \tag{32}$$

Notes and Comments

$K(\cdot,\cdot)$ of section A has been established as the Kullback–Leibler number in many cases, e.g., finite state space Markov chains, iid Banach space valued random variables, etc. (see Bahadur et al. [1980]). Therefore, we may view the first theorem in this chapter as an extension of the lower bound formulation of Sanov's theorem.

The work in section B showing that $b(\varepsilon, \gamma)$, for small ε is attained by the maximum likelihood estimates in the iid case was performed originally by Bahadur [1960, 1967] and Fu [1975]. The final theorem quoted in this section departs from our usual practice in that we did not specify all of the technical conditions needed for the theorem statement. We refer the reader to the original papers, where there are extensive discussion of the necessary topological and measure theoretic conditions needed to prove the result rigorously. It seems that except for the conditions mentioned explicitly in the text, these conditions hold quite generally. These papers consider a more general case: estimating $g(\gamma)$ where $g(\cdot)$ is a function of the parameter controlling the distribution of the observations. More recently, investigations (see Kester and Kallenberg [1986]) have proceeded for fixed (i.e., not asymptotically small) $\varepsilon > 0$. These investigations have shown that in many cases (e.g., convex parameter spaces and exponential families) the maximum likelihood estimator is still the best.

EXERCISES

1. For finite-dimensional Markov chains, show that $K(q, p)$ is the large deviation rate function.

2. Let $\{\mathbf{x}_i\}$ be a sequence of iid Gaussian, mean value of γ, variance 1, random variables. What is $b(\varepsilon, \gamma)$ in this case? Does the sample mean attain this performance bound?

3. Let $\{\mathbf{x}_i\}$ be a sequence of iid exponential random variables with density $\gamma \exp[-\gamma x]$ $(x \geqslant 0)$. What is the maximum likelihood estimate of γ? Do the small ε asymptotics of this estimate equal the bound given in the text?

4. Show directly (for the iid case) via Taylor's theorem that $K(p(\cdot, \gamma + \varepsilon), p(\cdot, \gamma)) \approx \varepsilon^2 k(\gamma)/2$, where $K(\cdot, \cdot)$ is the Kullback–Leibler number and $k(\cdot)$ is the Fisher information at the point γ.

CHAPTER X

Applications to Information Theory

A. ERROR EXPONENTS OF SHANNON'S FIRST CODING THEOREM

Consider an information source that emits an independent identically distributed random variable every T seconds. Such a source is said to be *memoryless* (present outputs have no "memory" of what past outputs were). If the random variables emitted only take on values in a finite set, we say the *source alphabet is finite* (*or discrete finite*). In this chapter we consider only discrete finite memoryless sources. Without loss of generality, we can always take the source alphabet to be the set $\{1, 2, \ldots, d < \infty\}$. Consider dividing up the source output stream in blocks n random variables long; i.e., we equivalently consider the source to be emitting an n-dimensional random vector every nT seconds. Each one of these random vectors is called a *sourceword*. In the data compression problem, one desires to represent each sourceword with a *codeword* of blocklength r from a code alphabet of size A. Without loss of generality we can always assume that $A \equiv 2$, which implies that there are at most 2^r-codewords. The natural logarithm of this number divided by n is called the *coderate* $R \equiv r \log 2/n$. Since there are d^n number of sourcewords, data compression is achieved only if $2^r < d^n$. It is assumed that $2^r - 1$ of the sourcewords will be given unique codewords. The remaining sourcewords will be assigned to a default codeword that essentially will signal a decoding failure.

Since the source outputs are independent, identically distributed random variables, we can completely characterize the source by specifying the probability distribution of the first source output. Hence $P(\mathbf{x}_1 = i) = p_i$, $i = 1, \ldots, d$, specifies the source P; source Q will be specified by $Q(\mathbf{x}_1 = 1) = q_i$, $i = 1, \ldots, d$; and the uniform source U is specified by

$U(\mathbf{x}_1 = i) = d^{-1}, i = 1, \ldots, d$. The *entropy* of the source P, $H(P)$, is defined by

$$H(P) \equiv -\sum_{i=1}^{d} p_i \log p_i$$

It is easy to check that the entropy of U is $\log d$.

Shannon's first coding theorem established that entropy is the basic measure of the information content of a source. This theorem states that given any $\varepsilon > 0$, there exists a code mapping sourcewords of blocklength n into binary codewords of blocklength r such that the probability of decoding failure can be made less than ε (provided that $R > H$ and n is sufficiently large). In other words, as n approaches infinity, the probability of a decoding failure can be made to approach zero. In this subsection we show that this convergence is exponentially fast and derive the rate via large deviation theory techniques.

Suppose that $H < R < \log d$. Consider a sourceword $x \equiv (\mathbf{x}_1, \mathbf{x}_2, \ldots, \mathbf{x}_n)$. One can create an empirical distribution

$$L_{n,\mathbf{x}}(j) \equiv \frac{1}{n} \sum_{i=1}^{n} 1_j(\mathbf{x}_i) \equiv \frac{\#\text{ of } \mathbf{x}_i\text{'s that equal } j}{n}$$

Associated with this empirical distribution is an entropy

$$H(L_{n,\mathbf{x}}) \equiv -\sum_{j=1}^{d} L_{n,\mathbf{x}}(j) \log L_{n,\mathbf{x}}(j)$$

Note also that $H(\cdot)$ may be thought of as a continuous mapping from the space of probability measures to the real line $\mathbb{R}$. How many of the d^n sourcewords give rise to empirical distributions with entropy less than R? Let us denote this set as

$$S_n \equiv \{x \in \mathbb{R}^n : H(L_{n,x}) < R\}$$

and the number of elements in this set (or the cardinality of the set) as $|S_n|$. Note that S_n is an open set (actually, since the space is discrete, all sets are open). Now it is easy to see that

$$|S_n| = d^n U(S_n) \tag{1}$$

(since the U source puts equal weight to every possible sourceword). Hence, by Sanov's theorem, we have

$$\overline{\lim_{n\to\infty}} \frac{1}{n} \log U(S_n) \leqslant \overline{\lim_{n\to\infty}} \frac{1}{n} \log U(\{x : H(L_{n,x}) \leqslant R\})$$

$$\leqslant - \inf_{Q:H(Q)\leqslant R} \left[\sum_{i=1}^{d} q_i \log \frac{q_i}{d^{-1}} \right]$$

$$= -\log d + \sup_{Q:H(Q)\leqslant R} H(Q) = R - \log d \qquad (2)$$

Hence, from (1)–(2),

$$\overline{\lim_{n\to\infty}} \frac{1}{n} \log |S_n| \leqslant R$$

Now let us consider a coding scheme whereby we represent every sourceword in S_n by a unique codeword. We can thus code every sourceword in S_n with

$$m_n \equiv \left[\frac{\log(|S_n| + 1)}{\log 2} \right]$$

binary digits, where $[x]$ indicates the greatest integer less than or equal to x. We use one more codeword to code all of the sequences that lie outside S_n. Hence the probability that a sequence is coded incorrectly is merely $P(H(L_{n,x}) \geqslant R)$. The asymptotics of this probability are then obtained via Sanov's theorem to be

$$\overline{\lim_{n\to\infty}} \frac{1}{n} \log P(H(L_{n,x}) \geqslant R) \leqslant - \inf_{Q:H(Q)\geqslant R} \left[\sum_{i=1}^{d} q_i \log \frac{q_i}{p_i} \right]$$

Hence we have proved the following theorem.

Theorem (Jelinek–Csiszar and Longo–Anantharam)
For any R such that $0 \leqslant R \leqslant \log d$, there exists a sequence of n to m_n binary block codes such that $\overline{\lim}_{n\to\infty} m_n/n \leqslant R/\log 2$ with

$$\overline{\lim_{n\to\infty}} \frac{1}{n} \log P(\text{block decoding failure}) \leqslant - \inf_{Q:H(Q)\geqslant R} \left[\sum_{i=1}^{d} q_i \log \frac{q_i}{p_i} \right]$$

This theorem is also optimal in that this is the best rate available. Consider any other n to m_n code such that $\overline{\lim}_{n\to\infty} m_n/n \leqslant R/\log 2$. Now not only is $H(\cdot)$ a continuous mapping from the space of distributions to the real line but for a fixed P, so is

$$H(\cdot, P) \equiv \text{Kullback–Leibler information} \equiv \sum_{i=1}^{d} q_i \log \frac{q_i}{p_i}$$

Consider some Q_0 such that $H(Q_0) > R$. Then by continuity, for any $\varepsilon > 0$, we can find an open set of distributions G such that $Q_0 \in G$ and $|H(Q) - H(Q_0)| < \varepsilon$ and $|H(Q, P) - H(Q_0, P)| < \varepsilon$ for all $Q \in G$.

How many sourcewords give rise to empirical distributions that lie in G? Denote this set as

$$S_{G,n} \equiv \{x : L_{n,x}(\cdot) \in G\}$$

Then mimicking a similar argument given above, we have

$$\begin{aligned}\underline{\lim}_{n\to\infty} \frac{1}{n} \log |S_{G,n}| &= \underline{\lim}_{n\to\infty} \frac{1}{n} \log d^n U(S_{G,n}) \geqslant \log d - \inf_{Q\in G} H(Q, U) \\ &= \log d - \inf_{Q\in G} \sum_{i=1}^{d} q_i \log \frac{q_i}{d^{-1}} = \sup_{Q\in G} H(Q) \\ &\geqslant H(Q_0) - \varepsilon > R\end{aligned}$$

(if $\varepsilon > 0$ is chosen small enough). Now let S_n denote the set of sourcewords that are given a uniquely decodable codeword. Then necessarily

$$|S_n| \leqslant 2^{m_n}$$

or, equivalently,

$$\overline{\lim}_{n\to\infty} \frac{1}{n} \log |S_n| \leqslant \overline{\lim}_{n\to\infty} \frac{m_n}{n} \log 2 \leqslant R$$

Hence for n large enough at least $\frac{1}{2}$ (actually any fraction strictly between 0 and 1 will suffice) of the sourcewords in $S_{G,n}$ lie outside S_n.

Furthermore, suppose that x is a sourceword such that

$$L_{n,x}(\cdot) \equiv Q \in G \qquad (\text{i.e., } x \in S_{G,n})$$

This means that out of n independent source letters from the source P exactly nq_i of them are equal to i. The probability of this particular sourceword is bounded below as

$$\begin{aligned}
\frac{1}{n}\log P(x) &= \frac{1}{n}\log \prod_{i=1}^{d} p_i^{nq_i} = \sum_{i=1}^{d} q_i \log p_i \\
&= \sum_{i=1}^{d}\left[q_i \log q_i - q_i \log \frac{q_i}{p_i}\right] = -H(Q) - H(Q, P) \\
&\geqslant -\sup_{Q \in S_{G,n}} [H(Q) + H(Q, P)] \\
&\geqslant -\sup_{Q \in S_{G,n}} [H(Q_0) + H(Q_0, P) + 2\varepsilon] \\
&= -[H(Q_0) + H(Q_0, P) + 2\varepsilon]
\end{aligned}$$

Therefore,

$$\begin{aligned}
\underline{\lim}_{n\to\infty} \frac{1}{n}\log P(\text{block decoding failure}) &\geqslant \underline{\lim}_{n\to\infty} \frac{1}{n}\log P(x \notin S_n \text{ and } x \in S_{G,n}) \\
&\geqslant \underline{\lim}_{n\to\infty} \frac{1}{n}\log[\tfrac{1}{2}|S_{G,n}|\, P(x \mid x \in S_{G,n})] \\
&\geqslant H(Q_0) - \varepsilon - [H(Q_0) + H(Q_0, P) + 2\varepsilon] \\
&= -H(Q_0, P) - 3\varepsilon
\end{aligned}$$

Letting $\varepsilon \to 0$ then gives the converse to the first theorem statement.

Theorem

Let any sequence of n to m_n binary codes be given such that $\overline{\lim}_{n\to\infty} m_n/n \leqslant R/\log 2$. Then

$$\underline{\lim}_{n\to\infty} \frac{1}{n}\log P(\text{block decoding failure}) \geqslant -\inf_{Q: H(Q) \geqslant R} H(Q, P)$$

B. SOURCE CODING WITH RESPECT TO A FIDELITY CRITERION

Information theory rests on two key results, the channel coding theorem and the source coding with respect to a fidelity criterion theorem. The channel coding theorem [Shannon, 1948] establishes a quantity known as the *channel capacity* as the fundamental limitation to reliable communication over an unreliable signaling medium. The bulk of Shannon's 1948 paper deals with this theorem and calculations related to it.

The source coding with respect to a fidelity criterion theorem established a quantity known as the *rate distortion function* as the fundamental limitation to the performance of data compression schemes. The question considered here is: If one is constrained to coding at information rates below the inherent information rate of the source, how much distortion must one suffer for a given coding rate? This and the related question of what information should be transmitted to attain this performance was answered in a landmark paper by Shannon [1959].

We consider the case of *single-letter distortion measures*, $\rho(j, k)$, as being the distortion (a positive number) produced when the source alphabet letter j is represented by the reproducing alphabet letter k. In general, we will assume that there are M source letters and N reproducing letters (M need not necessarily equal N). From this single-letter distortion measure we may create a distortion measure for sourcewords mapped into codewords as

$$\rho_n(x, y) \equiv \frac{1}{n} \sum_{i=1}^{n} \rho(x_i, y_i)$$

Let $P(j)$ be defined to be the probability that the source letter from the discrete memoryless source is j, $j = 1, \ldots, M$. Let $Q(k \mid j)$ denote a conditional probability assignment between the source alphabet and the reproducing alphabet. Any such conditional probability assignment may be thought of as a communications channel linking the source and reproducing alphabets. $Q(\cdot \mid \cdot)$ along with $P(\cdot)$ generates a marginal distribution on the reproducing alphabet as

$$Q(k) \equiv \sum_{j=1}^{M} P(j)Q(k \mid j)$$

Associated with every $Q(\cdot \mid \cdot)$, we can associate a number

$$d(Q) \equiv \text{average distortion} = \sum_{j,k} P(j)Q(k \mid j)\rho(j, k)$$

A conditional probability assignment is said to be *D-admissible* if and only if $d(Q) \leqslant D$. We denote the set of all such D-admissible conditional probability assignments as Q_D. For each conditional probability assignment we may also define a number called the *average mutual information*

$$I(Q) \equiv \sum_{j,k} P(j)Q(k \mid j) \log \frac{Q(k \mid j)}{Q(k)}$$

The average mutual information is a measure of how much information about the source can be gleaned from observing the sequence of reproducing letters at the channel output. Finally, we may now define the quantity

$$R(D) \equiv \min_{Q \in \mathcal{Q}_D} I(Q)$$

the *rate distortion function*. Let Q^* denote the minimizing $Q \in \mathcal{Q}_D$. It is a simple exercise in the variational calculus to establish (see exercise 2) the following lemma.

Lemma (Gallagher)

$$R(D) \geqslant \sup_{\theta \leqslant 0} [\theta D - f(\theta)]$$

where

$$f(\theta) \equiv \sum_j P(j) \log \left[\sum_k \exp[\theta \rho(j, k)] Q(k) \right]$$

and $Q(\cdot)$ is derived from a conditional probability assignment in $\mathcal{Q}_D$.

We are now ready to state and prove the source coding with respect to a fidelity criterion theorem for discrete memoryless sources. Any set $B \equiv \{y_1, \ldots, y_k\}$ of n-dimensional vectors whose components lie in the reproducing alphabet is called a *code of size K and blocklength n.* The elements of B are called *codewords.* Each sourceword $\mathbf{x}$ is mapped into whichever codeword $y \in B$ minimizes $\rho_n(\mathbf{x}, y)$, with resulting distortion denoted by

$$\rho_n(\mathbf{x} \mid B) \equiv \min_{y \in B} \rho_n(\mathbf{x}, y)$$

The *average distortion of the code* is defined to be

$$E\{\rho_n(\mathbf{x} \mid B)\} \equiv \sum_x P(x) \rho_n(x \mid B)$$

If $E\{\rho_n(\mathbf{x} \mid B)\} \leqslant D$, we say that the *code is D-admissible.* The *coderate* is defined to be

$$R \equiv \frac{1}{n} \log K$$

If the logarithm were base 2, then nR would be the number of binary digits needed to assign a unique binary number to each codeword in B. R would then be the number of binary digits per source output that is used to code the source. Our logarithms are always to the natural base, but this does not invalidate our representation. (Instead of "bits" of information, information theorists use natural logarithms and have "nats" of information.)

Theorem (Shannon)
Let a discrete memoryless source and single-letter distortion measure be given. Then, given any $\varepsilon > 0$ and any $D \geqslant 0$, an integer n can be found such that there exists a $(D + \varepsilon)$ admissible code of blocklength n with rate $R < R(D) + \varepsilon$.

Proof: The proof proceeds by a random coding argument. We will choose our K codewords as independent samples from some distribution. We will then show that on the average, this procedure generates D-admissible codes of the desired rate. If this is true on the average, there must exist at least one code that attains the average performance. The theorem would then be proved.

We choose the K codewords to be independent samples from the distribution

$$Q(y) \equiv \prod_{i=1}^{n} Q(y_i)$$

where the y_i's are the components of the vector y and $Q(\cdot)$ is some distribution on the reproducing alphabet to be chosen later. Let

$$S(x) \equiv \{y : \rho_n(x, y) \leqslant D + \delta\}$$

the "sphere of radius $D + \delta$ and center x." Furthermore, let

$$Q(S(x)) \equiv \sum_{y \in S(x)} Q(y)$$

the probability that a codeword chosen at random will lie in $S(x)$. Hence the probability that no codewords of the randomly chosen code of size K will lie in $S(x)$ is merely $[1 - Q(S(x))]^K$. Let $\rho_{\max}$ denote the largest value of $\rho(\cdot, \cdot)$. Let us now consider developing an upper bound to the average distortion produced by a randomly chosen code (denoted as $\bar{\rho}$).

Therefore,

$$\bar{\rho} \leqslant \sum_x P(x)[(D + \delta)(1 - [1 - Q(S(x))]^K]) + \rho_{\max}[1 - Q(S(x))]^K]$$
$$\leqslant D + \delta + \rho_{\max} \sum_x P(x)[1 - Q(S(x))]^K$$
$$\leqslant D + \delta + \rho_{\max} \sum_x P(x) \exp[-KQ(S(x))]$$

(note that for $0 \leqslant x \leqslant 1$, $(1 - x)^K \leqslant \exp[-Kx]$). In general we will be allowing K to approach infinity as n does (in fact, we will take $K \equiv [\exp[nR(D) + 2\delta]]$). The problem is to determine how fast $Q(S(x))$ approaches zero. If it approaches zero at an exponential rate less than $nR(D) + 2\delta$, then $\exp[-KQ(S(x))]$ approaches zero, and consequently for n large enough, the quantity

$$\rho_{\max} \sum_x P(x) \exp[-KQ(S(x))]$$

can be made less than δ. This is enough to assert the theorem statement. Therefore, we are interested in the large deviations of the quantity $Q(S(x))$ for a fixed x.

Now let us choose

$$Q(y) \equiv \prod_{i=1}^{n} Q^*(y_i)$$

where we recall that Q^* is the "minimizing" distribution in Q_D. Then

$$Q(S(x)) = Q^*\left(\frac{1}{n}\sum_{i=1}^{n} \rho(x_i, \mathbf{y}_i) \leqslant D + \delta\right)$$

Hence we are interested in the large deviations of a sum of independent but not identically distributed random variables $\rho(x_i, \mathbf{y}_i)$. To invoke the Gärtner–Ellis result we need to compute

$$\varphi(\theta) \equiv \lim_{n\to\infty} \frac{1}{n} \log E_{Q^*}\left\{\exp\left[\theta \sum_{i=1}^{n} \rho(x_i, \mathbf{y}_i)\right]\right\}$$
$$= \lim_{n\to\infty} \frac{1}{n} \sum_{i=1}^{n} \log[E_{Q^*}\{\exp[\theta\rho(x_i, \mathbf{y}_i)]\}]$$
$$= E_P\{\log[E_{Q^*}\{\exp[\theta\rho(\mathbf{x}_1, \mathbf{y})]\}]\}$$
$$= \sum_j P(j) \log\left[\sum_k \exp[\theta\rho(j, k)]Q^*(k)\right] = f(\theta)$$

where we have used independence of the summand elements and the law of large numbers, respectively, to obtain these last expressions. Hence, by the Gärtner–Ellis theorem,

$$\underline{\lim}_{n\to\infty} \frac{1}{n} \log Q^*\left(\frac{1}{n}\sum_{i=1}^{n} \rho(x_i, \mathbf{y}_i) \leqslant D + \delta\right) \geqslant -\sup_{\theta}\,[\theta D - f(\theta)]$$

$$= -\sup_{\theta \leqslant 0}\,[\theta D - f(\theta)] \geqslant -R(D)$$

The last result follows from the lemma. The preceding inequality follows from identical arguments made in the proof of Cramér's theorem about how the supremum can be taken over negative θ if the mean value of the sum is greater than the right endpoint of the interval under consideration. The theorem is proved. □

Notes and Comments

Shannon presented the foundations of information theory in his famous paper Shannon [1948]. His first coding theorem (and its converse) established that entropy is the true measure of the information content of a source. Information theoretic proofs of the error exponent were studied by Jelinek [1968] and by Csiszar and Longo [1971] for the case of memoryless sources. An elegant treatment of the material in section A via these methods is given by Blahut [1987] who also discusses related work and generalizations. The methodology and proof given here is due to Anantharam [1990], who generalizes the result to include finite state space Markov chains.

The source coding with respect to a fidelity criterion theorem has a converse in that no code can give better performance than promised by $R(D)$, but large deviation theory is not required to prove it and so is not given here. The notation and preliminary setup of results of section B is due to Berger [1971] in his classic text on rate distortion theory. He gives a more general version of this theorem for arbitrary alphabets and ergodic sources. In Bucklew [1988] can be found a large deviations proof along the lines taken in the text for this more general case. Another proof for the discrete memoryless source case using Sanov's theorem is given in Bucklew [1987].

EXERCISES

1. For a given P, find an expression for the Q that attains $\inf_{Q:H(Q)\geqslant R} H(Q, P)$.

2. Prove the lemma in section B.

3. Show a lower bound to the convergence rate of $Q^*(S(x))$ via Sanov's theorem.

APPENDIX A

Set Theory and Topology

Definition. A *topology* for a set X is a family $\mathscr{T}$ of subsets of X satisfying the following three properties:

a. The set X and the empty set $\varnothing$ are in T.

b. The union of any collection of members of $\mathscr{T}$ is in $\mathscr{T}$.

c. The intersection of any finite family of members of $\mathscr{T}$ is in $\mathscr{T}$. □

The members of $\mathscr{T}$ are called *open sets*. For example, the usual open sets on the real line form the topology (called the *Euclidean topology*) with which we are most familiar. A *topological space* or simply *space* is a pair $(X, \mathscr{T})$, where X is a set and $\mathscr{T}$ is a topology for X. A subset of X is said to be *closed* if its complement is an open set. A *neighborhood* of a point $x \in X$ is an open set containing x. A point is a *limit point* of a subset A of X if every neighborhood of x contains a point of A distinct from x. The *closure* of a set A (denoted $\bar{A}$) is the union of A with its set of limit points. The *boundary* of A (denoted ∂A) is the intersection of $\bar{A}$ with $\overline{X \backslash A}$. The *interior* of A (denoted $\mathring{A}$) is the largest open set contained in A. The *complement* of A, $X \backslash A$, is denoted A^c.

The point of these definitions is to emphasize that many of our concepts, such as limit point, closure, boundary, etc., may all be defined once we have defined the topology.

Theorem

A subset A of a space X is closed if and only if A contains all of its limit points. A subset O of X is open if and only if O contains a neighborhood of each of its points.

Definition. A space X is a *Hausdorff space* provided that for each pair x_1, x_2 of distinct points of X there exist disjoint neighborhoods O_1 and O_2 of x_1 and x_2, respectively. □

Without exception, in this book, all spaces that are considered are Hausdorff.

Definition. A *covering* C of a subset Y contained in a space X is a family of subsets of X whose union contains Y. A covering each of whose members is an open set is called an *open covering*. □

Definition. A subset A of X is called *compact* provided that every open covering of A has a finite subcovering. Let us consider for a moment what this definition means. Suppose that $C = \{A_i \mid i \in I\}$ is an open covering of A, a compact set, where I is some index set possibly uncountably infinite. Then there exists some collection of indices $i_1, i_2, \ldots, i_n$ $(n < \infty)$ such that $A \subset \bigcup_{j=1}^{n} A_{i_j}$. □

Theorem
In a Hausdorff space, compact sets are closed.

Theorem (Heine–Borel)
A subset of $\mathbb{R}^d$ is compact if and only if it is closed and bounded.

Definition. A *metric* or *distance function* for a set X is a function d mapping from $X \times X$ into the nonnegative real numbers such that for all x, y, z in X,

a. $d(x, y) = d(y, x)$.
b. $d(x, y) = 0$ if and only if $x = y$.
c. $d(x, y) + d(y, z) \leqslant d(x, z)$ (triangle inequality). □

If a metric has been defined on a set X, we may generate a topology for the set by taking our open sets to be all possible unions of open balls. The *open ball with center x and radius r* is defined as $B_x(r) \equiv \{y \in X : d(x, y) < r\}$. X with a topology given by a metric is called a *metric space*. For example, we define Euclidean n-dimensional space $\mathbb{R}^n$ (n a positive integer) as the set of all real n-tuples with the topology defined by the Euclidean metric: $d(x, y) = [\Sigma_{i=1}^{n} (x_i - y_i)^2]^{1/2}$.

Definition. A function $f: X \to Y$ from a space X to a space Y is *continuous* provided that for each open set U in Y the *inverse image* $f^{-1}(U) \equiv \{x \in X : f(x) \in U\}$ is open in X. As a consequence of this definition, one also has that if f is continuous, then for each closed set C of Y, $f^{-1}(C)$ is closed in X; i.e., inverse images of closed sets are closed. □

APPENDIX B

Elementary Analysis

1. LIMITS

If S is a set of real numbers, we say that u is an *upper bound* for S if for each $x \in S$, $x \leqslant u$. A number b is the *least upper bound* or *supremum* for S if it is an upper bound for S and $b \leqslant u$ for every upper bound u of S. We define *lower bound* and *greatest lower bound* or *infimum* in an exactly analogous fashion. We shall denote the least upper bound of S by $\sup S$ or $\sup_{x \in S} x$ or by $\sup\{x : x \in S\}$. Similarly, we denote the greatest lower bound of a set S by $\inf S$ or $\inf_{x \in S} x$ or by $\inf\{x : x \in S\}$. Note that

$$\inf_{x \in S} x = -\sup_{x \in S} (-x)$$

Let $\{x_n\}$ be a sequence of real numbers. ℓ is the *limit of the sequence* if given any $\varepsilon > 0$, all but a finite number of terms of the sequence are within ε of ℓ. c is a *cluster point* of the sequence if infinitely many terms of the sequence are within ε of c for any $\varepsilon > 0$. The limit (if it exists) is clearly the unique cluster point of the sequence. Consider the sequence

$$x_n \equiv (-1)^n$$

1 and -1 are the cluster points of this sequence and no limit exists. Clearly, if there is more than one cluster point, a sequence does not converge to any limit. The *limit superior* (denoted $\overline{\lim}$ or lim sup) of a sequence is defined to be the supremum of the cluster points of the sequence. More precisely,

$$\overline{\lim}\, x_n \equiv \inf_n \sup_{k \geqslant n} x_k$$

Similarly, the *limit inferior* (denoted $\underline{\lim}$ or lim inf) of a sequence is defined to be the infimum of the cluster points of the sequence, or

$$\underline{\lim}\, x_n \equiv \sup_n \inf_{k \geqslant n} x_k$$

If we allow $\pm\infty$ as possible cluster points, the lim sup and the lim inf of a sequence always exist. Not having to worry about existence is the primary reason these quantities are used instead of the more desirable limit. If the lim sup equals the lim inf, the *limit* exists and is equal to these.

For further discussion of these and other concepts, see Royden [1968].

2. RIEMANN–STIELTJES INTEGRAL

A function $g(x)$ is of *bounded variation* on $[a, b]$ if it can be written in the form $g(x) = g_1(x) - g_2(x)$, where the functions $g_1(\cdot)$ and $g_2(\cdot)$ are bounded and nondecreasing on $[a, b]$. Hence a probability distribution function (always a nondecreasing function) is a function of bounded variation.

The Riemann–Stieltjes integral of the function $f(x)$ with respect to the function $g(x)$ over the finite interval $[a, b]$ is defined as (if the limit exists)

$$\int_a^b f(x)\, dg(x) \equiv \lim_{\max|x_i - x_{i-1}| \to 0} \sum_{i=1}^{m} f(\xi_i)[g(x_i) - g(x_{i-1})]$$

for any sequence of partitions of the form

$$a = x_0 < \xi_1 < x_1 < \xi_2 < x_2 < \cdots < \xi_m < x_m = b$$

A sufficient condition for the limit to exist is if f is of bounded variation and g is continuous, or vice versa. One may extend the definition to infinite intervals by taking limits of the finite interval integrals.

The main application of these integrals to us is in defining the expectation of a random variable. For a more in-depth discussion and proofs of these statements, see Rudin [1974].

3. CONVEX SETS AND FUNCTIONS

A subset C of $\mathbb{R}^d$ is said to be *convex* if $\lambda x + (1 - \lambda)y$ is in C for all x and y in C and for $0 < \lambda < 1$. We say that a function f is *convex* or *proper convex* on a subset C of $\mathbb{R}^d$ if C is convex, $f(x)$ is finite for at least one $x \in C$, f does not take the value $-\infty$, and

$$f(\lambda x + (1-\lambda)y) \leqslant \lambda f(x) + (1-\lambda)f(y)$$

for every x and y in C and $0 < \lambda < 1$. A function is *strictly convex* if the inequality above is a strict inequality for every $x, y \in C$ with $x \neq y$. The *domain* of f is denoted

$$D_f \equiv \{x \in \mathbb{R}^d : f(x) < \infty\}$$

f is a *closed convex function* if for any sequence $\{x_n\}$ such that $x_n \to x$, we have $\underline{\lim}_{n\to\infty} f(x_n) \geqslant f(x)$.

Theorem
A convex function f on $\mathbb{R}^1$ is continuous on $\mathring{D}_f$.

Recall that the interior of a set A, $\mathring{A}$, is the largest open set contained in the set A. The theorem above then implies that for a convex function to be closed it is sufficient to have the function be continuous up to *and* including its boundary points.

For a function f on $\mathbb{R}^d$ (not necessarily convex), the *convex conjugate function* is defined as

$$f^*(y) \equiv \sup_{x \in \mathbb{R}^d} \{\langle x, y\rangle - f(x)\}$$

for $y \in \mathbb{R}^d$. f^* is also known as the *Legendre–Fenchel transform* of f.

Theorem
Let f be a closed convex function on $\mathbb{R}^d$. Then

a. f^* is a closed convex function on $\mathbb{R}^d$.
b. $\langle x, y\rangle \leqslant f(x) + f^*(y)$ for all $x, y \in \mathbb{R}^d$.
c. $f^{**} = f$, i.e., $f(x) = \sup_{y\in\mathbb{R}^d} \{(x, y) - f^*(y)\}$.

Property (c), especially, has important applications for the determination of large deviation rate functionals. The classic text for convex analysis is Rockafellar [1970]. A very readable and less exhaustive treatment appears in Van Tiel [1984].

APPENDIX C

Matrix Results

1. DIAGONALIZATION OF COVARIANCE MATRICES

Let K be an $n \times n$ positive definite symmetric matrix (i.e., for any vector $x \neq 0$, $x^{\mathrm{T}}Kx > 0$ and $K^{\mathrm{T}} = K$, where T indicates the transpose operation). Consider the eigenvector problem

$$\lambda\varphi = K\varphi$$

where λ is a complex number called the eigenvalue and φ is an n-tuple called the eigenvector.

Theorem

There exist n linearly independent solutions $\varphi_1, \varphi_2, \ldots, \varphi_n$ and $\lambda_1, \lambda_2, \ldots, \lambda_n$ to the eigenvector problem given above, where the φ_i are orthogonal and can be normalized to have unit length, i.e., $\langle\varphi_i, \varphi_j\rangle = \delta_{ij}$.

Let Φ be the matrix consisting of all these eigenvectors arranged in columns, and let Λ be the diagonal matrix of the eigenvalues. It is easy to check that $\Phi^{\mathrm{T}} = \Phi^{-1}$ and $(\Phi^{\mathrm{T}})^{\mathrm{T}} = \Phi$. Then we have the representation

$$\Lambda = \Phi^{\mathrm{T}}K\Phi$$

Note that if K is the covariance matrix of the random vector $\mathbf{x}$, the covariance matrix of the random vector $y = \Phi\mathbf{x}$ is just $\Phi^{\mathrm{T}}K\Phi = \Lambda$, which is a diagonal matrix. This means that with this transformation we have removed the correlation structure of the original random vector. From the representation above we have also that

$$K = \Phi\Lambda\Phi^{\mathrm{T}} = \sum_{i=1}^{n} \lambda_i\varphi_i\varphi_i^{\mathrm{T}}$$

A good introduction to various matrix techniques useful in statistics is given in Rao [1973].

2. KARHUNEN–LOÈVE EXPANSION

The Karhunen–Loève expansion is a generalization of the results of section 1 to covariance functions defined for continuous-time random processes. One obtains in an exactly analogous fashion a diagonalization of the covariance function in terms of eigenvalues and eigenfunctions of an integral operator arising from that covariance function.

Let $R(s, t)$ be the covariance function of a zero mean Gaussian random process. Consider solutions to the integral equation

$$\lambda_i \varphi_i(t) = \int_0^T R(s, t)\varphi_i(s)\, ds$$

The $\{\lambda_i\}$ and the $\{\varphi_i(\cdot)\}$ are the eigenvalues and eigenfunctions of the covariance operator, respectively. We will always normalize the eigenfunctions to have unit energy (i.e., $\int \varphi_i^2 = 1$).

Theorem (Mercer)
Suppose that $R(s, t)$ is continuous for $0 \leqslant s,\ t \leqslant T$. Solutions to the integral equation above exist and there are at most a countable number of distinct solutions. Then $R(s, t)$ may be expanded in the series

$$R(s, t) = \sum_{i=1}^{\infty} \lambda_i \varphi_i(s)\varphi_i(t)$$

for $0 \leqslant s,\ t \leqslant T$, where the infinite series is absolutely and uniformly convergent in that rectangle.

Theorem
If $R(s, t)$ is strictly positive definite, the eigenfunctions can be chosen to form a complete orthonormal set (in the L_2 sense) over the interval $[0, T]$.

We note that it is fairly easy to show that any two eigenfunctions belonging to different eigenvalues have to be orthogonal to each other. Furthermore, an eigenfunction may always be normalized so that it has unit energy on the interval. The important point about this last theorem is that the set of eigenfunctions is complete; i.e., one may expand any square integrable function on the interval in terms of a linear combination of eigenfunctions. If a covariance function is not strictly positive definite, its set of eigenfunctions will not be complete. However, one may augment this set with enough additional orthonormal functions to obtain a complete set. These "extra" eigenfunctions are sometimes called the eigenfunctions associated with the zero eigenvalue.

Theorem (Karhunen–Loève)
Let $\mathbf{x}(t)$ be a zero mean Gaussian random process with continuous covariance function $R(s, t)$. Then on the interval $[0, T]$, we have the expansion

$$\mathbf{x}(t) = \sum_{i=1}^{\infty} \mathbf{x}_i \varphi_i(t)$$

where

$$\mathbf{x}_i \equiv \int_0^T \mathbf{x}(t)\varphi_i(t)\, dt$$

where the convergence of the infinite sum is to be understood as mean square convergence, i.e.,

$$\lim_{n \to \infty} E\left\{\left[\mathbf{x}(t) - \sum_{i=1}^{n} \mathbf{x}_i \varphi_i(t)\right]^2\right\} = 0$$

for all $t \in [0, T]$. Furthermore, the $\{\mathbf{x}_i\}$ are zero mean independent Gaussian random variables with

$$E\{\mathbf{x}_k^2\} = \lambda_k \qquad k = 1, 2, \ldots$$

For a discussion of Mercer's theorem and related topics, see Riesz and Sz.-Nagy [1955]. For a more in-depth discussion of these and other results dealing with the eigenfunctions of random processes, see Van Trees [1968].

3. TOEPLITZ DISTRIBUTION THEOREM

A *Toeplitz matrix* $V \equiv (v_{ij})$ is one whose ijth element depends only on the difference between i and j, i.e.,

$$v_{ij} = v_{i-j}$$

The kth *diagonal* is the collection of all (i, j) positions in the matrix where $i - j = k$.

Consider the case where we have an infinite rowed and columned Toeplitz matrix. Let t_k be the matrix entry on the kth diagonal. The function

$$f(\omega) \equiv \sum_{k=-\infty}^{\infty} t_k \exp[ik\omega]$$

is called the *spectrum of the matrix*. Note that $f(\cdot)$ is periodic (with period 2π) in ω so that it need only be specified over a period (typically, $[-\pi, \pi]$).

Suppose that one considers sampling a stationary random process at evenly spaced time instants. The covariance matrix $V^{(n)}$ of the first n samples is a Toeplitz matrix. n can, in fact, be infinity, in which case the power spectrum of the discrete random process corresponds exactly to the spectrum of the covariance matrix. It is this observation that makes Toeplitz matrices so important in the study of stationary Gaussian random processes. The eigenvalue distribution of Toeplitz matrices in many cases approach certain limiting densities as the matrix size approaches infinity. The Toeplitz distribution theorem quantifies this relationship.

Theorem (Toeplitz Distribution Theorem)
Let $V^{(\infty)}$ be an infinite Toeplitz matrix with entry t_k on the kth diagonal and corresponding spectrum $f(\omega)$. Suppose that $|f(\omega)|$ is a bounded function of ω. Then if $h(\cdot)$ is a continuous function over the range of $f(\cdot)$, we have

$$\lim_{n \to \infty} \frac{1}{n} \sum_{k=1}^{n} h(\lambda_k^{(n)}) = \frac{1}{2\pi} \int_{-\pi}^{\pi} h(f(\omega))\, d\omega$$

where the $\lambda_k^{(n)}$ are the eigenvalues of the nth-order matrix $V^{(n)}$ centered about the main diagonal of $V^{(\infty)}$.

To prove this theorem, one first constructs a circulant matrix $V_c^{(n)}$ approximation to the Toeplitz $V^{(n)}$. The definition of a finite-dimensional *circulant matrix* $V_c^{(n)}$ is that the lth row is obtained by performing a cyclic right shift on the $(l-1)$th row (the first row is the cyclic right shift of the nth row). The eigenvectors of any circulant matrix are the columns from the discrete Fourier transform (DFT) matrix; i.e., the jth eigenvector of a circulant matrix is given by

$$\psi^{(j)} = (\psi_{0j} \quad \psi_{1j} \quad \psi_{(n-1)j})^{\mathrm{T}}$$

where

$$\psi_{mj} \equiv \frac{\exp[2\pi ijm/n]}{\sqrt{n}}$$

for $j = 0, 1, \ldots, n-1$ and $i \equiv \sqrt{-1}$. The eigenvalues of any circulant matrix are given by the DFT of the first row. One proves the Toeplitz distribution theorem by showing that $V^{(n)} - V_c^{(n)}$ converges (in the weak matrix norm) to zero as n approaches infinity. This then is enough to assert that the eigenvalues are acting in a distributional fashion like each other.

Intuitively, one would like to say that since $V^{(n)}$ converges to $V_c^{(n)}$, the eigenvectors should be converging also; i.e., the eigenvectors should be converging to complex exponentials. This is true but only in a distributional sense. Regardless, as long as one is working with sums of functions of the eigenvalues, one can obtain the correct limit results by assuming that the eigenvectors of $V^{(n)}$ are converging (strongly) to complex exponentials. We employ this methodology in chapter VI to obtain large deviation principles for Gaussian processes into quadratic form detectors. A good introductory treatment of this material may be found in the tutorial paper by Gray [1972]. A more advanced approach is taken in Grenander and Szego [1958].

4. PERRON–FROBENIUS THEOREM FOR POSITIVE MATRICES

Suppose that $A = (a_{ij})$ is an $n \times n$ square matrix. If every a_{ij} is nonnegative, we write $A \geqslant 0$; if $A \geqslant 0$ and at least one a_{ij} is positive, we write $A > 0$ and call A a *positive matrix*. If *all* the a_{ij} are positive, we write $A \gg 0$. We also use the same notation for a vector $x \equiv (x_1 \quad x_2 \quad \cdots \quad x_n)$; i.e., $x \geqslant 0$ requires that $x_i \geqslant 0$ for $i = 1, \ldots, n$ and $x > 0$ implies that $x \geqslant 0$ and at least one x_i is positive; $x \gg 0$ means all of the x_i are positive. Let $A \geqslant 0$, and let Λ be the set consisting of all real numbers λ for which there corresponds a vector x such that $\Sigma_{i=1}^{n} x_i = 1$, $x > 0$, and $Ax \geqslant \lambda x$. Let $\lambda_0 \equiv \sup_{\Lambda} \lambda$.

Theorem (Perron)
If $A \gg 0$:

a. There exists an $x_0 \gg 0$ such that $Ax_0 = \lambda_0 x_0$.

b. If $\lambda \neq \lambda_0$ is any other eigenvalue of A, then $|\lambda| < \lambda_0$.

c. There is only one eigenvector with eigenvalue λ_0.

Theorem (Frobenius)
If $A > 0$ and $A^m \gg 0$ for some integer $m > 0$, the assertions of the preceding theorem hold.

Theorem (Frobenius)
Assume that $A > 0$, and let λ_0 be defined as above. Then λ_0 is an eigenvalue of A with eigenvector $x_0 > 0$. Furthermore, if λ is any other eigenvalue of A, then $|\lambda| \leqslant \lambda_0$.

Perron gave the proof of his theorem for positive matrices in 1907. Frobenius, in subsequent papers (published in 1909 and 1912), greatly extended the usefulness of this first result.

It is actually the second theorem that will be of greatest interest to us in our study of large deviation results for Markov chains. The transition probability matrix of all finite-dimensional aperiodic irreducible Markov chains has the quoted property for some m.

A good text for these and other matrix results is Gantmacher [1960].

5. PERTURBATION THEORY

Suppose that $A(\alpha)$ is a collection of $N \times N$ real valued matrices indexed by a parameter α. Suppose that we have available a power series representation for $A(\alpha)$:

$$A(\alpha) = A + \alpha A^{(1)} + \alpha^2 A^{(2)} + \cdots$$

where $A \equiv A(0)$ and the $A^{(i)}$ are $N \times N$ matrices forming the "coefficients" of the power series. Let λ be a particular eigenvalue of interest of the unperturbed matrix A. In our applications we are usually interested in the largest eigenvalue. As α increases from zero, this eigenvalue also changes, to $\lambda(\alpha)$, say. If λ has multiplicity $m > 1$, the eigenvalue can split (bifurcate) into m different eigenvalues, following independent trajectories as α varies.

For simplicity, suppose that the multiplicity of λ is 1. Associated with this eigenvalue is a (right) eigenvector φ, i.e.,

$$A\varphi = \lambda\varphi$$

The transpose matrix A^{T} will also have the eigenvalue λ but with a different eigenvector ψ, i.e.,

$$A^{\mathrm{T}}\psi = \lambda\psi$$

or, equivalently, $\psi^{\mathrm{T}} A = \lambda \psi^{\mathrm{T}}$. ψ can then be viewed as the left eigenvector of A for λ.

Theorem

With the assumptions above, $\lambda(\alpha)$ has a power series representation as

$$\lambda(\alpha) = \lambda + \alpha\lambda^{(1)} + \alpha^2\lambda^{(2)} + \cdots$$

where

$$\lambda^{(1)} = \psi A^{(1)} \varphi$$

and (if A can be diagonalized)

$$\lambda^{(2)} = \psi A^{(2)} \varphi - \sum_j (\mu_j - \lambda)^{-1} (\psi_j A^{(1)} \varphi)(\psi A^{(1)} \varphi_j)$$

The $\{\mu_j\}$, $\{\psi_j\}$, and $\{\varphi_j\}$ are the other eigenvalues, left eigenvectors, and right eigenvectors, respectively, of A.

Our main interest for applications will be in the first order perturbation term $\lambda^{(1)}$. An exhaustive treatment of these and other results in perturbation theory may be found in Kato [1976].

APPENDIX D

Topics in Probability

0. FUNDAMENTALS OF MEASURE THEORY AND PROBABILITY

Let Ω be an abstract space and $\mathscr{B}$ be a collection of subsets of Ω satisfying the following conditions:

(i) $\Omega \in \mathscr{B}$.
(ii) $A \in \mathscr{B}$ implies that $A^c \in \mathscr{B}$.
(iii) $A_i \in \mathscr{B}$, $i \geqslant 1$ implies that $\bigcup_{i=1}^{\infty} A_i \in \mathscr{B}$.

where A^c denotes the complement of the set A. $\mathscr{B}$ is said to be a *σ-field* of subsets of Ω. The pair $(\Omega, \mathscr{B})$ is called a *measurable space.* Let $\mathscr{C}$ be a collection of subsets of Ω. The *σ-field generated by* $\mathscr{C}$ is defined as the smallest σ-field of Ω that contains $\mathscr{C}$. If Ω is a metric space, then the σ-field generated by the open balls is called the *Borel σ-field.*

Let μ be a set function defined on $(\Omega, \mathscr{B})$ with the following properties:

(i) $0 \leqslant \mu(A)$ for every $A \in \mathscr{B}$.
(ii) $\mu(\varnothing) = 0$.
(iii) $A_i \in \mathscr{B}$, $A_i \cap A_j = \varnothing$, $i, j \geqslant 1$ implies that $\mu(\bigcup_{i=1}^{\infty} A_i) = \Sigma_{i=1}^{\infty} \mu(A_i)$.

μ is called a *measure* defined on $(\Omega, \mathscr{B})$ and the triple $(\Omega, \mathscr{B}, \mu)$ is called a *measure space.* A measure space $(\Omega, \mathscr{B}, \mu)$ is said to be *complete* if $\mathscr{B}$ contains all subsets of sets of measure zero, that is, if $A \in \mathscr{B}$, $\mu(A) = 0$, and $C \subset A$ implies that $C \in \mathscr{B}$. One may always complete any measure space by the addition of subsets of sets of measure zero. A measure μ is called *finite* if $\mu(\Omega) < \infty$. It is called *σ-finite* if there is a sequence $\{\Omega_n\}$ of sets in $\mathscr{B}$ such that

$$\Omega = \bigcup_{n=1}^{\infty} \Omega_n$$

and $\mu(\Omega_n) < \infty$. There is a very important measure for the measurable space $(\mathbb{R}^d, \mathscr{L})$, where $\mathscr{L}$ is the completion of the σ-field generated by the open sets (i.e., $\mathscr{L}$ is the completion of the Borel σ-field of $\mathbb{R}^d$). The measure $m(\cdot)$, called *Lebesgue measure*, assigns the volume (length in $\mathbb{R}$ or area in $\mathbb{R}^2$) of a set in $\mathscr{L}$ as its measure, e.g., in $\mathbb{R}$ $m([a, b]) \equiv b - a$ for $b > a$.

We may now define a particular class of finite measures that will be of great interest to us, the probability measures. Let P be a set function defined on $(\Omega, \mathscr{B})$ with the following properties:

(i) $0 \leqslant P(A) \leqslant 1$ for every $A \in \mathscr{B}$.
(ii) $P(\Omega) = 1$.
(iii) $A_i \in \mathscr{B}$, $A_i \cap A_j = \phi$, $i, j \geqslant 1$ implies that $P(\bigcup_{i=1}^{\infty} A_i) = \Sigma_{i=1}^{\infty} P(A_i)$.

P is called a *probability measure* defined on $(\Omega, \mathscr{B})$ and the triple $(\Omega, \mathscr{B}, P)$ is called a *probability space*.

Any set $A \in \mathscr{B}$ in a measure space is called a *measurable set*. Any set $A \in \mathscr{B}$ in a probability space is called an *event*. Two events are said to be *independent* if $P(A \cap B) = P(A)P(B)$. A collection of events A_i, $1 \leqslant i \leqslant k$, are said to be *independent* if for every finite subset $1 \leqslant i_m \leqslant k$, $m = 1, 2, \ldots, r$, $P(A_{i_1} \cap A_{i_2} \cap \cdots \cap A_{i_r}) = \Pi_{m=1}^{r} P(A_{i_m})$. In general, a family of events $\{A_\alpha, \alpha \in I\}$ is said to be independent if every finite subset of the family is independent.

Let $(\Omega_1, \mathscr{B}_1)$ and $(\Omega_2, \mathscr{B}_2)$ be two measurable spaces. A mapping $\mathbf{X}$ from Ω_1 into Ω_2 is said to be a *random variable* if $\mathbf{X}^{-1}(B) \in \mathscr{B}_1$ for every $B \in \mathscr{B}_2$. Such a mapping is also called an Ω_2 *valued random variable* or a *random variable taking values in* Ω_2. Ω_2 is said to be the *state space* of the random variable $\mathbf{X}$. Usually, in the study of real analysis when no particular emphasis is being placed on probability constructs, a random variable is called a *measurable function.*

Let us now define an integral on a measure space. First we define a simple function $\varphi: \Omega \to \mathbb{R}$ as a finite linear combination of indicator functions of measurable sets A_i, i.e.,

$$\varphi(x) = \sum_{i=1}^{n} c_i \cdot 1_{A_i}(x)$$

where $c_i \in \mathbb{R}$, $i = 1, \ldots, n$. We define the integral for simple functions over a measurable set A as

$$\int_A \varphi \, d\mu \equiv \sum_{i=1}^{n} c_i \mu(A_i \cap A)$$

[If it is desired that the independent variable be made explicit, common notation for the integral is $\int \varphi \, d\mu \equiv \int \varphi(x) \, d\mu(x) \equiv \int \varphi(x)\mu(dx)$.] Now let f be a nonnegative real valued measurable function on $(\Omega, \mathscr{B}, \mu)$. Then

$$\int f \, d\mu \equiv \sup \left\{ \int \varphi \, d\mu : 0 \leqslant \varphi \leqslant f \right\}$$

One may then define the integral for a general real valued measurable function by piecing together the integrals over the positive and negative portions of the function.

When $\mu = m$ (Lebesgue measure) and f is continuous except on a set of measure zero, it can be shown that the integral above (then called the *Lebesgue integral*) is equal to the usual Riemann integral of f on $\mathbb{R}^d$ studied in calculus, i.e., $\int f \, d\mu = \int f \, dm = \int f(x) \, dx$. The Lebesgue integral strictly generalizes the Riemann integral. That is, if a function is Riemann integrable, it is Lebesgue integrable and the two have the same value; but there are functions that are Lebesgue integrable that are not Riemann integrable. When $\mu = m$ we will frequently write our integral as $\int f(x) \, dx$, in keeping with calculus notation.

Most of the time we are concerned with random variables taking values in $\mathbb{R}$ or $\mathbb{R}^d$. Let $(\Omega, \mathscr{B}, P)$ be a probability space and $\mathbf{X}$ be an $\mathbb{R}^d$ valued random variable defined on it. Denote the components of $\mathbf{X}$ as $\mathbf{X} = (\mathbf{X}_1, \mathbf{X}_2, \ldots, \mathbf{X}_d)$ and define the *joint distribution function* of $\mathbf{X}$, $F(\underline{x})$ $(\underline{x} \in \mathbb{R}^d)$, as

$$F(\underline{x}) = P(\mathbf{X}_1 \leqslant x_1, \mathbf{X}_2 \leqslant x_2, \ldots, X_d \leqslant x_d)$$

If the context requires it, we will sometimes denote the distribution function as $F_{\mathbf{X}}(\cdot)$ to make clear of which random variable we are speaking. For the $\mathbb{R}$ valued case, and for some function $g: \mathbb{R} \to \mathbb{R}$, denote the integral (if it exists)

$$\int g(x) \, dF(x)$$

as $E\{g(\mathbf{X})\}$. $E\{g(\mathbf{X})\}$ is called the *mean* or *expected value* of $g(\mathbf{X})$. If $g(x) \equiv x$, then $E\{\mathbf{X}\}$ is often just called the *mean*. The mean value of an $\mathbb{R}^d$ valued random variable is defined to be the vector of the mean values of each component. One may show that the definition of mean value above is equivalent to

$$E\{g(\mathbf{X})\} = \int g \, dP$$

which is in terms of our integral defined above.

A collection of real valued random variables $\{\mathbf{X}_i\}_{i=1}^k$ is said to be *independent* if $F_{\mathbf{X}}(\underline{x}) = F_{\mathbf{X}_1}(x_1)F_{\mathbf{X}_2}(x_2)\cdots F_{\mathbf{X}_k}(x_k)$, where $F_{\mathbf{X}}(\cdot)$ is the joint distribution of the collection.

1. MOMENT GENERATING FUNCTIONS

The *moment generating function* of a real valued random variable **X** is defined as

$$M(\theta) \equiv E\{\exp[\theta \mathbf{X}]\} = \int \exp[\theta x]\, dF(x)$$

where $F(\cdot)$ is the distribution function. For $\theta = 0$ this integral exists (and equals 1), but for other values of θ it may or may not exist, depending on the distribution. Let us substitute the Taylor series expansion for the exponential into the integral and integrate term by term. [We may do this provided that $M(\theta)$ is finite in a neighborhood about the origin.] Hence

$$M(\theta) = \int \sum_0^\infty \frac{(\theta x)^k}{k!}\, dF(x) = \sum_0^\infty \frac{\theta^k}{k!}\left(\int x^k\, dF(x)\right) = \sum_0^\infty E\{\mathbf{X}^k\}\frac{\theta^k}{k!}$$

Now in a Taylor series expansion about zero, the coefficient of $\theta^k/k!$ is the kth derivative at zero. Hence

$$E\{\mathbf{X}^k\} = M^{(k)}(0)$$

Thus by successive differentiation of the moment generating function, we can obtain all of the moments of **X**.

If **X** and **Y** are independent random variables, the moment generating function of the sum $\mathbf{X} + \mathbf{Y}$ is a particularly simple combination of the individual moment generating functions, namely their product:

$$\begin{aligned} M_{\mathbf{X}+\mathbf{Y}}(\theta) &= E\{\exp[\theta(\mathbf{X} + \mathbf{Y})]\} = E\{\exp[\theta\mathbf{X}]\exp[\theta\mathbf{Y}]\} \\ &= E\{\exp[\theta\mathbf{X}]\}E\{\exp[\theta\mathbf{Y}]\} = M_{\mathbf{X}}(\theta)M_{\mathbf{Y}}(\theta) \end{aligned}$$

2. PROBABILITY INEQUALITIES

Theorem (Chebyshev's Inequality)
Let **X** be a random variable and f be a nonnegative, nondecreasing function.

Then for any real number c for which $f(c) > 0$, we have

$$P\{\mathbf{X} \geqslant c\} \leqslant \frac{E\{f(\mathbf{X})\}}{f(c)}$$

This theorem is usually stated in probability and statistics books for the case $f(x) = x^2$ or $f(x) = |x|$.

Theorem (Jensen's Inequality)
Let f be a finite convex function on an open interval A of $\mathbb{R}$. Let $\mathbf{X}$ be a random variable such that $P\{X \in A\} = 1$. Then if $E\{\mathbf{X}\}$ is finite,

$$f(E\{\mathbf{X}\}) \leqslant E\{f(\mathbf{X})\}$$

Suppose that $f(x) = x^2$. Then $(E\{\mathbf{X}\})^2 \leqslant E\{\mathbf{X}^2\}$, by Jensen's inequality. We already knew that this was true since the variance of $\mathbf{X}$ is just the difference between these two quantities and is of course nonnegative.

3. NOTIONS OF CONVERGENCE

A sequence of probability measures or random variables may converge in a variety of ways. We wish to consider the case of convergence on a metric space. In this section we let Ω be a metric space and let $\mathscr{B}$ denote the Borel σ-field generated by the open balls. We now give the definitions of the major forms of convergence for this setting.

Definition. Let f be a bounded continuous function mapping from a probability space $(\Omega, \mathscr{B}, P)$ into $\mathbb{R}$. Then *a sequence of random variables* $\{\mathbf{x}_i\}$ defined on the probability space *converges in distribution* or *weakly* to another random variable $\mathbf{x}$ on the probability space if

$$\lim_{n \to \infty} E\{f(\mathbf{x}_n)\} = E\{f(\mathbf{x})\}$$

for all bounded continuous $f(\cdot)$.

We say that *a sequence of probability measures* $\{P_n\}$ *converges in distribution* or *weakly* to P if

$$\lim_{n \to \infty} \int f\, dP_n = \int f\, dP$$

for all bounded continuous f. □

For the probability spaces that we consider in this book, we can further restrict the class of functions of interest to be bounded continuous functions of compact support [i.e., $f(x) = 0$ for x outside some sufficiently large compact set]. If we restrict ourselves to real valued random variables, it can be shown that this definition is equivalent to

$$\lim_{n\to\infty} P\{\mathbf{x}_i \leqslant y\} = P\{\mathbf{x} \leqslant y\}$$

for all y at which the right-hand side of the equality is continuous. This is an important point: the distribution functions have to converge only at the points of continuity of the limiting distribution [see part (e) of the following theorem].

An extremely useful result that characterizes weak convergence is the portmanteau theorem given in Billingsley [1968].

Theorem

The following statements are equivalent:

a. P_n converges in distribution to P.

b. $\lim_{n\to\infty} \int f(x)\,dP_n \equiv \lim_{n\to\infty} E\{f(\mathbf{x}_n)\} = \int f(x)\,dP \equiv E\{f(\mathbf{x})\}$ for all bounded continuous functions $f(\cdot)$.

c. $\overline{\lim}\, P_n(F) \leqslant P(F)$ for all closed F.

d. $\underline{\lim}\, P_n(G) \geqslant P(G)$ for all open F.

e. $\lim_{n\to\infty} P_n(A) = P(A)$ for all sets of P-continuity. A is a set of P-continuity if $P(\text{boundary of } A) \equiv P(\delta A) = 0$.

Definition. Let $d(\cdot,\cdot)$ be a metric that defines the topology of the probability space $(\Omega, \mathscr{B}, P)$. Let $\{\mathbf{x}_n\}$ be a sequence of random variables defined on the space. Then if for all $\varepsilon > 0$,

$$\lim_{n\to\infty} P\{d(\mathbf{x}_n, \mathbf{x}) > \varepsilon\} = 0$$

we say that $\{\mathbf{x}_i\}$ *converges in probability to* $\mathbf{x}$. □

Definition. Let $d(\cdot,\cdot)$ and $\{\mathbf{x}_n\}$ be as above. Then if

$$\lim_{n\to\infty} E\{d(\mathbf{x}_n, \mathbf{x})^2\} = 0$$

we say that $\{\mathbf{x}_i\}$ *converges in mean square* to $\mathbf{x}$. □

Definition. Let $\{\mathbf{x}_n\}$ denote a sequence of random variables defined on $(\Omega, \mathscr{B}, P)$. Let $N \equiv \{\omega \in \Omega: \lim_{n\to\infty} \mathbf{x}_n(\omega) = \mathbf{x}(\omega)\}$. If $P(N) = 1$, we say that $\{\mathbf{x}_i\}$ *converges to* $\mathbf{x}$ *with probability* 1. □

Theorem
Probability 1 convergence implies convergence in probability, which in turn implies convergence in distribution. Mean square convergence implies convergence in probability. Probability 1 convergence does not imply, nor is it implied by, mean square convergence.

4. LIKELIHOOD RATIOS

Let P and Q be two measures (not necessarily probability measures) defined on a measurable space $(\Omega, \mathscr{B})$. P is said to be *absolutely continuous* with respect to Q (denoted $P \ll Q$) if $P(A) = 0$ for each set $A \in \mathscr{B}$ for which $Q(A) = 0$.

Theorem (Radon–Nikodym)
Let $(\Omega, \mathscr{B}, \mu)$ be a σ-finite measure space and let v be a measure defined on $\mathscr{B}$ such that $v \ll \mu$. Then there exists a nonnegative measurable function g such that for each set $E \in \mathscr{B}$ we have

$$v(E) = \int_E g(x)\, d\mu(x)$$

g is called the *Radon–Nikodym derivative* of the measure v with respect to μ (and is usually denoted $dv/d\mu$). If P is a probability measure on $\mathbb{R}^d$ that is absolutely continuous with respect to Lebesgue measure, then by the theorem above, dP/dm exists. Integration with respect to Lebesgue measure is just our normal integration on $\mathbb{R}^d$. Hence

$$P(A) = \int_A \frac{dP}{dm}\, dm = \int_A \frac{dP}{dm}(x)\, dx$$

However, if the integration of a nonnegative function over a set A gives the probability of A, we say that that function is the *probability density function* $p(\cdot)$ of the probability measure P. Hence if $P \ll m$, the probability density $p(\cdot)$ exists and we have

$$\frac{dP}{dm}(x) = p(x)$$

Suppose that for two probability measures P and Q with densities $p(\cdot)$ and $q(\cdot)$ on $\mathbb{R}^d$, we have $P \ll Q$. Then it is easy to see that

$$\frac{dP}{dQ}(x) \equiv \frac{p(x)}{q(x)}$$

$dP(x)/dQ$ in this case is frequently called the *likelihood ratio* and plays an important role in maximum likelihood estimation theory and in detection theory. Royden [1968] is a good reference for the Radon–Nikodym theorem. Any textbook dealing with probability theory will discuss densities and their relationship to the Radon–Nikodym theorem.

5. MARKOV CHAINS

Let $\{\mathbf{x}_n\}$ denote a sequence of random variables whose state space is a finite set $S \equiv \{1, 2, \ldots, N\}$. Suppose that

$$P\{\mathbf{x}_{n+1} = i_{n+1} \mid \mathbf{x}_0 = i_0, \ldots, \mathbf{x}_n = i_n\} = P\{\mathbf{x}_{n+1} = i_{n+1} \mid \mathbf{x}_n = i_n\}$$

for every n and every collection $\{i_{n+1}, i_n, i_{n-1}, \ldots, i_0\}$ such that $i_j \in S$ for each j. Then $\{\mathbf{x}_n\}$ is called a *discrete-time Markov chain with (finite) state space S*. The conditional probabilities

$$P\{\mathbf{x}_{n+1} = y \mid \mathbf{x}_n = x\} \equiv p(y \mid x)$$

are called the *transition probabilities* of the chain. If $P\{\mathbf{x}_{n+1} = y \mid \mathbf{x}_n = x\}$ does not depend on n, we say that the chain has *stationary transition probabilities*. In this book we always assume stationary transition probabilities. We generalize our notation and write

$$p^{(m)}(y \mid x) \equiv P\{\mathbf{x}_{n+m} = y \mid \mathbf{x}_n = x\}$$

which are called the *m-step transition probabilities* and again for stationary transition probabilities are independent of n. Let us denote the *initial distribution* of the chain as

$$\mu_0(x) \equiv P\{\mathbf{x}_0 = x\}$$

(in general, the notation will be $\mu_n(x) \equiv P\{\mathbf{x}_n = x\}$). It is easy to verify by

Bayes' rule and the Markov property that

$$P\{\mathbf{x}_0 = i_0, \mathbf{x}_1 = i_1, \ldots, \mathbf{x}_n = i_n\} = \mu_0(x_0)p(i_1 \mid i_0)p(i_2 \mid i_1) \cdots p(i_n \mid i_{n-1})$$

Frequently, the transition probabilities are arranged into a matrix called the *transition probability matrix*,

$$P \equiv (p(j \mid i))$$

i.e., P is an $N \times N$ matrix with the ith row and jth column position occupied by $p(j \mid i)$. Note also that $\mu_n(x)$, $x = 1, 2, \ldots, N$, can be arranged into a vector (denoted as μ_n). It is easy to verify that

$$\mu_n = \mu_{n-1} P$$

and hence

$$\mu_n = \mu_0 P^n$$

We say that a *state* y *leads to* x if and only if $p^{(m)}(x \mid y) > 0$ for a positive integer m. A collection of states is said to be *irreducible* if x leads to y for all choices x and y in S. An *irreducible Markov chain* is a chain whose state space is irreducible. A probability vector π for which

$$\pi = \pi P$$

is called a *stationary distribution.*

Theorem
Every irreducible finite state space Markov chain has a unique stationary distribution.

In some sense (as indicated in the following theorem) the μ_n converge to the stationary distribution.

Theorem
Let $f(x)$ be some function defined on the state space of a finite state space irreducible Markov chain. Then $(1/n)\,\Sigma_{i=0}^{n-1} f(\mathbf{x}_i)$ converges with probability 1 to $\Sigma_{i=1}^{N} f(i)\pi(i)$, where π is the unique stationary distribution.

Another way of saying the theorem above is that a finite state space irreducible Markov chain is *ergodic*.

Most of the foregoing discussion goes over into finite state space continuous-time Markov processes (pure jump processes) with little or no modification. The books by Hoel et al. [1972] or Karlin and Taylor [1975] are good introductions to Markov processes in general.

6. STOCHASTIC INTEGRALS

Stochastic differential equations are defined by certain stochastic integrals which in their turn need to be defined. The most commonly utilized stochastic integral is the one due to Itô and is called the Itô integral. Itô presented his work on integration as a method to study the class of stochastic Markov processes known as diffusions. Essentially, these processes are those resulting from driving differential equations with "white" Gaussian noise.

Let $\mathbf{W}(t)$ be a standard Wiener process (also called Brownian motion), i.e., $\mathbf{W}(t)$ is a mean zero nonstationary Gaussian random process for $t \geqslant 0$, with covariance function given by

$$E\{\mathbf{W}(t)\mathbf{W}(s)\} = \min(t, s)$$

$\mathbf{W}(t)$ is in some sense the integral of white Gaussian noise. White Gaussian noise formally must have infinite variance, which provides the impetus to work with it through an integral. Now, we would like to be able to make sense of integrals like

$$\int_a^b d\mathbf{W}(t)$$

or

$$\int_a^b \mathbf{y}(t)\, d\mathbf{W}(t)$$

where $\mathbf{y}(t)$ is some other random process. This of course is pure symbolism based on the Stieltjes integral, but the first integral should be $\mathbf{W}(b) - \mathbf{W}(a)$.

As with the Lebesgue integral we first define the integral for "simple function like" random processes called step function processes. A step function random process $\varphi(t)$ is a random process for which there exist nonrandom times $t_0, t_1, \ldots, t_n$ such that

$$a = t_0 < t_1 < \cdots < t_n$$

and

$$\boldsymbol{\varphi}(t) = \boldsymbol{\varphi}_i$$

for $t_i \leqslant t < t_{i+1}, i = 0, 1, \ldots, n-1$. Let us now give a formal definition of the stochastic integral for such processes.

Definition. Let $\boldsymbol{\varphi}(t)$ be a step function random process. We then define the stochastic integral by

$$\int_a^b \boldsymbol{\varphi}(t)\, d\mathbf{W}(t) = \sum_{i=1}^{n-1} \varphi_i[\mathbf{W}(t_{i+1}) - \mathbf{W}(t_i)] \qquad \square$$

It turns out that many stochastic processes of interest besides the step function processes can have a stochastic integral defined for them. The result is essentially that as long as the stochastic process $\mathbf{y}(t)$ depends only on the past up to time t of $\mathbf{W}(t)$ (mathematically, they are called nonanticipating random processes), we can approximate $\mathbf{y}(t)$ arbitrarily closely with a sequence of step function random processes $\{\boldsymbol{\varphi}_n(t)\}_{n=1}^{\infty}$. The integrals associated with these approximating random processes can then be shown to converge. The stochastic integral for $\mathbf{y}(t)$ is then defined to be the limit of these approximating integrals.

We often wish to make sense of a differential equation driven by white noise which formally would be written as

$$\mathbf{x}_t' = m(\mathbf{x}_t, t) + \sigma(\mathbf{x}_t, t)\boldsymbol{\psi}_t$$

where $\boldsymbol{\psi}_t$ is a Gaussian white noise driver. This equation should be equivalent to the integral equation

$$\mathbf{x}_t = \mathbf{x}_a + \int_a^t m(\mathbf{x}_s, s)\, ds + \int_a^t \sigma(\mathbf{x}_s, s)\, d\mathbf{W}(s)$$

Under certain conditions on the growth rates and regularity of $m(\cdot,\cdot)$ and $\sigma(\cdot,\cdot)$, one can show that solutions to this system exist where the first integral can be thought of as the normal Riemann integral and the second integral is the stochastic integral.

Stochastic differential equations and integrals is a vast subject. A good introduction for systems scientists is given in Wong and Hajek [1985]. A more advanced introduction is contained in Ethier and Kurtz [1986].

APPENDIX E

Hypothesis Testing

In digital communication systems at the end of every signaling interval a decision must be made as to which symbol in the signaling set had been transmitted. In this text we will concern ourselves only with the simplest case, that of binary signaling sets. This digital communications theory problem is what statisticians call *hypothesis testing*. Consider the following binary hypothesis testing problem:

$$H_0\colon \mathbf{X} \text{ has distribution } P$$

$$H_1\colon \mathbf{X} \text{ has distribution } Q$$

where $\mathbf{X}$ is a collection of observations and P and Q are probability measures on the space of observations. Let π_i denote the *a priori* probability of H_i $(i = 0, 1)$ being true. *A priori* means before we have seen any data. There are two possible ways that an error can be made. One can announce H_0 when H_1 is really true (called *miss* or *type II* error) or one can announce H_1 when H_0 is true (called *false-alarm* or *type I* error). The probability of false alarm is usually denoted α and is called the *size* of the test. The probability of miss is usually denoted $1-\beta$, where β is called the *power* of the test. Obviously, by Bayes' theorem we would have the total probability of error given by $\pi_0\alpha + \pi_1(1-\beta)$. It is easy to show (see Van Trees [1968] or Poor [1988]) that the following test minimizes the probability of error:

$$\frac{dQ}{dP}(\mathbf{X}) \underset{H_0}{\overset{H_1}{\gtrless}} \frac{\pi_0}{\pi_1}$$

where dQ/dP is the Radon–Nikodym derivative of Q with respect to P (if it exists). In the case of probability densities existing, dQ/dP is merely the ratio of the densities. In any case, the test above is called the *likelihood ratio test*

(LRT). The symbolism is to be interpreted as: If $[dQ/dP](\mathbf{X})$ is less than 1, then announce H_0 as true; otherwise, announce H_1. If $[dQ/dP](\mathbf{X}) = \pi_0/\pi_1$, then announce either, as it will not effect the probability of error. Since the logarithm is monotone, one frequently states the LRT as

$$\log\left[\frac{dQ}{dP}(\mathbf{X})\right] \underset{H_0}{\overset{H_1}{\gtrless}} \log\frac{\pi_0}{\pi_1}$$

In some cases one does not know the a priori probabilities π_0 and π_1. One can then make use of the *Neyman–Pearson criterion*, which tells how to find the test which for a fixed size (fixed false-alarm probability) maximizes β (or minimizes the miss probability). The Neyman–Pearson lemma (see Van Trees [1968] or Poor [1988]) essentially states that the LRT again is the optimal test but with a threshold chosen so that the size is the prescribed value. In the case of discrete random variables one may have to randomize in a certain fashion at the threshold value to be able to obtain a given α. We will attempt to avoid such considerations and the resulting necessary technical details in the text by assuming that densities exist whenever we need.

APPENDIX F

Estimation Theory

Suppose ζ is an unknown real parameter that needs to be estimated from a random observation $\mathbf{X}$. Typically, one can parameterize the distribution (density if it exists) by ζ as $p(\mathbf{X}, \zeta)$. This parameterized density is sometimes called the *likelihood function.*

The *maximum likelihood estimate* is to choose that value of ζ which maximizes $p(\mathbf{X}, \zeta)$ for a given observed $\mathbf{X}$:

$$\hat{\zeta}_{\mathrm{ml}}(\mathbf{X}) \equiv \text{maximum likelihood estimate based on the observation } X$$

is defined by

$$p(\mathbf{X}, \hat{\zeta}_{\mathrm{ml}}(\mathbf{X})) = \max p(\mathbf{X}, \zeta)$$

It is this estimate which is of greatest interest for the material of chapter IX.

An estimate is said to be *consistent* if as the number of observations approaches infinity the estimate approaches the true value (in probability or with probability 1). An estimate is said to be *unbiased* if $E\{\hat{\zeta}_{\mathrm{ml}}(\mathbf{X})\} = \zeta$.

A common measure of goodness for an unbiased estimate $\hat{\zeta}$ of a parameter is its variance $\sigma_{\hat{\zeta}}^2$. Obviously, the lower the variance, the better the estimate. The *Cramér–Rao inequality* sets a lower bound on the estimator variance for any unbiased estimate. This bound is

$$\sigma_{\hat{\zeta}}^2 \geqslant E\left\{\left(\frac{\partial}{\partial \zeta} \log p(\mathbf{X}, \zeta)\right)^2\right\}^{-1} = -E\left\{\left(\frac{\partial^2}{\partial \zeta^2} \log p(\mathbf{X}, \zeta)\right)\right\}^{-1}$$

Simple derivations of this bound are given in Van Trees [1968] or Blahut [1987]. The inverse of the quantity making up the lower bound is called the

Fisher information at point ζ, i.e.,

$$k(\zeta) \equiv \text{Fisher information at point } \zeta = E\left\{\left(\frac{\partial}{\partial\zeta}\log p(\mathbf{X}, \zeta)\right)^2\right\}$$

When ζ is a parameter vector instead of a simple scalar, $k(\zeta)$ becomes a matrix called the *Fisher information matrix.*

There is an interesting relationship between the Kullback–Leibler information numbers and the Fisher information. Essentially, under some mild technical assumptions, one can show that $K(p(\cdot, \zeta + \varepsilon), p(\cdot, \zeta)) = \frac{1}{2}\varepsilon^2 k(\zeta) + o(\varepsilon^2)$ (see exercise 4 of chapter IX). Basically, this relationship is indicating that $k(\zeta)$ is the second derivative of the Kullback–Leibler information number. A rigorous proof and statement of these results may be found in Blahut [1987].

APPENDIX G

Calculus of Variations

The classical calculus of variations deals with the study of how to extremize the value of various integral functionals. Specifically, one of the simplest cases is to consider a function $F(x, y, z)$ with continuous first and second partial derivatives with respect to all of its arguments. Denote F_x, F_y, and F_z as the partial derivatives of F with respect to x, y, and z, respectively. Then among all functions $y(x)$ continuously differentiable on $x \in [a, b]$ and satisfying boundary/initial conditions $y(a) = A$, $y(b) = B$, find the function for which the functional

$$J(y) \equiv \int_a^b F(x, y, y')\, dx$$

is extremized (minimized or maximized). One can then show that the following theorem supplies a necessary condition:

Theorem
A necessary condition for $J(y)$ to have an extremum for a given function $y(x)$ is that $y(x)$ satisfy *Euler's equation*:

$$F_y - \frac{d}{dx} F_{y'} \equiv F_y(x, y(x), y'(x)) - \frac{d}{dx} F_{y'}(x, y(x), y'(x)) = 0$$

Euler's equation is merely a necessary condition and it is not in general sufficient by itself to specify the optimizing trajectory to extremize $J(y)$. However, in many cases, Euler's equation will have only one trajectory satisfying it. In such cases, of course, it is sufficient, which will be the situation for the material in this book.

Euler's equation is in general a second order differential equation. In certain special cases it can be reduced to a first order equation. For example, suppose that $F(x, y, y')$ does not depend on x directly, i.e.,

$$J(y) \equiv \int_a^b F(y, y')\, dx$$

Then Euler's equation becomes

$$F_y - \frac{d}{dx} F_{y'} = F_y - F_{y'y} y' - F_{y'y'} y'' = 0$$

Multiply both sides by y' to obtain

$$F_y y' - F_{y'y} y'^2 - F_{y'y'} y' y'' = \frac{d}{dx}(F - y' F_{y'}) = 0$$

We may now integrate with respect to x to find

$$F - y' F_{y'} = c \equiv \text{constant}$$

When the endpoints of the trajectory are not fixed, one must introduce further conditions in addition to Euler's equation. For example, suppose that we wish to solve the problem: Among all smooth curves whose endpoints (x_0, y_0) and (x_1, y_1) lie on two given curves $y = \varphi(x)$ and $y = \psi(x)$, find the curve for which the functional

$$J(y) = \int_{x_0}^{x_1} F(x, y, y')\, dx$$

has an extremum. The first necessary condition for this problem is that Euler's equation be satisfied. This has to be true since once the endpoints are known for the optimal trajectory, that same trajectory must also satisfy the fixed endpoint problem considered above. One may show that the additional conditions necessary are

$$F + (\varphi' - y') F_{y'}|_{x = x_0} = 0$$

and

$$F + (\psi' - y') F_{y'}|_{x = x_1} = 0$$

the *transversality conditions*, which come about due to the variable endpoints. A good introductory textbook containing these and other results is Gelfand and Fomin [1963].

Solutions to Exercises

CHAPTER II

1. **(a)** $p(x) = \lambda \exp[-\lambda x]$ for $x \geqslant 0$. Hence $M(\theta) = \lambda/(\lambda - \theta)$ for $\theta < \lambda$. $I(x)$ is then straightforwardly found to be $\lambda x - 1 - \log[\lambda x]$ for $x > 0$ and ∞ otherwise.

(b) $P\{\mathbf{x}_1 = 1\} = p = 1 - P\{\mathbf{x}_1 = 0\}$. Hence $M(\theta) = p\exp[\theta] + (1-p)$. Therefore, $I(x) = x\log(x/p) + (1-x)\log[(1-x)/(1-p)]$ for $0 < x < 1$ and ∞ otherwise.

(c) $p(x) = 1/[\pi(1+x^2)]$. $M(\theta) = 1$ for $\theta = 0$ and equals ∞ for all other values of θ. $I(x) = 0$ for all x.

2. **(a)**
$$\lim_{\theta_n \to \lambda} \frac{M'(\theta_n)}{M(\theta_n)} = \lim_{\theta_n \to \lambda} \frac{1}{\lambda - \theta_n} = \infty$$

Hence $\log M(\cdot)$ is steep.

(b) $D_M = \mathbb{R}$. Hence there is no finite boundary to the domain and $\log M(\cdot)$ is steep.

(c) $D_M = \{0\}$. Log $M(\cdot)$ is not differentiable on its domain and hence $\log M(\cdot)$ is not steep.

3. Let $x \in (b, \infty)$. Note that $M(\theta) \equiv E\{\exp[\theta \mathbf{x}]\} \leqslant \exp[\theta b]$ for $\theta > 0$. Hence $I(x) = \sup_\theta \{\theta x - \log M(\theta)\} = \sup_{\theta > 0}\{\theta x - \log M(\theta)\} \geqslant \sup_{\theta > 0}\{\theta x - \theta b\} = (x - b)\sup_{\theta > 0}\{\theta\} = \infty$.

4.
$$E\{\exp[\theta \mathbf{Y}_n]\} = .5[E\{\exp[\theta \mathbf{S}_n^+] \mid \mathbf{Y}_n = \mathbf{S}_n^+\} + E\{\exp[-\theta \mathbf{S}_n^-] \mid \mathbf{Y}_n = \mathbf{S}_n^-\}]$$
$$= .5\left[\sum_{j=0}^{n} \exp[\theta j]\binom{n}{j}.5^n + \exp[-\theta(n-j)]\binom{n}{n-j}.5^n\right]$$
$$= .5^{n+1}[(1+\exp[\theta])^n + (1+\exp[-\theta])^n]$$

Hence

$$\lim_{n\to\infty} \frac{1}{n} \log E\{\exp[\theta \mathbf{Y}_n]\} = \log \frac{1 + \exp[|\theta|]}{2}$$

The lower bounds hypotheses do not apply since this limiting function is not differentiable at zero. The upper bounds hypotheses are, however, still valid.

5. $E\{\exp[\theta \mathbf{Y}_n]\} = .5^{n+1}[\exp[n\theta] + \exp[-n\theta]] + 1 - .5^n$

If $\exp[|\theta|]/2 \leqslant 1$. then $\lim_{n\to\infty} (1/n) \log E\{\exp[\theta \mathbf{Y}_n]\} = 0$. If $\exp[|\theta|]/2 \geqslant 1$, then $\lim_{n\to\infty} (1/n) \log E\{\exp[\theta \mathbf{Y}_n]\} = |\theta| - \log 2$. Hence $\varphi(\theta)$ is differentiable at zero but not everywhere. Upper bound will be correct but lower bound fails. $I(x) = |x| \log 2$ for $|x| < 1$ and ∞ elsewhere.

6. $\varphi_w(\theta) = \exp[.5\theta^{\mathrm{T}} K\theta]$. $\varphi(\theta) = .5\theta^{\mathrm{T}}[I - A]^{-1}\Gamma K \Gamma^{\mathrm{T}}[I - A]^{-1\mathrm{T}}\theta \equiv .5\theta^{\mathrm{T}} B\theta$ (note that B is symmetric and hence B^{-1} will be symmetric).

$$I(x) \equiv \sup_{\theta} \{\langle \theta, x\rangle - .5\theta^{\mathrm{T}} B\theta\} = .5x^{\mathrm{T}}(B^{-1})^{\mathrm{T}} x = .5x^{\mathrm{T}} B^{-1} x$$

7. (a) $$\binom{n}{k} = \frac{n\Gamma(n)}{(n-k)k!\,\Gamma(n-k)} \approx \left[\frac{n}{n-k}\right]^{.5} \frac{\exp[-k]k^k}{k!} \left[\frac{k}{n}\right]^{-k} \left[1 - \frac{k}{n}\right]^{-(n-k)}$$

Note that for $k = \alpha n$ for $0 < \alpha < 1$ we have

$$\frac{\exp[-k]k^k}{k!} \approx k^k \exp[-k] \frac{\sqrt{k}}{\sqrt{2\pi}\,(k/e)^k} = \frac{\sqrt{k}}{\sqrt{2\pi}} = \frac{\sqrt{\alpha n}}{\sqrt{2\pi}}$$

Therefore,

$$\lim_{n\to\infty} \frac{1}{n} \log \frac{\exp[-k]k^k}{k!} = 0$$

Also note that

$$\lim_{n\to\infty} \frac{1}{n} \log \left[\frac{n}{n-k}\right]^{.5} = \lim_{n\to\infty} \frac{1}{n} \log \left[\frac{1}{1-\alpha}\right]^{.5} = 0$$

Hence

$$\frac{1}{n} \log \binom{n}{k} \approx \frac{k}{n} \log \frac{k}{n} - \left(1 - \frac{k}{n}\right) \log\left(1 - \frac{k}{n}\right) \qquad \text{for } k = \alpha n \ (0 < \alpha < 1).$$

(b) $\mathbf{S}_n \equiv (1/n)\,\Sigma_{i=1}^n \mathbf{x}_i$ and $\{\mathbf{x}_i\}$ are iid with $P\{\mathbf{x}_1 = 0\} = P\{\mathbf{x}_1 = 1\} = .5$. Let us just consider the upper bound: $P\{\mathbf{S}_n \geqslant x\}$ for $.5 < x < 1$. $P\{\mathbf{S}_n \geqslant x\} = \binom{n}{[nx]}.5^n$, where $[a]$ is the smallest integer greater than or equal to a. Therefore,

$$\frac{1}{n}\log P\{\mathbf{S}_n \geqslant x\} = \log .5 + \frac{1}{n}\log\binom{n}{[nx]}$$

$$\approx -\log 2 - x\log x - (1-x)\log(1-x)$$

8. To show hint: Integrate $\int_w^\infty (3/x^4)\exp[-x^2/2]\,dx$ by parts. Let $u = \exp[-x^2/2]$ $dv = (3/x^4)\,dx$ and apply $\int u\,dv = uv - \int v\,du$. Now

$$\int_w^\infty \exp\left[-\frac{x^2}{2}\right]dx \geqslant \int_w^\infty \left(1 - \frac{3}{x^4}\right)\exp\left[-\frac{x^2}{2}\right]dx$$

$$= \left(\frac{1}{w} - \frac{1}{w^3}\right)\exp\left[-\frac{w^2}{2}\right] \geqslant \left(\frac{1}{w} - \frac{1}{w^2}\right)\exp\left[-\frac{w^2}{2}\right]$$

which gives the lower bound. Now note that

$$\int_w^\infty \exp\left[-\frac{x^2}{2}\right]dx \leqslant \int_w^\infty \frac{x}{w}\exp\left[-\frac{x^2}{2}\right]dx = \frac{1}{w}\exp\left[-\frac{w^2}{2}\right]$$

which is the upper bound.

Now $\mathbf{S}_n \equiv (1/n)\,\Sigma_{i=1}^n \mathbf{x}_i$ for $\{\mathbf{x}_i\}$ iid and $\mathbf{x}_1$ a unit variance zero mean Gaussian random variable. Then $\mathbf{S}_n$ is a zero mean variance $(1/n)$ Gaussian random variable. Then for $x > 0$ we have

$$P\{\mathbf{S}_n \geqslant x\} = P\left\{\frac{\mathbf{x}_1}{\sqrt{n}} \geqslant x\right\} = P\{\mathbf{x}_1 \geqslant x\sqrt{n}\} = \frac{\int_{x\sqrt{n}}^\infty \exp[-y^2/2]\,dy}{\sqrt{2\pi}}$$

Use our upper and lower bounds and it is easy to obtain that $\lim_{n\to\infty}(1/n)$ $\log P\{\mathbf{S}_n \geqslant x\} = -x^2/2$.

CHAPTER III

1. Let P and Q be two probability measures.

Claim: If $\alpha, \beta > 0$ and if $P(F) \leqslant Q(F^\alpha) + \beta$ (for all closed F), then $Q(F) \leqslant P(F^\alpha) + \beta$ (for all closed F).

Proof: Let F_1 be an arbitrary closed set and define $F_2 = (F_1^\alpha)^c$, where the superscript c indicates complementation. F_2 is closed and $F_1 \subset (F_2^\alpha)^c$.

Then $P(F_1^\alpha) = 1 - P(F_2) \geqslant 1 - Q(F_2^\alpha) - \beta \geqslant Q(F_1) - \beta$. Hence $Q(F_1) \leqslant P(F_1^\alpha) + \beta$.

Now the claim immediately implies that $\rho(P, Q) = \rho(Q, P)$. If $\rho(P, Q) = 0$, then $P(F) = Q(F)$ for all closed F. Equality on the closed sets (and hence on the open sets) is enough to state that $P = Q$. Finally, if P, Q, and R are probability measures with $\rho(P, Q) < \delta$ and $\rho(Q, R) < \varepsilon$, then $P(F) \leqslant Q(F^\delta) + \delta \leqslant Q(\overline{F^\delta}) + \delta \leqslant R((\overline{F^\delta})^\varepsilon) + \delta + \varepsilon = R(F^{\delta+\varepsilon}) + \delta + \varepsilon$ for all closed F. Hence $\rho(P, R) \leqslant \delta + \varepsilon$, which gives the triangle inequality.

2. Suppose $\mathbf{S}_n \to S \equiv (\pi_1, \pi_2, \ldots, \pi_d)$ (i.e., $\lim_{n \to \infty} \mathbf{S}_{ni} = \pi_i$ for $i = 1, \ldots, d$). Consider some bounded continuous function $g(\cdot)$ (since this is a discrete state space, all functions are continuous). Then

$$\lim_{n \to \infty} E_{\mathbf{L}_n}\{g\} = \lim_{n \to \infty} \sum_{i=1}^{d} g(\alpha_i)\mathbf{S}_{ni} = \sum_{i=1}^{d} g(\alpha_i)\pi_i$$

Hence $\mathbf{L}_n$ converges in distribution to π (see appendix D.3). The same argument going backward gives that convergence in distribution of $\mathbf{L}_n$ to π must imply that $\mathbf{S}_n$ converges as a vector to $(\pi_1, \pi_2, \ldots, \pi_d)$.

3. Suppose that P_n converges to P in Prohorov metric (i.e., $\lim_{n \to \infty} \rho(P_n, P) = 0$). Let $\varepsilon_n = \rho(P_n, P) + 1/n$. Let f be any bounded continuous positive function of compact support. Denote the distribution function of $f(\mathbf{x}_n)$ as $F_n(\cdot)$ (where $\mathbf{x}_n$ is the random variable that has distribution P_n). Let $\|f\|$ denote the maximum value of f. Then

$$\int f\, dP_n = \int_0^{\|f\|} t\, dF_n(t) = F_n(t)t\big|_0^{\|f\|} - \int_0^{\|f\|} F_n(t)\, dt = \|f\| - \int_0^{\|f\|} F_n(t)\, dt$$

$$= \int_0^{\|f\|} [1 - F_n(t)]\, dt = \int_0^{\|f\|} P_n(\{f > t\})\, dt = \int_0^{\|f\|} P_n(\{f \geqslant t\})\, dt$$

Now since $f(\cdot)$ is continuous, $\{f \geqslant t\}$ is closed. Hence $P_n(\{f \geqslant t\}) \leqslant P(\{f \geqslant t\}^{\varepsilon_n}) + \rho(P_n, P) + 1/n$ (where F^ε denotes the "ε-fattening" of the set F) $= P(\{f \geqslant t\}^{\varepsilon_n}) + \varepsilon_n$. Hence

$$\int f\, dP_n \leqslant \int_0^{\|f\|} P(\{f \geqslant t\}^{\varepsilon_n})\, dt + \varepsilon_n \|f\|$$

Therefore,

$$\overline{\lim_{n \to \infty}} \int f\, dP_n \leqslant \lim_{n \to \infty} \int_0^{\|f\|} P(\{f \geqslant t\}^{\varepsilon_n})\, dt = \int_0^{\|f\|} P(\{f \geqslant t\})\, dt = \int f\, dP$$

Now let f be any bounded continuous function of compact support. Then $\|f\| + f$ and $\|f\| - f$ are nonnegative bounded continuous functions of compact support. Hence

$$\overline{\lim_{n\to\infty}} \int (\|f\| + f)\, dP_n \leqslant \int (\|f\| + f)\, dP = \|f\| + \int f\, dP$$

which implies that

$$\overline{\lim_{n\to\infty}} \int f\, dP_n \leqslant \int f\, dP$$

Similarly, using $\|f\| - f$ we find that

$$\underline{\lim}_{n\to\infty} \int f\, dP_n \geqslant \int f\, dP$$

Hence

$$\lim_{n\to\infty} \int f\, dP_n = \int f\, dP$$

for all bounded continuous f of compact support, which is equivalent to convergence in distribution.

Suppose now that we have convergence in distribution. We will use the portmanteau theorem (see appendix D.3) to show convergence in the Prohorov metric. Let $\varepsilon > 0$ be given. Let $E_1, E_2, \ldots$ be a partition of the real line into intervals of length less than $\varepsilon/2$. Let N be the smallest integer n so that $P(\bigcup_{i=1}^{n} E_i) > 1 - \varepsilon/2$. Now consider the "$\varepsilon$-fattening" of this set $(\bigcup_{i=1}^{n} E_i)^{\varepsilon/2}$. This is an open set on the real line, which implies that it is composed of a disjoint collection of open intervals G. It is easy to see that the number of disjoint open intervals in G must be finite. Now from the portmanteau theorem we have that convergence in distribution is equivalent to $\underline{\lim}_{n\to\infty} P_n(O) \geqslant P(O)$ for all open sets O. Hence for all $n \geqslant n_0$ and for all $G_i \subset G$, we must have $P(G_i) \leqslant P_n(G_i) + \varepsilon/2$. Let F be any closed set. Then define $F_0 \equiv \bigcup \{E_i : 1 \leqslant i \leqslant N,\ E_i \cap F \neq \varnothing\}$. It then must be true that $F_0 \subset G$ and $P(F) \leqslant P(F_0^{\varepsilon/2}) + \varepsilon/2 \leqslant P_n(F_0^{\varepsilon/2}) + \varepsilon \leqslant P_n(F^{\varepsilon}) + \varepsilon$ for all $n \geqslant n_0$. Hence $\rho(P_n, P) \leqslant \varepsilon$ for all $n \geqslant n_0$. This implies convergence in Prohorov metric.

4. Let $g(\cdot)$ be a bounded continuous function. Since $f(\cdot)$ and $g(\cdot)$ are bounded continuous functions, so is $g(f(\cdot))$. Hence $\lim_{n\to\infty} E\{g(f(\mathbf{x}_n))\} = E\{g(f(\mathbf{x}))\}$ for all such $g(\cdot)$. Hence $f(\mathbf{x}_n)$ converges in distribution to $f(\mathbf{x})$.

Let $\mathbf{x}_n$ be converging in distribution to $\mathbf{x}$. Let $f(\cdot)$ be a continuous function. Let $g(\cdot)$ be a bounded continuous function. Then $g(f(\cdot))$ is a bounded continuous function. Hence $\lim_{n\to\infty} E\{g(f(\mathbf{x}_n))\} = E\{g(f\mathbf{x}))\}$. Therefore, $f(\mathbf{x}_n)$ is converging in distribution to $f(x)$ and no counterexample exists.

Let $\mathbf{x}_n$ be uniformly distributed on the interval $(-1/n, 1/n)$. $\mathbf{x}_n$ is converging in distribution to the degenerate random variable $\mathbf{x}$, where $P(x = 0) = 1$. Define $f(x) = 1$ if $x \geqslant 0$ and $f(x) = 0$ otherwise. f is bounded but not continuous. $P(f(\mathbf{x}_n) = 1) = .5 = P(f(\mathbf{x}_n) = 0)$ for every n. Hence $f(\mathbf{x}_n)$ is converging in distribution but not to $f(\mathbf{x})$, the degenerate random variable where $P(f(\mathbf{x}) = 1) = 1$.

5. (a) $$\left[\int f(y)^{1/\lambda}\,dy\right]^{\lambda}\left[\int g(y)^{1/(1-\lambda)}\,dy\right]^{1-\lambda} \geqslant \int f(y)g(y)\,dy \qquad (*)$$

where $\lambda \in [0, 1]$ for nonnegative g, f by Holder's inequality. Let $f(y) \equiv [dP_2/dW]^{\lambda\alpha_1}[dP_1/dW]^{\lambda(1-\alpha_1)}$ and $g(y) \equiv [dP_2/dW]^{(1-\lambda)\alpha_2} \times [dP_1/dW]^{(1-\lambda)(1-\alpha_2)}$. Now apply (*) and take logarithms of both sides of inequality to obtain convexity.

(b) $\log H_1 = \log H_0 = 0$. Then, by convexity, the minimum exists and must lie in the interval $[0, 1]$.

(c) If probabilities exist, $dW(y)$ can be replaced by the density $w(y)$, which then just gets canceled out of the integrand. The general case is just a measure theoretic extension of that argument; i.e., one shows $H_\alpha = \int [dP_1(y)/dP_2(y)]^\alpha\, dP_2(y)$, which of course has no W dependence.

6. $$H_\alpha = \int \left[(2\pi)^{-.5}\exp\left[-\frac{x^2}{2}\right]\right]^\alpha \left[(2\pi)^{-.5}\exp\left[-\left(\frac{(x-s)^2}{2}\right)\right]\right]^{1-\alpha} dx$$

$$= \exp\left[\left(-\frac{s^2}{2}\right)(\alpha - \alpha^2)\right]$$

Hence

$$\min H_\alpha = H_{.5} = \exp\left[-\frac{s^2}{8}\right]$$

and

$$\lim_{n\to\infty} \frac{1}{n}\log P(\text{error}) = -\frac{s^2}{8}$$

7. We heuristically set up a Lagrange multiplier minimization. Minimize over the class of probability densities $r(y)$ the quantity $\int \log[r(y)/p(y)]r(y)\,dy$ subject to $\int h_i(y)r(y)\,dy = \beta_i$, $i = 1, \ldots, k$, where $p(y)$ is a given probability density and the $\beta_1, \beta_2, \ldots, \beta_d$ are fixed constants. We proceed by a perturbation argument to find a necessary condition: Perturb $r(\cdot)$ to $r(\cdot) + \varepsilon d(\cdot)$, where $d(\cdot)$ must satisfy $\int d(y)\,dy = 0$ and $r(y) + \varepsilon d(y) \geqslant 0$. We ignore the inequality constraint for now, for if we can find a valid relative minimum without the inequality constraint, then we have found a valid relative minimum with the constraint. Hence

$$\frac{\delta}{\delta\varepsilon}\left[\int \log\frac{r(y)+\varepsilon d(y)}{p(y)}[r(y)+\varepsilon d(y)]\,dy - \sum_{i=1}^{k}\theta_i \int h_i(y)[r(y)+\varepsilon d(y)]\,dy - (\theta_0+1)\int [r(y)+\varepsilon d(y)]\,dy\right] = 0$$

$(\theta_0 + 1)$ is the Lagrange multiplier that represents our constraint that $r(y) + \varepsilon d(y)$ must integrate to 1 and $\theta_1, \theta_2, \ldots, \theta_d$ are the multipliers that represent the given constraints. Formally taking the derivative, we find that

$$\int d(y)\left[1 + \log\frac{r(y)}{p(y)} - \sum_{i=1}^{k}\theta_i h_i(y) - (\theta_0+1)\right]dy = 0$$

for all allowable perturbations $d(y)$. The only way that the integral can be zero for all such perturbations is for the term in brackets to be zero for all y. Hence

$$\log\frac{r(y)}{p(y)} = \theta_0 + \sum_{i=1}^{k}\theta_i h_i(y)$$

8. (a) $\log\left[\dfrac{x^{a-1}(1-x)^{b-1}}{\Gamma(a)\Gamma(b)/\Gamma(a+b)}\right] = \theta_0 + \theta_1 h_1(x) + \theta_2 h_2(x)$. Equating coefficients we find that $\theta_0 = \log(\Gamma(a+b)/\Gamma(a)\Gamma(b))$; $\theta_1 = a-1$, $h_1(x) = \log(x)$; $\theta_2 = b-1$, $h_2(x) = \log(1-x)$. $\beta_1 = E\{\log(\mathbf{x})\}$, $\beta_2 = E\{\log(1-\mathbf{x})\}$.

(b) $\theta_0 = \log(a)$; $\theta_1 = -a$, $h_1(x) = x$. $\beta_1 = 1/a$.

(c) $\theta_0 = b\log(a) - \log(\Gamma(b))$; $\theta_1 = b-1$, $h_1(x) = \log(x)$; $\theta_2 = -a$, $h_2(x) = x$. $\beta_1 = E\{\log(\mathbf{x})\}$, $\beta_2 = E\{\mathbf{x}\} = \Gamma(b+1)/a\Gamma(b) = b/a$.

(d) $\theta_0 = -.5\log(2\pi\sigma^2) - m^2/2\sigma^2$; $\theta_1 = -1/2\sigma^2$, $h_1(x) = x^2$; $\theta_2 = 1/\sigma^2$, $h_2(x) = x$. $\beta_1 = \sigma^2$, $\beta_2 = m$.

(e) $\theta_0 = \log(.5a)$; $\theta_1 = -a$, $h(x) = |x|$. $\beta_1 = 2/a$.

9. $$I_\mu(\lambda P_1 + (1-\lambda)P_2) = \int \log\left[\frac{d(\lambda P_1 + (1-\lambda)P_2)}{d\mu}\right] d(\lambda P_1 + (1-\lambda)P_2)$$
$$= \lambda \int \log\left[\frac{d(\lambda P_1 + (1-\lambda)P_2)}{d\mu}\right] dP_1$$
$$+ (1-\lambda) \int \log \frac{d(\lambda P_1 + (1-\lambda)P_2)}{d\mu} dP_2$$

Aside: Now consider

$$\int \log \frac{dP_i}{d\mu} dP_i - \int \log \frac{dQ}{d\mu} dP_i = \int \log \frac{dP_i}{dQ} dP_i \geqslant \int \left[1 - \frac{dQ}{dP_i}\right] dP_i$$

[since $\log(x) \geqslant 1 - x$]

$$= \int dP_i - \int dQ = 1 - 1 = 0$$

Hence

$$\int \log \frac{dP_i}{d\mu} dP_i \geqslant \int \log \frac{dQ}{d\mu} dP_i$$

for all probability measures Q. Therefore,

$$I_\mu(\lambda P_1 + (1-\lambda)P_2) \leqslant \lambda \int \log \frac{dP_1}{d\mu} dP_1 + (1-\lambda) \int \log \frac{dP_2}{d\mu} dP_2$$
$$= \lambda I_\mu(P_1) + (1-\lambda) I_\mu(P_2)$$

Hence $I_\mu(v)$ is convex in the space of probability measures.

CHAPTER IV

1. $$\int_0^T \int_0^T K^{-1}(s,t)(y(s)-m(s))(y(t)-m(t))\, ds\, dt$$
$$= \int_0^T \int_0^T \left[\sum_{k=0}^{\infty} \lambda_k^{-1} \varphi_k(s)\varphi_k(t)\right] (y(s)-m(s))(y(t)-m(t))\, ds\, dt$$
$$= \sum_{k=0}^{\infty} \lambda_k^{-1} \int_0^T \int_0^T \varphi_k(s)\varphi_k(t)(y(s)-m(s))(y(t)-m(t))\, ds\, dt$$
$$= \sum_{k=0}^{\infty} \lambda_k^{-1} (y_k - m_k)^2$$

2. **(a)** Consider $f_n(x)=\int_{-\infty}^{x}\exp[-nt^2/2]/\sqrt{2\pi/n}\,dt$. $f_n(x)$ is obviously differentiable, but its limit is given by

$$f(x)=\begin{cases}1 & \text{if } x>0\\ 0 & \text{if } x<0\\ .5 & \text{if } x=0\end{cases}$$

which is clearly not differentiable. It is easy, however, to verify that

$$\lim_{n\to\infty}\int_{-\infty}^{\infty}(f_n(x)-f(x))^2\,dx=0$$

(b) Let $f_n(x)=1$ for $x\in[-1/n,1/n]$ and be zero elsewhere. $f_n(x)$ is converging to zero in mean square [pointwise it is converging to $f(x)=1$ if $x=0$ and zero elsewhere]. Note that

$$\lim_{n\to\infty}\int_{-\infty}^{\infty}(f_n(x)-0)^2\,dx=\lim_{n\to\infty}\frac{1}{n}=0$$

(c) In part (a) we have a sequence of differentiable functions converging in the L_2 norm to a nondifferentiable function. Hence the set of all differentiable functions does not contain all of its limit points and hence it is not closed. In part (b) $f_n(x)$ is not differentiable for any n but for large n becomes arbitrarily close to the zero function, which is differentiable. This implies that there exist nondifferentiable functions arbitrarily close in L_2 norm to a differentiable function. This implies, then, that the set of differentiable functions is not open.

3. We first need to establish a variant of a martingale inequality due to Kolmogorov. Our proof will follow that of Hall and Heyde [1980]. $\{\mathbf{S}_i\}$ is a martingale if

$$E\{\mathbf{S}_n\,|\,\mathbf{S}_{n-1},\mathbf{S}_{n-2},\ldots\}=\mathbf{S}_{n-1}$$

This immediately implies by successive conditioning that $E\{\mathbf{S}_n\,|\,\mathbf{S}_{n-i},\mathbf{S}_{n-2},\ldots\}=\mathbf{S}_{n-i}$ for $i\geqslant 0$.

Claim 1: $\lambda P(\max_{i\leqslant n}\mathbf{S}_i>\lambda)\leqslant E\{|\mathbf{S}_n|\}$.

Proof of Claim 1: Define $E\equiv\{\max_{i\leqslant n}\mathbf{S}_i>\lambda\}=\bigcup_{i=1}^{n}\{\mathbf{S}_i>\lambda;\ \max_{1\leqslant j<i}\mathbf{S}_j\leqslant\lambda\}\equiv\bigcup_{i=1}^{n}E_i$. Therefore,

$$\lambda P(E)=\lambda\sum_{i=1}^{n}P(E_i)=\lambda\sum_{i=1}^{n}E\{\mathbf{1}_{E_i}\}\leqslant\sum_{i=1}^{n}E\{\mathbf{S}_i\mathbf{1}_{E_i}\}$$

$$=\sum_{i=1}^{n}E\{E\{\mathbf{S}_n\,|\,\mathbf{S}_i,\mathbf{S}_{i-1},\ldots\}\mathbf{1}_{E_i}\}$$

$$= \sum_{i=1}^{n} E\{E\{\mathbf{S}_n \mathbf{1}_{E_i} \mid \mathbf{S}_i, \mathbf{S}_{i-1}, \ldots\}\} = \sum_{i=1}^{n} E\{\mathbf{S}_n \mathbf{1}_{E_i}\}$$

$$= E\{\mathbf{S}_n \mathbf{1}_E\} \leqslant E\{|\mathbf{S}_n|\}$$

Now consider the slow Markov random process model $\mathbf{x}_{k+1} = \mathbf{x}_k + (1/n)V(\mathbf{x}_k, \xi_k)$ and $\mathbf{x}_0 \equiv 0$. We assume that $\sigma_x^2 \equiv E\{(V(x, \xi_1) - m(x))^2\} \leqslant \sigma^2 < \infty$. $\{\mathbf{x}_k\}$ is obviously a Markov chain. Interestingly enough, for any Markov chain we have the following assertion:

Claim 2: $\displaystyle\prod_{j=0}^{k-1} \frac{f(\mathbf{x}_{j+1})}{E\{f(\mathbf{x}_{j+1})|\mathbf{x}_j\}}$ is a martingale.

Proof of Claim 2: $$E\left\{\prod_{j=0}^{k-1} \frac{f(\mathbf{x}_{j+1})}{E\{f(\mathbf{x}_{j+1})|\mathbf{x}_j\}} |\mathbf{x}_{k-1}, \mathbf{x}_{k-2}, \ldots\right\}$$

$$= \left[\prod_{j=0}^{k-2} \frac{f(\mathbf{x}_{j+1})}{E\{f(\mathbf{x}_{j+1}) \mid \mathbf{x}_j\}}\right] \frac{E\{f(\mathbf{x}_k) \mid \mathbf{x}_{k-1}, \mathbf{x}_{k-2}, \ldots\}}{E\{f(\mathbf{x}_k) \mid \mathbf{x}_{k-1}\}}$$

$$= \left[\prod_{j=0}^{k-2} \frac{f(\mathbf{x}_{j+1})}{E\{f(\mathbf{x}_{j+1})|\mathbf{x}_j\}}\right]$$

Now in our problem we wish to take $f(x) \equiv \exp[\theta x]$. Then

$$\prod_{j=0}^{k-1} \frac{\exp[\theta \mathbf{x}_{j+1}]}{E\{\exp[\theta \mathbf{x}_j + (\theta/n)V(\mathbf{x}_j, \xi_j)] \mid \mathbf{x}_j\}}$$

$$= \prod_{j=0}^{k-1} \frac{\exp[\theta \mathbf{x}_{j+1}]}{\exp[\theta \mathbf{x}_j] E\{\exp[(\theta/n)V(\mathbf{x}_j, \xi_j)] \mid \mathbf{x}_j\}}$$

$$= \frac{\exp[\theta \mathbf{x}_k]}{\Pi_{j=0}^{k-1} E\{\exp[(\theta/n)V(\mathbf{x}_j, \xi_j)] \mid \mathbf{x}_j\}}$$

is a martingale. Define $m(x_j) \equiv E\{V(x_j, \xi_j)\}$. Then $\exp[\theta \mathbf{x}_k]/\Pi_{j=0}^{k-1} \times \exp[(\theta/n)m(\mathbf{x}_j)]\mathbf{r}_j$ is a martingale where $\mathbf{r}_j \equiv E\{\exp[(\theta/n)V(\mathbf{x}_j, \xi_j) - (\theta/n)m(\mathbf{x}_j)] \mid \mathbf{x}_j\}$. Now for large n,

$$\mathbf{r}_j = E\left\{1 + \frac{\theta}{n}(V(\mathbf{x}_j, \xi_j) - m(\mathbf{x}_j)) + \frac{\theta^2}{n^2}(V(\mathbf{x}_j, \xi_j) - m(\mathbf{x}_j))^2 + O\left(\frac{1}{n^3}\right)\right\}$$

$$= 1 + \frac{\theta^2}{n^2}\sigma_{\mathbf{x}_j}^2 + O\left(\frac{1}{n^3}\right) \leqslant 1 + \frac{\theta^2}{n^2}\sigma^2$$

Therefore, $\exp[\theta \mathbf{x}_k - (\theta/n)\,\Sigma_{j=0}^{k-1} m(\mathbf{x}_j)]/\Pi_{j=0}^{k-1} \mathbf{r}_j$ is a martingale and $\Pi_{j=0}^{k-1} \mathbf{r}_j \leqslant (1 + [\theta^2/n^2]\sigma^2)^k \leqslant (1 + [\theta^2/n^2]\sigma^2)^n$ for all $k \leqslant n$.

Now consider for $\theta > 0$ and $n \gg \theta$

$$P\left(\sup_{0 \leqslant k \leqslant n}\left(\mathbf{x}_k - \frac{1}{n}\sum_{j=0}^{k-1} m(\mathbf{x}_j)\right) \geqslant \varepsilon\right)$$

$$= P\left(\sup_{0 \leqslant k \leqslant n} \exp\left[\theta\left(\mathbf{x}_k - \frac{1}{n}\sum_{j=0}^{k-1} m(\mathbf{x}_j)\right] \geqslant \exp[\theta\varepsilon]\right)$$

$$\leqslant P\left(\sup_{0 \leqslant k \leqslant n} \exp\left[\theta\left(\mathbf{x}_k - \frac{1}{n}\sum_{j=0}^{k-1} m(\mathbf{x}_j)\right)\right]\Big/\prod_{j=0}^{k-1} \mathbf{r}_j \geqslant \exp[\theta\varepsilon]\left(1 + \frac{\theta^2}{n^2}\sigma^2\right)^{-n}\right)$$

Now we wish to invoke the martingale inequality at this point. Note that the martingale is positive, hence $E\{|\mathbf{S}_n|\} = E\{\mathbf{S}_n\}$. But it is easy to show that martingales are mean stationary, i.e., $E\{\mathbf{S}_n\} = E\{\mathbf{S}_0\} = 1$ for all n. Hence

$$P\left(\sup_{0 \leqslant k \leqslant n}\left(\mathbf{x}_k - \frac{1}{n}\sum_{j=0}^{k-1} m(\mathbf{x}_j)\right) \geqslant \varepsilon\right) \leqslant \exp[-\theta\varepsilon]\left(1 + \frac{\theta^2}{n^2}\sigma^2\right)^n$$

Now for θ and n large enough, we can make the term on the right-hand side of the inequality arbitrarily small. One can then go through this whole construction for the Markov chain defined by $\{-\mathbf{x}_k\}$ to find that $P(\sup_{0 \leqslant k \leqslant n} |(\mathbf{x}_k - (1/n)\sum_{j=0}^{k-1} m(\mathbf{x}_j))| \geqslant \varepsilon)$ can be made arbitrarily small for large n. Now we may form a continuous-time function from $\{\mathbf{x}_k\}$ by linearly interpolating between the defined points $\mathbf{x}^n(k/n) \equiv \mathbf{x}_k$. Then we immediately have that $P(\sup_{0 \leqslant t \leqslant 1} |\mathbf{x}^n(t) - \int_0^t m(\mathbf{x}^n(t))\, dt| \geqslant \varepsilon)$ can be made arbitrarily small for n large enough.

We now invoke a lemma from the theory of ordinary differential equations.

Lemma (Gronwall's Inequality)
Suppose that $|F(x) - F(y)| \leqslant L|x - y|$ for all x and y and some constant L. Let $X(s)$ denote the solution of $dx(s)/ds = F(X(s))$ for all $s \leqslant t$ and $X(0) = 0$. Then

$$\sup_{s \leqslant t}\left|X_n(s) - \int_0^s F_n(X_n(u))\, du\right| \leqslant \delta$$

implies that

$$\sup_{s \leqslant t} |X_n(s) - X(s)| \leqslant \left(\delta + t\sup_x |F(x) - F_n(x)|\right)\exp[Lt]$$

Now in our setting ($F_n(\cdot) \equiv F(\cdot) \equiv m(\cdot)$, $X_n(t) = \mathbf{x}^n(t)$, and $X(s) \equiv x(s)$, where $x(s)$ is the solution to the differential equation $dx(s)/ds = m(x(s))$ with $x(0) = 0$. Fix $\varepsilon > 0$ and choose δ so that $\varepsilon \exp[L] < \delta$. Then since

$$\left\{ \sup_{0 \leqslant t \leqslant 1} \left| \mathbf{x}^n(s) - \int_0^s m(\mathbf{x}^n(s))\, ds \right| \leqslant \varepsilon \right\}$$

$$\subset \left\{ \sup_{0 \leqslant t \leqslant 1} |\mathbf{x}^n(s) - x(s)| \leqslant \varepsilon e^L < \delta \right\}$$

We have

$$P\left(\sup_{0 \leqslant t \leqslant 1} |\mathbf{x}^n(t) - x(s)| > \delta \right) \leqslant P\left(\sup_{0 \leqslant t \leqslant 1} \left| \mathbf{x}^n(t) - \int_0^t m(\mathbf{x}^n(t))\, dt \right| \geqslant \varepsilon \right)$$

which gets arbitrarily small as n get large. The proof is completed. See Kurtz [1971] for a proof that the convergence is in fact exponentially fast.

4. $f^*(\theta) = \sup_x \{\theta x - f(x)\}$. Now the optimum choice for x is to solve the equation $\theta = f'(x)$ or $x_{\text{opt}} = f'^{-1}(\theta)$. Hence $f^*(\theta) = \theta f'^{-1}(\theta) - f(f'^{-1}(\theta))$.
Now

$$f^{**}(x) = \sup_\theta \{\theta x - f^*(\theta)\} = \sup_\theta \{\theta x - \theta f'^{-1}(\theta) + f(f'^{-1}(\theta))\}$$

Now we wish to solve for the optimum θ by considering the equation

$$\begin{aligned} x - f'^{-1}(\theta) - \theta[f'^{-1}(\theta)]' + f'(f'^{-1}(\theta))[f'^{-1}(\theta)]' \\ = x - f'^{-1}(\theta) - \theta[f'^{-1}(\theta)]' + \theta[f'^{-1}(\theta)]' \\ = x - f'^{-1}(\theta) = 0 \end{aligned}$$

This implies that $\theta_{\text{opt}} = f'(x)$. Therefore,

$$\begin{aligned} f^{**}(x) &= x\theta_{\text{opt}} - f^*(\theta_{\text{opt}}) = x\theta_{\text{opt}} - \theta_{\text{opt}} f'^{-1}(\theta_{\text{opt}}) + f(f'^{-1}(\theta_{\text{opt}})) \\ &= x\theta_{\text{opt}} - \theta_{\text{opt}} x + f(x) = f(x) \end{aligned}$$

5. Now from the previous solution we have that $f^*(\theta) = \theta f'^{-1}(\theta) - f(f'^{-1}(\theta))$. Therefore,

$$f^{*\prime}(\theta) = \theta[f'^{-1}(\theta)]' + f'^{-1}(\theta) - f'(f'^{-1}(\theta))[f'^{-1}(\theta)]'$$

Now in general for a function $g(\cdot)$ and its inverse $g^{-1}(\cdot)$, we have the relationship $[g^{-1}(y)]' = 1/[g'(g^{-1}(y))]$. For example, if $g(x) = \exp[x]$ and $g^{-1}(y) = \log[y]$, we see that $[g^{-1}(y)]' = [\log[y]]' = 1/y = 1/\exp[\log[y]] = 1/[g'(g^{-1}(y))]$. Therefore,

$$f^{*\prime}(\theta) = \frac{\theta}{f''(f'^{-1}(\theta))} + f'^{-1}(\theta) - \frac{f'(f'^{-1}(\theta))}{f''(f'^{-1}(\theta))}$$

Evaluate this at $\theta = f'(x)$ to obtain

$$f^{*\prime}(f'(x)) = \frac{f'(x)}{f''(x)} + x - \frac{f'(x)}{f''(x)} = x$$

Since $\log M_x(\cdot)$ and $I_x(\cdot)$ are convex duals of each other, we have thus demonstrated both derivative relationships.

6. One way to view the situation (in discrete time) is that at every position point x along the optimal trajectory, we twist the distribution $F_x(\cdot)$ of $V(x, \xi_1)$ to $dF_x^*(z) = \exp[\theta_x z]\, dF_x(z)/M_x(\theta_x)$. The twisted jump distribution $F_x^*(\cdot)$ will have mean value denoted by $m^*(x)$. We choose θ_x so that the twisted jump distribution's mean will track the optimal trajectory, i.e., θ_x is selected so that $f'_{\text{opt}}(t) = m^*(f_{\text{opt}}(t))$. Note that

$$m^*(x) = \int z\, dF_x^*(z) = \frac{d/d\theta_x \int \exp[\theta_x z]\, dF_x(z)}{M_x(\theta_x)} = [\log M_x(\theta_x)]'$$

Recall from the optimization derivation given in the text that the slope parameter v was given parametrically by the optimum trajectory as $f'_{\text{opt}}(t) = v = [\log M_x(I'_x(v))]'$. But from the above we also have $f'_{\text{opt}}(t) = m^*(f_{\text{opt}}(t)) = [\log M_x(\theta_x)]'$. Therefore, $\theta_x = I'_x(v) = -C'_S(x)$ is the correct interpretation of the twisting parameter at position x.

In the case of the discrete time M/M/1 queue (example 2), $\theta_x = \log[\bar{\lambda}/\lambda]$. Hence the twisted jump distribution is changed to

$$P(V^*(x, \xi_1) = 1) = \exp\left[\log\frac{\bar{\lambda}}{\lambda}\right] \frac{\lambda}{\exp[\log(\bar{\lambda}/\lambda)]\lambda + \exp[-\log(\bar{\lambda}/\lambda)]\bar{\lambda}} = \bar{\lambda}$$

and similarly,

$$P(V^*(x, \xi_1) = -1) = \lambda$$

Hence the jump rates have been switched, which corresponds in a queue to the interchange of customer arrival and service rates. For the continuous-time case, one can make limiting arguments from the discrete-time case of do a similar construction directly. One can also make use of the fact that an M/M/1 queue is reversible; i.e., when looking at the number of customers in the system it is impossible to tell if it is looking at the system forwardly or backwardly in time. In that case the service process becomes the arrival process, and vice versa.

7. For queue 1 to blow up to n, the arrival rate should be μ_1 and its service rate should be ν over a time interval like $n/(\mu_1 - \nu)$. Note that the arrival rate for queue 2 is μ_1 also for this period.

Suppose that $\mu_1 > \mu_2$. Here the arrival rate is greater than the service rate for queue 2, so the queue builds up linearly at rate $\mu_1 - \mu_2$. Hence

$$E\{\mathbf{x}_2 \,|\, \mathbf{x}_1 = n\} \approx (\mu_1 - \mu_2)\frac{n}{\mu_1 - \nu} \qquad \text{or} \qquad \lim_{n \to \infty} \frac{1}{n} E\{\mathbf{x}_2 \,|\, \mathbf{x}_1 = n\} = \frac{\mu_1 - \mu_2}{\mu_1 - \nu}$$

Suppose that $\mu_1 < \mu_2$. Here queue 2 is stable with arrival rate μ_1 and service rate μ_2. The steady state number in the queue is given by the classical formula $\rho/(1 - \rho)$, where $\rho \equiv \mu_1/\mu_2$ is called the loading factor. Therefore,

$$\lim_{n \to \infty} E\{\mathbf{x}_2 \,|\, \mathbf{x}_1 = n\} = \frac{\rho}{1 - \rho}$$

We note that the case $\mu_1 = \mu_2$ can be handled similarly (the expectation blows up like $\sqrt{n}$ in this case).

8. Queue 1 has a combined arrival rate of $\lambda_1 + \nu$. This will change to a rate of μ_1 as the queue size gets large. The question that must be answered is: How do the subsidiary streams change? Recall (from exercise 6) that the jump distribution get multiplied by $\exp[\theta_x z]$ and normalized. Hence the ratio of the jump rates should remain the same. Therefore, ν becomes $\nu\mu_1/(\nu + \lambda_1)$ and λ_1 becomes $\lambda_1\mu_1/(\nu + \lambda_1)$. [Note that the sum rate is $\nu\mu_1/(\nu + \lambda_1) + \lambda_1\mu_1/(\nu + \lambda_1) = \mu_1$ as needed.] The time interval needed for queue 1 to achieve level n is like $n/(\mu_1 - \lambda_1 - \nu)$. Hence queue 2 in this time interval is behaving like an M/M/1 queue with arrival rate $\lambda_2 + \nu\mu_1/(\nu + \lambda_1)$ and service rate μ_2. The solutions then follows as in exercise 7. Define

$$\rho \equiv \frac{\lambda_2 + \nu\mu_1/(\nu + \lambda_1)}{\mu_2}$$

If $\rho > 1$, we have

$$\lim_{n\to\infty} \frac{1}{n} E\{\mathbf{x}_2 \mid \mathbf{x}_1 = n\} = \frac{\rho\mu_2 - \mu_2}{\mu_1 - \lambda_1 - \nu}$$

If $\rho < 1$, we have

$$\lim_{n\to\infty} E\{\mathbf{x}_2 \mid \mathbf{x}_1 = n\} = \frac{\rho}{1-\rho}$$

CHAPTER V

1. $\begin{vmatrix} 1-\lambda & c \\ 1 & -\lambda \end{vmatrix} = c + \lambda - \lambda^2 = 0$

or

$$\lambda = \frac{-1 \pm \sqrt{1+4c}}{-2} \qquad \text{and} \qquad \lambda_{\max} = \frac{1+\sqrt{1+4c}}{2}$$

The associated eigenvector is $[\varphi_1 \quad \varphi_1]^{\mathrm{T}}$, where $\varphi_1 = 1$ and $\varphi_2 = 1/\lambda_{\max}$. The matrix adjoint will have the eigenvector $\varphi_1 = 1$, $\varphi_2 = c/\lambda_{\max}$.

2. The memoryless property of the exponential distribution may be formulated as follows: Let τ be an exponentially distributed random variable with mean value μ^{-1}. What is the distribution of τ given that τ is greater than $a \in \mathbb{R}^+$? Note that for $x \geqslant a$, we have

$$P(\tau \leqslant x \mid \tau \geqslant a) = \frac{P(a \leqslant \tau \leqslant x)}{P(\tau \geqslant a)} = \frac{\int_a^x \mu \exp[-\mu y]\, dy}{\int_a^\infty \mu \exp[-\mu y]\, dy}$$
$$= \frac{\exp[-\mu a] - \exp[-\mu x]}{\exp[-\mu a]} = 1 - \exp[-\mu(x-a)]$$

Hence $\tau - a$ given that $\tau \geqslant a$ is again an exponentially distributed random variable with mean value μ^{-1}. In other words, given that we are at state 1 and that we have waited a seconds for the random telegraph to flip states gives us no more predictive information other than that we are at state 1. We still have an exponential waiting time with the same parameter ahead of us.

3. Define $J(f) \equiv \Sigma_{i=1}^{k} f(i)v(i) - \log \Sigma_{i=1}^{k} \exp[f(i)]\alpha(i)$. We desire to find the vector (function actually) that maximizes $J(f)$. Note that $J(f(\cdot)+c) = J(f(\cdot))$ for any constant c. Hence we may constrain $f(\cdot)$ so that $\Sigma_{i=1}^{k} \exp[f(i)]\alpha(i) = 1$. Consider now the Lagrange multiplier problem

$$J(f) - \lambda \sum_{i=1}^{k} \exp[f(i)]\alpha(i)$$

Take the derivative with respect to $f(j)$ to find

$$v(j) - \frac{\exp[f(j)]\alpha(j)}{\Sigma_{i=1}^{k} \exp[f(i)]\alpha(i)} - \lambda \exp[f(j)]\alpha(j) = 0$$

where λ is chosen to meet the constraint. Hence we must have

$$f(j) = \log \frac{v(j)}{\alpha(j)} - \log[1 + \lambda]$$

Choosing λ to meet the constraint gives $\lambda = 0$ and hence

$$f(j) = \log \frac{v(j)}{\alpha(j)}$$

is the optimum choice. Hence

$$J\left(\log \frac{v(\cdot)}{\alpha(\cdot)}\right) = \sum_{i=1}^{k} v(i) \log \frac{v(i)}{\alpha(i)} - \log \sum_{i=1}^{k} v(i) = \sum_{i=1}^{k} v(i) \log \frac{v(i)}{\alpha(i)}$$

which is the rate function from Sanov's theorem.

4. This exercise follows Kim and David [1979].

$$P_n \equiv P\left(\frac{1}{n} \sum_{i=1}^{n} a(\mathbf{x}_i, \mathbf{x}_{i-1}) > 0\right)$$

Now we know that

$$\lim_{n \to \infty} \frac{1}{n} \log P_n = \inf_{\theta} \log \lambda(\theta) \equiv \log \lambda(\theta_0)$$

where $\theta_0 \in (0, \infty)$ and where $\lambda(\theta)$ is the largest eigenvalue of the operator:

$$\sum_x \exp[\theta a(x, y)]p(x \mid y)f(x) \equiv T_\theta(f)(y)$$

Denote the left (right) eigenvector associated with this eigenvalue as $r_\theta^*(x)$ $(r_\theta(x))$. First note that for any function $s: \mathbb{R} \to \mathbb{R}$, we have

$$\sum_x \sum_y s(x)f(x, y) = \sum_x \sum_y s(y)f(x, y)$$

for all $f(\cdot, \cdot) \in F$. Then for positive θ, $f(\cdot, \cdot) \in F_a$

$$I(f, p) \geqslant \sum_x \sum_y \log \frac{f(x \mid y)}{p(x \mid y)} f(x, y) - \sum_x \sum_y \log \frac{r_\theta(x)}{r_\theta(y)} f(x, y) - \theta \sum_x \sum_y a(x, y) f(x, y)$$

$$= \sum_x \sum_y -\log\left[p(x \mid y) r_\theta(x) \frac{\exp[\theta a(x, y)]}{f(x \mid y) r_\theta(y)}\right] f(x, y)$$

$$\geqslant -\log \sum_x \sum_y p(x \mid y) f(y) r_\theta(x) \frac{\exp[\theta a(x, y)]}{r_\theta(y)}$$

$$= -\log \sum_y \lambda(\theta) f(y) \frac{r_\theta(y)}{r_\theta(y)} = -\log \lambda(\theta)$$

Now take

$$f_0(y) \equiv r_{\theta_0}(y) r_{\theta_0}^*(y)$$

and

$$f_0(x \mid y) \equiv \exp[\theta_0 a(x, y)] p(x \mid y) \frac{r_{\theta_0}(x)}{\lambda(\theta_0) r_{\theta_0}(y)}$$

Define

$$f_0(x, y) \equiv f_0(x \mid y) f_0(y)$$

It is easy to check (by differentiation of a moment generating function) that

$$\sum_x \sum_y a(x, y) f_0(x, y) = 0$$

In fact, it is easy to check that $f_0(\cdot, \cdot) \in F_a$ and $I(f_0, p) = -\log(\theta_0)$. Hence we have that

$$\inf_{f \in F_a} I(f, p) = I(f_0, p) = -\log \lambda(\theta_0) = -\inf_{\theta \in (0,\infty)} \log \lambda(\theta)$$

Of course what we have shown here is that $I(f, p)$ for $f \in F$ is the Sanov theorem rate function analog for the bivariate empirical distributions of Markov chains. We could have proceeded in this fashion to derive the rate function for the one-dimensional distributions considered in section B of this chapter. Why is the rate function constrained to be only for bivariate distributions in F? Let us consider how bivariate empirical distributions would be obtained. Consider a vector $x = (x_1 \quad x_2 \quad \cdots \quad x_n)$. Let $n_{ij}(x) \equiv$ number of transitions from i to j in x with the cyclic convention that x_1 follows x_n. It is easy to verify that $\Sigma_{i,j} n_{ij}/n = 1$ and $\Sigma_j n_{ij} = \Sigma_j n_{ji}$. n_{ij}/n is obviously then an element of the class F.

5. We may write the transition probabilities for this chain as $p(x_n \mid x_{n-1}) = \delta(x_n - x_{n-1})(1 - p) + v(x_n)p$. We are interested then in the eigenvector/eigenvalue problem for the operator

$$\begin{aligned}\sum_x \exp[\theta f(x)]p(x \mid y)\psi(x) &= \exp[\theta f(y)]\psi(y)(1 - p) + p\sum_x \exp[\theta f(x)]v(x)\psi(x)\\ &= \lambda(\theta)\psi(y)\end{aligned}$$

Now let us scale the eigenvector $\psi(\cdot)$ so that

$$p\sum_x \exp[\theta f(x)v(x)\psi(x) = 1$$

In the eigenvector problem this then implies that

$$\exp[\theta f(y)]\psi(y)(1 - p) + 1 = \lambda(\theta)\psi(y)$$

or

$$\psi(y) = \frac{1}{\lambda(\theta) - \exp[\theta f(y)](1 - p)}$$

with condition implied by the scale factor that

$$p\sum_x \frac{\exp[\theta f(x)]v(x)}{\lambda(\theta) - \exp[\theta f(x)](1 - p)} = 1$$

This then implies that $\lambda(\theta)$ is found as being the largest solution to the equation above. Depending on the function $f(\cdot)$ of interest, a variety of techniques can be used to solve the foregoing equation numerically (Newton–Raphson, steepest descent, etc.).

6. The generator may be computed by taking the limit as Δ approaches zero of

$$\frac{g(x)\exp[-\Delta\mu] + [\Sigma_y g(y)v(y)](1 - \exp[-\Delta\mu]) - g(x)}{\Delta}$$

$$\approx \frac{g(x)(1 - \Delta\mu) + \Sigma_y g(y)v(y)\Delta\mu - g(x)}{\Delta}$$

$$= \frac{\Sigma_y g(y)v(y)\Delta\mu - g(x)\Delta\mu}{\Delta} = \mu\left[\sum_y g(y)v(y) - g(x)\right]$$

$$= \mu\left[\sum_y [g(y) - g(x)]v(y)\right]$$

The eigenvector problem is thus

$$\theta f(x)\psi(x) + \mu\left[\sum_y [\psi(y) - \psi(x)]v(y)\right] = \lambda(\theta)\psi(x)$$

This is solved as in exercise 5 to give

$$\psi(x) = \frac{\mu}{\lambda(\theta) + \mu - \theta f(x)}$$

and where $\lambda(\theta)$ is chosen to be the maximum solution to

$$\sum_x \frac{\mu}{\lambda(\theta) + \mu - \theta f(x)} v(x) = 1$$

7. Using similar arguments as in the preceding problem, one can show that the infinitesmal generator is given by

$$\mu\left[\sum_y (g(y) - g(x))v(y, x)\right]$$

Even more generally, one can have the rate be state dependent and have the generator be given by

$$\mu(x)\left[\sum_{y}(g(y)-g(x))\nu(y,x)\right]$$

Closed form solutions to the eigenvector/eigenvalue problem do not appear to be available for these more general cases.

CHAPTER VI

1. Now it is desired that the type I error remain fixed, i.e.,

$$P_0\left(\frac{1}{n}\sum_{i=1}^{n}\log\frac{p(\mathbf{x}_i\mid\mathbf{x}_{i-1})}{q(\mathbf{x}_i\mid\mathbf{x}_{i-1})}>u_n\right)=\alpha$$

where $0<\alpha<1$. Now (under P_0)

$$\lim_{n\to\infty}\frac{1}{n}\sum_{i=1}^{n}\log\frac{p(\mathbf{x}_i\mid\mathbf{x}_{i-1})}{q(\mathbf{x}_i\mid\mathbf{x}_{i-1})}=\sum_{x}\sum_{y}p(x,y)\log\frac{p(x\mid y)}{q(x\mid y)}\equiv K(p,q)\equiv K$$

Therefore,

$$\begin{aligned}&\lim_{n\to\infty}\frac{1}{n}\log P_1\left(\frac{1}{n}\sum_{i=1}^{n}\log\frac{p(\mathbf{x}_i\mid\mathbf{x}_{i-1})}{q(\mathbf{x}_i\mid\mathbf{x}_{i-1})}>u_n\right)\\&\quad=\lim_{n\to\infty}\frac{1}{n}\log P_1\left(\frac{1}{n}\sum_{i=1}^{n}\log\frac{p(\mathbf{x}_i\mid\mathbf{x}_{i-1})}{q(\mathbf{x}_i\mid\mathbf{x}_{i-1})}>K\right)\\&\quad=\lim_{n\to\infty}\frac{1}{n}\log P_1\left(\frac{1}{n}\sum_{i=1}^{n}\left[\log\frac{p(\mathbf{x}_i\mid\mathbf{x}_{i-1})}{q(\mathbf{x}_i\mid\mathbf{x}_{i-1})}-K\right]>0\right)\\&\quad=\inf_{\theta}\log\lambda(\theta)\equiv\log\lambda(\theta_0)\end{aligned}$$

The operator of interest to go with this eigenvalue is

$$\begin{aligned}T_{\theta,q}(g)(y)&\equiv\sum_{x}\exp\left[\theta\left[\log\frac{p(x\mid y)}{q(x\mid y)}-K\right]\right]q(x\mid y)g(x)\\&=\exp[-\theta K]\sum_{x}p(x\mid y)^{\theta}q(x\mid y)^{1-\theta}g(x)\end{aligned}$$

θ_0 is such that (exercise 4 of chapter 5 implies $\Sigma_{x,y}\, a(x,y) f_0(x,y) = 0$)

$$\sum_x \sum_y p(x \mid y)^{\theta_0} q(x \mid y)^{1-\theta_0} r_{\theta_0}(x) r^*_{\theta_0}(y) \frac{\log[p(x \mid y)/q(x \mid y)] - K}{\lambda(\theta_0)} = 0$$

Claim: $\theta_0 = 1$ is the only solution.

Proof: Note that $T_{1,q}(g)(y) = \exp[-K]\, \Sigma_x\, p(x \mid y) g(x)$. This operator has eigenvalue $\lambda(1)$ given by $\exp[-K]$ and eigenvector (r_1), consisting of all 1's. The adjoint operator $T^*_{1,q}(g)(y) = \exp[-K]\, \Sigma_x\, p(y \mid x) g(x)$ has eigenvalue $\lambda(1)$, $\exp[-K]$ and eigenvector (r^*_1), $p(x)$, the stationary distribution of the chain with transition probabilities $p(x \mid y)$.

Hence

$$\sum_y p(x \mid y) r^*_1(y) \log \frac{p(x \mid y)}{q(x \mid y)} = \sum_y p(x \mid y) p(y) \log \frac{p(x \mid y)}{q(x \mid y)} = K$$

Hence

$$\sum_x \sum_y p(x \mid y)^{\theta_0} q(x \mid y)^{1-\theta_0} r_{\theta_0}(x) r^*_{\theta_0}(y) \frac{\log[p(x \mid y)/q(x \mid y)] - K}{\lambda(\theta_0)} \bigg|_{\theta_0 = 1}$$

$$= \sum_x \sum_y p(x \mid y) r_1(x) r^*_1(y) \frac{\log[p(x \mid y)/q(x \mid y)] - K}{\lambda(\theta_1)}$$

$$= \sum_x \sum_y p(x \mid y) p(y) \frac{\log[p(x \mid y)/q(x \mid y)] - K}{\lambda(\theta_1)} = 0$$

Therefore, the rate of the type II error to zero is merely

$$\lim_{n \to \infty} \frac{1}{n} \log P_1 \left(\frac{1}{n} \sum_{i=1}^{n} \log \frac{p(\mathbf{x}_i \mid \mathbf{x}_{i-1})}{q(\mathbf{x}_i \mid \mathbf{x}_{i-1})} > u_n \right) = \log \lambda(1) = -K$$

2. The optimal nonlinearity is given by $\log[p_1(y \mid x)/p_0(y \mid x) \equiv g(x, y)$. Hence $g(0,0) = g(1,1) = \log[3/7]$ and $g(1,0) = g(0,1) = \log[7/3]$. We now need to find the largest eigenvalue of the operator

$$\sum_x p_0(x \mid y)^{-\theta} p_1(x \mid y)^{1+\theta} g(x)$$

In matrix form we need to find the largest eigenvalue of the matrix

$$A \equiv \begin{bmatrix} .3[3/7]^{\theta} & .7[7/3]^{\theta} \\ .7[7/3]^{\theta} & .3[3/7]^{\theta} \end{bmatrix}$$

Solving the resulting quadratic form for $|\lambda I - A| = 0$ obtains

$$\lambda(\theta) = .3[3/7]^{\theta} + .7[7/3]^{\theta}$$

The minimizing θ, θ_0, is obtained by simple differentiation and is $\theta_0 = -.5$. The rate to zero of the probability of error of the optimal detector is thus

$$\log \lambda(\theta_0) = \log\left[.3\sqrt{\frac{7}{3}} + .7\sqrt{\frac{3}{7}}\right]$$

3. Suppose that two solutions $g(\cdot)$ and $g'(\cdot)$ exist. Consider the operators

$$J_i^{\varepsilon}(\cdot)(y) \equiv \sum_x \exp[\theta_i(g(x) + \varepsilon(g'(x) - g(x)))](\cdot)p_i(x \mid y)$$

where $p_i(\cdot \mid \cdot)$ is the transition probability for the ith chain ($i = 0, 1$). Now as ε varies, the largest eigenvalue of each of these operators graphs a convex function. Now at $\varepsilon = 0$ and $\varepsilon = 1$ these graphs must cross each other. The slopes of the graphs at the crossing points ($D_{m,0}$ and $D_{m,1}$) must be of opposite sign. However, it is impossible for two convex functions to have this behavior at two separate points. Hence we have a contradiction and the solution for each θ must be unique.

4. It is immediate from the definitions of the operators that

$$\lambda_{f+c,i} = \exp[\theta_i c]\lambda_{f,i}$$

Since $\theta_0 - \theta_1 = 1$, it is clear that

$$c_f \equiv \log\frac{\lambda_{f,1}}{\lambda_{f,0}}$$

Now we make use of our perturbation formulas. When $f(\cdot)$ is perturbed to $f(\cdot) + \varepsilon m(\cdot)$,

$$\begin{aligned} c_{f+\varepsilon m} &= \log \lambda_{f+\varepsilon m,1} - \log \lambda_{f+\varepsilon m,0} \\ &\approx \log\left[\lambda_{f,1} + \varepsilon\theta_1\lambda_{f,1}\sum_x m(x) r_{f,1}(x) r_{f,1}^*(x)\right] \\ &\qquad - \log\left[\lambda_{f,0} + \varepsilon\theta_0\lambda_{f,0}\sum_x m(x) r_{f,0}(x) r_{f,0}^*(x)\right] \end{aligned}$$

$$\approx c_f + \varepsilon\theta_1 \sum_x m(x) r_{f,1}(x) r^*_{f,1}(x) - \varepsilon\theta_0 \sum_x m(x) r_{f,0}(x) r^*_{f,0}(x)$$

Therefore,

$$\lambda_{f+\varepsilon m} = \lambda_{f+\varepsilon m + c_{f+\varepsilon m},1}$$

$$= \lambda_{f+\varepsilon m,1} \exp[\theta_1 c_{f+\varepsilon m}]$$

$$\approx \left[\lambda_{f,1} + \varepsilon\theta_1 \lambda_{f,1} \sum_x m(x) r_{f,1}(x) r^*_{f,1}(x)\right]$$

$$\cdot \exp\left[\theta_1 \left[c_f + \varepsilon\theta_1 \sum_x m(x) r_{f,1}(x) r^*_{f,1}(x) - \varepsilon\theta_0 \sum_x m(x) r_{f,0}(x) r^*_{f,0}(x)\right]\right]$$

$$\approx \left[\lambda_{f,1} + \varepsilon\theta_1 \lambda_{f,1} \sum_x m(x) r_{f,1}(x) r^*_{f,1}(x)\right] \exp[\theta_1 c_f]$$

$$\cdot \left[1 - \varepsilon\theta_0\theta_1 \sum_x m(x) r_{f,0}(x) r^*_{f,0}(x) + \varepsilon\theta_1^2 \sum_x m(x) r_{f,1}(x) r^*_{f,1}(x)\right]$$

$$\approx \left[\lambda_f + \varepsilon\theta_1 \lambda_f \sum_x m(x) r_{f,1}(x) r^*_{f,1}(x)\right]$$

$$\cdot \left[1 - \varepsilon\theta_0\theta_1 \sum_x m(x) r_{f,0}(x) r^*_{f,0}(x) + \varepsilon\theta_1^2 \sum_x m(x) r_{f,1}(x) r^*_{f,1}(x)\right]$$

$$\approx \lambda_f + \varepsilon\theta_1 \lambda_f \sum_x m(x) r_{f,1}(x) r^*_{f,1}(x) - \varepsilon\theta_0\theta_1 \lambda_f \sum_x m(x) r_{f,0}(x) r^*_{f,0}(x)$$

$$+ \varepsilon\theta_1^2 \lambda_f \sum_x m(x) r_{f,1}(x) r^*_{f,1}(x)$$

$$= \lambda_f + \varepsilon\theta_1 \lambda_f \left[-\theta_0 \sum_x m(x) r_{f,0}(x) r^*_{f,0}(x) + (1+\theta_1) \sum_x m(x) r_{f,1}(x) r^*_{f,1}(x)\right]$$

$$= \lambda_f + \varepsilon\theta_1\theta_0 \lambda_f \left[\sum_x m(x)[r_{f,1}(x) r^*_{f,1}(x) - r_{f,0}(x) r^*_{f,0}(x)]\right]$$

$$= \lambda_f + \lambda_f \theta_1 \theta_0 \varepsilon \langle m, r_{f,1} r^*_{f,1} - r_{f,0} r^*_{f,0} \rangle$$

The above is an explicit expression for the derivative taken in the $m(\cdot)$ direction. The critical point is when the derivative is zero for all directions $m(\cdot)$. From the expression above that is obviously given by the condition $[r_{f,1}(x) r^*_{f,1}(x) - r_{f,0}(x) r^*_{f,0}(x)] = 0$, which of course is equation (28) given in the text. The steepest descent direction from f to the critical point is simply to choose $m(x) = [r_{f,1}(x) r^*_{f,1}(x) - r_{f,0}(x) r^*_{f,0}(x)]$ and normalize it so that it has unit length.

Now suppose that for some θ_0, θ_1 pair ($R \equiv \theta_1/\theta_0$) we have found the critical point $f_R(\cdot)$ such that $r_{f_R,0}(x)r^*_{f_R,0}(x) = r_{f_R,1}(x)r^*_{f_R,1}(x)$. What direction should we move θ_0 in order to bring about a lowering of the eigenvalue λ_{f_R}? Consider a perturbation of the θ terms of the operators of interest:

$$T_{\theta_i}(g)(y) \equiv \sum_x \exp[(\theta_i + \Delta\theta) f_R(x)] p_i(x \mid y) g(x)$$

Note that $(\theta_0 + \Delta\theta) - (\theta_1 + \Delta\theta) = \theta_0 - \theta_1 = 1$. The perturbation series for the principal eigenvalue then appears as

$$\lambda_{f_R}(\theta_i + \Delta\theta) = \lambda_{f_R} + \lambda_{f_R}\Delta\theta \sum_x f_R(x) r_{f_R,i}(x) r^*_{f_R,i}(x) + O(\Delta\theta^2)$$

which is the same for $i = 0$ or 1. Hence choose the sign of $\Delta\theta$ so that $\lambda_{f_R}(\theta_i + \Delta\theta)$ is getting smaller, or

$$\operatorname{sgn}[\Delta\theta] = -\operatorname{sgn}\left[\sum_x f_R(x) r_{f_R,i}(x) r^*_{f_R,i}(x)\right]$$

Furthermore, we immediately have a condition for a critical point when

$$\sum_x f_R(x) r_{f_R,i}(x) r^*_{f_R,i}(x) = 0$$

We should finally note (by convexity arguments) that one can show that only one critical point may exist. Hence this last expression characterizes the optimal nonlinearity.

5. (a) Given an $f(\cdot)$, it must have a false-alarm rate α, which implies that for every n,

$$P_0\left(\frac{1}{n}\sum_{i=1}^{n} f(\mathbf{x}_i) > 0\right) = \alpha$$

This immediately implies that the expectation of $f(\cdot)$ under the stationary distribution of the hypothesis H_0 chain must equal zero or

$$\sum_x p_0(x) f(x) = 0$$

We desire to maximize the rate to zero of the type II error or maximize I_1, where

$$\lim_{n\to\infty}\frac{1}{n}\log P_1\left(\frac{1}{n}\sum_{i=1}^{n} f(\mathbf{x}_i) < 0\right) = -I_1 = \inf_\theta \log \lambda(f,\theta) = \log \lambda(f,\theta_f)$$

Suppose that f_{opt} exists and $\theta_{\text{opt}}(\equiv\theta_{f_{\text{opt}}})$ is the associated θ value. $\theta_{\text{opt}}\in(-\infty,0]$. f_{opt} must satisfy two criteria:

1. $\Sigma_x f_{\text{opt}}(x)p_0(x) = 0$.
2. $\lambda(f_{\text{opt}} + \varepsilon m, \theta_{\text{opt}}) > \lambda(f_{\text{opt}}, \theta_{\text{opt}})$ for all perturbations m such that $\Sigma_x m(x)p_0(x) = 0$.

Consider the perturbation series for the perturbed eigenvalue

$$\lambda(f_{\text{opt}}+\varepsilon m,\theta_{\text{opt}})=\lambda(f_{\text{opt}},\theta_{\text{opt}})$$
$$+\varepsilon\theta_{\text{opt}}\lambda(f_{\text{opt}},\theta_{\text{opt}})\sum_x m(x)r_{f_{\text{opt}},1}(x)r^*_{f_{\text{opt}},1}(x)+O(\varepsilon^2)$$

Condition 2 implies that the term multiplying ε had better be zero for all m satisfying $\Sigma_x m(x)p_0(x)=0$; otherwise, m or $-m$ will cause a contradiction to it. The only possible solution is thus

$$p_0(x) = r_{f_{\text{opt}},1}(x)r^*_{f_{\text{opt}},1}(x) \qquad (*)$$

With (*) replacing (28) as our new condition, the presentation in the text is unchanged.

(b) In this setting the perturbation analysis is even simpler than in the discrete-time case. $\lambda(f,\theta)$ is the largest eigenvalue of the operator $\theta f(\cdot)+G$, where G is the generator of the Markov process. The perturbed operator is thus $\theta[f + \varepsilon m] + G$. The perturbation series has exactly the same form as in the discrete-time case and equation (28) is unchanged.

(c) Define the vector valued random process $\mathbf{Y}_0=(\mathbf{x}_0\quad \mathbf{x}_1\quad \cdots\quad \mathbf{x}_{m-1})$, $\mathbf{Y}_1=(\mathbf{x}_m\quad \mathbf{x}_{m+1}\quad \cdots\quad \mathbf{x}_{2m-1}),\ldots$. Now it is easy to see that $\{\mathbf{Y}_i\}$ is a finite state space Markov chain with transition function $p_i(y\,|\,x)$. Hence our theory will go through unchanged when we define the relevant operators to be

$$T_{\theta,i}(g)(x) \equiv \sum_{y\in S^m}\exp\left[\frac{1}{m}\sum_{y_i} f(y_i)\right]g(x)p_i(y\,|\,x) \qquad i = \{0,1\}$$

where $x \in S^m$; $g(\cdot)$ is a mapping from S^m to $\mathbb{R}$; and $f(\cdot)$ is the memoryless nonlinearity, which is a mapping from S to $\mathbb{R}$.

6. A computer program was written (by A. Hillery of the University of Wisconsin) to do the search for the optimum memoryless nonlinearity:

```
Program main
{the optimal memoryless nonlinearity f[·] is computed}
P0[x|y], P1[x|y], B; n×n array of reals
leftev0, leftev1, rightev0, rightev1, f; n×1 vector of reals
x, y {1≤x, y≤n}; integers
θ, Δθ, threshold, c; real
Initialize
          Read P0[x|y]
          Read P1[x|y]
          leftev0[x]:= leftev1[x] = 1 (for 1≤x≤n)
          rightev0[x]:= rightev1[x] = 1 (for 1≤x≤n)
          f[x]:= 1 (for 1≤x≤n)
          threshold:= 10E—8
          Δθ:= 10E—4
          θ:= .5
Repeat
          Repeat
                    B[x,y]:= exp[θ·f[x]]·P0[x|y] (for 1≤x,y≤n)
                    eigenvalue0:= eigen(rightev0,B)
                    transpose(B)
                    eigenvalue0:= eigen(leftev0,B)
                    normalize(rightev0,leftev0)
                    B[x,y]:= exp[(θ−1)·f[x]]·P1[x|y] (for 1≤x,y≤n)
                    eigenvalue1:= eigen(rightev1,B)
                    transpose(B)
                    eigenvalue1:= eigen(leftev1,B)
                    normalize(rightev1,leftev1)
                    grad f[x]:= rightev1[x]·leftev1[x]
                                   − rightev0[x]·leftev0[x] (for 1≤x≤n)
                    c:= log[eigenvalue0/eigenvalue1]
                    f[x]:= f[x] + c − grad f[x] (for 1≤x≤n)
          until ||grad f|| ≤ threshold
          θ:= θ − Δθ·f[x]·rightev1[x]·leftev1[x] (for 1≤x≤n)
until ||f·rightev1·leftev1|| ≤ threshold
end {main}

Procedure Normalize[v1,v2]
{normalizes two vectors so that their inner product is one}
          t:= 0
          t:= t + v1[x]·v2[x] (for 1≤x≤n)
          v1[x]:= v1[x]/√t
          v2[x]:= v2[x]/√t
end {normalize}
```

```
Function Eigen[ev,B]
{returns max eigenvalue of B in eigen}
{using power method, returns associated eigenvector in ev}
          A; n × n matrix of reals
          tv, lastv; n × 1 vector of reals
          A[x,y]:= B[x,y] (for 1⩽x,y⩽n)
          A[x,x]:= A[x,x] + 1 (for 1⩽x⩽n)
          Initialize
                    tv[x]:= 0 (for 1⩽x⩽n)
          Repeat
                    lastv[x]:= tv[x] (for 1⩽x⩽n)
                    (for 1⩽x⩽n)
                                 tv[x]:= A[x,y]·ev[y] (for 1⩽y⩽n)
                    end for x
                    ev[x]:= tv[x]/||tv||
          until ||lastv[x] − tv[x]|| ⩽ threshold
          eigen:= tv[1]/ev[1] − 1
end {eigen}
```

Note that in the subroutine eigenvalue, an identity matrix was added to the matrix representing the operators of interest. For the power method to be used to find the principal eigenvalue and eigenvector of a positive matrix that has zero entries, one should first add to the matrix (any) positive constant c times an identity matrix and power up the result. The resulting eigenvector is the principal eigenvector of the original matrix. The resulting eigenvalue should then have c subtracted back away. The result will be the principal real eigenvalue of the original matrix. The problem is that one may have in general complex eigenvalues of the same magnitude as the principal real eigenvalue when the matrix has zero entries. The resulting shifted matrix has a strictly largest positive eigenvalue and the power method will work fine for it.

Running this program on our problem gives the optimal nonlinearity is $[f[0] \quad f[1] \quad \cdots \quad f[5]]=[0 \quad -1 \quad 1 \quad -1 \quad 1]$. The value found for $\lambda(\theta_0)=.92563$ and $\theta_0=.4623$. The eigenvalue for the optimal detector (i.e., with memory) is the same!! Can you explain why?

7. The equation to be solved is

$$\gamma=\frac{1}{2\pi}\int_0^{2\pi}\frac{P(\omega)+(1-2\theta_0)}{(1-2\theta_0)^2}\,d\omega=\frac{P}{(1-2\theta_0)^2}+\frac{1}{1-2\theta_0}$$

or

$$P+(1-2\theta_0)-\gamma(1-2\theta_0)^2=0$$

This gives

$$\theta_0 = .5\left(1 - \frac{E}{2\gamma}\right)$$

where

$$E \equiv [1 + \sqrt{1 + 4\gamma P}]$$

Now substituting this value into the expression for the rate function gives

$$I(\gamma) = \frac{P}{2} + \frac{2\gamma - E}{4} - \frac{P\gamma}{E} + \frac{1}{2}\log\frac{E}{2\gamma}$$

8. The likelihood ratio test is

$$\log\frac{p_1(\mathbf{x})}{p_0(\mathbf{x})} = .5\log\frac{|\Sigma_0|}{|\Sigma_1|} - .5\mathbf{x}^{\mathrm{T}}[\Sigma_1^{-1} - \Sigma_0^{-1}]\mathbf{x} \underset{H_0}{\overset{H_1}{\gtrless}} 0$$

or, equivalently,

$$\mathbf{x}^{\mathrm{T}}[\Sigma_1^{-1} - \Sigma_0^{-1}]\mathbf{x} - \log\frac{|\Sigma_0|}{|\Sigma_1|} \underset{H_1}{\overset{H_0}{\gtrless}} 0$$

We have denoted the covariance matrix under the ith hypothesis as Σ_i. We will denote the power spectrum under the ith hypothesis as $S_i(\omega)$. Recall that for the optimal detector, we should have that the error rate for the type I error should equal the type II rate. Therefore, we will consider the situation for the type II error (i.e., when H_1 is true). First note that

$$\frac{1}{n}\log\left[\exp\left[-\theta\log\frac{|\Sigma_0|}{|\Sigma_1|}\right]\right] = \frac{1}{n}\log\prod_{i=1}^{n}\left[\frac{\lambda_{1i}}{\lambda_{0i}}\right]^{\theta}$$

$$= \frac{1}{n}\sum_{i=1}^{n}\theta\log\frac{\lambda_{1i}}{\lambda_{0i}} \xrightarrow[n\to\infty]{} \frac{\theta}{2\pi}\int_0^{2\pi}\log\frac{S_1(\omega)}{S_0(\omega)}\,d\omega$$

Now for large n, $\Sigma_1^{-1} \approx \Psi\Lambda_1^{-1}\Psi^{\mathrm{T}}$ and $\Sigma_0 \approx \Psi\Lambda_0\Psi^{\mathrm{T}}$, where Ψ is the DFT matrix and Λ_i is the diagonal eigenvalue matrix of Σ_i. Consequently, if $\mathbf{y} \equiv \Psi^{\mathrm{T}}\mathbf{x}$, we then have

$$\mathbf{x}^{\mathrm{T}}[\Sigma_1^{-1} - \Sigma_0^{-1}]\mathbf{x} = \mathbf{y}^{\mathrm{T}}[\Psi^{\mathrm{T}}\Sigma_1^{-1}\Psi - \Psi^{\mathrm{T}}\Sigma_0^{-1}\Psi]\mathbf{y}$$

$$\approx \mathbf{y}^{\mathrm{T}}\Lambda_1^{-1}\mathbf{y} - \mathbf{y}^{\mathrm{T}}\Lambda_0\mathbf{y} = \sum_{i=1}^{n}\left[\frac{\lambda_{0i} - \lambda_{1i}}{\lambda_{0i}\lambda_{1i}}\right]\mathbf{y}_i^2$$

Now under hypothesis H_1, the $\{\mathbf{y}_i\}$ are iid Gaussian, zero mean, variance $\{\lambda_{1i}\}$ random variables. Hence

$$\lim_{n\to\infty}\frac{1}{n}\log E_1\{\exp[\theta\mathbf{x}^{\mathrm{T}}[\Sigma_1^{-1} - \Sigma_0^{-1}]\mathbf{x}]\}$$

$$= \lim_{n\to\infty}\frac{1}{n}\log E_1\left\{\exp\left[\theta\sum_{i=1}^{n}\left[\frac{\lambda_{0i} - \lambda_{1i}}{\lambda_{0i}\lambda_{1i}}\right]\mathbf{y}_i^2\right]\right\}$$

$$= \lim_{n\to\infty}\frac{1}{n}\log\prod_{i=1}^{n}\left[1 - 2\theta\left[1 - \frac{\lambda_{1i}}{\lambda_{0i}}\right]\right]^{-.5}$$

$$= \lim_{n\to\infty} -\frac{1}{2n}\sum_{i=1}^{n}\log\left[1 - 2\theta\left[1 - \frac{\lambda_{1i}}{\lambda_{0i}}\right]\right]$$

$$= -\frac{1}{4\pi}\int_0^{2\pi}\log\left[1 - 2\theta\left(1 - \frac{S_1(\omega)}{S_0(\omega)}\right)\right]d\omega$$

Hence

$$\varphi(\theta) = \lim_{n\to\infty}\varphi_n(\theta)$$

$$= \lim_{n\to\infty}\frac{1}{n}\log E_1\left\{\exp\left[\theta\left[\mathbf{x}^{\mathrm{T}}[\Sigma_1^{-1} - \Sigma_0^{-1}]\mathbf{x} - \log\frac{|\Sigma_0|}{|\Sigma_1|}\right]\right]\right\}$$

$$= \frac{\theta}{2\pi}\int_0^{2\pi}\log\frac{S_1(\omega)}{S_0(\omega)}d\omega - \frac{1}{4\pi}\int_0^{2\pi}\log\left[1 - 2\theta\left(1 - \frac{S_1(\omega)}{S_0(\omega)}\right)\right]d\omega$$

The error rate is then

$$-\sup_{\theta}\{\theta\cdot 0 - \varphi(\theta)\} = \inf_{\theta}\varphi(\theta) \equiv \varphi(\theta_0)$$

By simple differentiation, one sees that the equation that the optimal θ, θ_0 must satisfy is

$$-\int_0^{2\pi}\log[R(\omega)]\,d\omega = \int_0^{2\pi}\frac{1 - R(\omega)}{1 - 2\theta_0(1 - R(\omega))}d\omega$$

where $R(\omega) \equiv S_1(\omega)/S_0(\omega)$.

9. We use the calculus of variations to evaluate

$$\inf_{f \in S} \int_0^{\lambda_0} \sup_\theta \left\{ \theta f'(t) - \log \int p(\xi)^\theta q(\xi)^{1-\theta} \, d\xi \right\} dt$$

This is a little different than some of the minimizations considered in chapter IV in that the endpoint of the optimal trajectory $f_{\mathrm{opt}}(\lambda_0)$ is constrained to be $g(\lambda_0)$. Euler's equation in this setting is $F - y' F_{y'} = c \equiv$ constant, where we are minimizing the functional $\int F(y, y')\, dx$ over the set of trajectories $y(x)$ such that $y(0) = 0$ and $y(\lambda_0) = g(\lambda_0)$. Writing out Euler's equation gives

$$I_x(f'(t)) - f'(t) I_x'(f'(t)) = c$$

where

$$I_x(u) \equiv \sup_\theta \left\{ \theta u - \log \int p(\xi)^\theta q(\xi)^{1-\theta} \, d\xi \right\}$$

Now since $I_x(u)$ doesn't depend on x and is a strictly convex function of u, the only possible solution to Euler's equation is if $f'(t) =$ constant. Hence $f_{\mathrm{opt}}'(t) = g(\lambda_0)/\lambda_0$.

10. Recall that we have replaced λ by $(1 - \lambda)$ in our original test statistic for convenience in calculating the type I error rate. Since there was no change, we could be a little careless in the ultimate meaning of this. Now we must interpret the revised test statistic in a different way. $\{\mathbf{x}_1, \mathbf{x}_2, \ldots, \mathbf{x}_{[n\lambda]}\}$ are iid distributed as $p(x)$ and $\{\mathbf{x}_{[n\lambda+1]}, \ldots, \mathbf{x}_n\}$ are iid distributed as $q(x)$. The type II error rate will be a function of $0 < \lambda < 1$, the point at which the change occurs. If the mean drift of the test before and after the change is such that the mean drift trajectory (which has zero cost) does not cross $g(t)$ for any $t \in [0, 1]$, the type II error rate must be zero. Obviously, if λ is close enough to zero, this must happen since there is simply not enough time for the change to be noticed. The mean drift before the change is $s^2/2$ and after the change is $-s^2/2$. The critical value of λ may then be found by finding the value of λ, call it λ_c, which solves the equation

$$\frac{s^2}{2} \lambda_c = g(\lambda_c) = \frac{-s^2}{2} \lambda_c + \sqrt{2\alpha\lambda_c}\, s$$

or

$$\lambda_c = \frac{2\alpha}{s^2}$$

For values of $\lambda > \lambda_c$ we expect exponential convergence of the type II error and desire to find its rate.

For $\lambda > \lambda_c$ we desire to consider paths starting from zero that do not cross the boundary function $g(\cdot)$ at any point. If the path can avoid crossing $g(\cdot)$ up to time λ, there will be no further cost since the mean drift after the change will pull it away from the boundary. A moment's thought will reveal that the optimum trajectory consists of two line segments. There is a straight line from the point $(0, 0)$ to the point $(\lambda, g(\lambda))$ followed by a line segment from $(\lambda, g(\lambda))$ to $(1, g(\lambda) - (s^2/2)(1 - \lambda))$. This second line segment is just following the mean drift and has no cost associated with it. The cost of the first line segment is

$$\begin{aligned} I(f_{\text{opt}}) &= \lambda \sup_{\theta} \left\{ \theta \frac{g(\lambda)}{\lambda} - \log\left[\exp\left[\frac{s^2(\theta + \theta^2)}{2} \right] \right] \right\} \\ &= \lambda \sup_{\theta} \left\{ \theta \frac{g(\lambda)}{\lambda} - \frac{s^2(\theta + \theta^2)}{2} \right\} \end{aligned}$$

By differentiation we find that the optimizing θ is

$$\theta_{\text{opt}} = \frac{g(\lambda)}{\lambda s^2} - \frac{1}{2}$$

Hence, substituting this value, we have

$$\begin{aligned} I(f_{\text{opt}}) &= \frac{g(\lambda)^2}{2\lambda s^2} - \frac{g(\lambda)}{2} + \frac{s^2}{8}\lambda \\ &= \frac{2s^2\alpha\lambda - s^2\lambda(2\alpha s^2\lambda)^{1/2} + s^4\lambda^2/4}{2\lambda s^2} + \frac{s^2\lambda}{4} - \frac{(2s^2\lambda\alpha)^{1/2}}{2} + \frac{s^2}{8}\lambda \\ &= \alpha + \frac{s^2\lambda}{2} - (2\alpha s^2\lambda)^{1/2} \\ &= \left(s\sqrt{\frac{\lambda}{2}} - \sqrt{\alpha} \right)^2 \qquad (\text{for } \lambda > \lambda_c) \end{aligned}$$

Now let us suppose λ is known and work out the performance of the optimal Neyman–Pearson detector. The threshold T on the detector must be set so that the false-alarm rate is α or

$$\begin{aligned}
&\lim_{n\to\infty} \frac{1}{n} \log P_q\left(\frac{1}{n}\sum_{i=1}^{n\lambda} \log \frac{p(\mathbf{x}_i)}{q(\mathbf{x}_i)} > T\right) \\
&= \lambda \lim_{n\to\infty} \frac{1}{n\lambda} \log P_q\left(\frac{1}{n\lambda}\sum_{i=1}^{n\lambda} \log \frac{p(\mathbf{x}_i)}{q(\mathbf{x}_i)} > \frac{T}{\lambda}\right) \\
&= -\lambda \sup_\theta \left\{\theta \frac{T}{\lambda} - \log \int p(\xi)^\theta q(\xi)^{1-\theta}\, d\xi\right\} \\
&= -\lambda \sup_\theta \left\{\theta \frac{T}{\lambda} + s^2 \frac{\theta - \theta^2}{2}\right\} = -\frac{[T + \lambda s^2/2]^2}{2\lambda s^2} = -\alpha
\end{aligned}$$

which implies that

$$T = -s^2 \frac{\lambda}{2} + \sqrt{2\alpha\lambda}\, s = g(\lambda)$$

Note that the changepoint detector operating on this data (for unknown λ) will reject the null hypothesis whenever the optimal Neyman–Pearson detector rejects. Hence it has at least as good type II error performance as the optimum detector, which means that its performance equals the optimal detector. This means that the changepoint detector is asymptotically uniformly most powerful.

Note that the principal results of this example hold in general for the iid setting. In other words, $T = g(\lambda)$ and (asymptotic) uniformly most powerful properties of the test follow for the design procedure presented in the chapter.

CHAPTER VII

1. Now we know that

$$\lim_{n\to\infty} E\{\mathbf{S}_n\} = a$$

where the expectation is taken over the twisted chains distribution. Define

$$\exp[n\psi_n(\theta)] = E\{\exp[\theta n \mathbf{S}_n]\}$$

Further define

$$\lim_{n \to \infty} \psi_n(\theta) \equiv \psi(\theta)$$

Then

$$\begin{aligned} n^2 E\{\mathbf{S}_n^2\} &= \frac{d^2}{d\theta^2} \exp[n\psi_n(\theta)]|_{\theta=0} \\ &= n^2 \psi_n'(\theta)^2 \exp[n\psi_n(\theta)] + n\psi_n''(\theta) \exp[n\psi_n(\theta)]|_{\theta=0} \\ &= n^2 \psi_n'(0)^2 + n\psi_n''(0) \end{aligned}$$

Also note that

$$E\{\mathbf{S}_n\} = \psi_n'(0)$$

which implies that

$$\lim_{n \to \infty} \psi_n'(0) \equiv \psi'(0) = a$$

(The limit of derivatives equaling the derivative of the limit requires some sort of analyticity argument.)

Now note that

$$\begin{aligned} \operatorname{var}\{\mathbf{S}_n \sqrt{n}\} &= \operatorname{var}\{[\mathbf{S}_n - E\{\mathbf{S}_n\}]\sqrt{n}\} = [E\{\mathbf{S}_n^2\} - E\{\mathbf{S}_n\}^2]n \\ &= n\psi_n'(0)^2 + \psi_n''(0) - n\psi_n'(0)^2 = \psi_n''(0) \end{aligned}$$

Now for the twisted chain, the relevant operator is

$$T_\theta(g)(y) \equiv \sum_x \exp[\theta f(x)] \exp[\theta_a f(x)] r_{\theta_a}(x) p(x \mid y) \frac{g(x)}{r_{\theta_a}(y)\lambda(\theta_a)}$$

It is simple to verify that the largest eigenvalue is $\lambda(\theta + \theta_a)/\lambda(\theta_a)$ with associated eigenvector $r_{\theta+\theta_a}(x)/r_{\theta_a}(x)$. Hence

$$\lim_{n \to \infty} \psi_n(\theta) \equiv \psi(\theta) = \log \frac{\lambda(\theta + \theta_a)}{\lambda(\theta_a)}$$

which implies that

$$\psi'(0) = \frac{\lambda'(\theta_a)}{\lambda(\theta_a)} = a$$

and

$$\psi''(0) = \frac{\lambda''(\theta_a)}{\lambda(\theta_a)} - a^2$$

Then

$$\lim_{n\to\infty} \operatorname{var}\{\mathbf{S}_n\sqrt{n}\} = \lim_{n\to\infty} \psi_n''(0) = \psi''(0) = \frac{\lambda''(\theta_a)}{\lambda(\theta_a)} - a^2$$

2. Since we know that

$$I(a) = \frac{a^2}{2}$$

for a Gaussian random variable we need only compute θ_a and σ^2. Now

$$\varphi(\theta) \equiv E\{\exp[\theta\mathbf{x}]\} = \exp\left[\frac{\theta^2}{2}\right]$$

θ_a is the value of θ that attains the supremum for $\{\theta a - \log\varphi(\theta)\} = \{\theta a - \theta^2/2\}$, which implies that

$$\theta_a = a$$

$$\varphi''(\theta) = \frac{d}{d\theta}\theta\exp\left[\frac{\theta^2}{2}\right] = \theta^2\exp\left[\frac{\theta^2}{2}\right] + \exp\left[\frac{\theta^2}{2}\right]$$

$$= \exp\left[\frac{\theta^2}{2}\right][1 + \theta^2] = [1 + \theta^2]\varphi(\theta)$$

Hence

$$\sigma^2 = \frac{\varphi''(\theta_a)}{\varphi(\theta_a)} - a^2 = [1 + \theta_a^2] - a^2 = 1$$

Therefore,

$$P\left(\frac{1}{n}\sum_{i=1}^{n}\mathbf{x}_i \geqslant a\right) \approx \frac{\exp[-nI(a)]}{\sqrt{2\pi n}\,\sigma\theta_a}$$

$$= \frac{\exp[-na^2/2]}{\sqrt{2\pi n}\,a}$$

This matches up with the exercise in chapter II since

$$P\left(\frac{1}{n}\sum_{i=1}^{n}\mathbf{x}_i \geqslant a\right) = P(x_1 \geqslant \sqrt{n}a) = \frac{\int_{\sqrt{n}a}^{\infty}\exp[-.5x^2]\,dx}{\sqrt{2\pi}}$$

3. Let the distribution function of $\mathbf{x}_1$ be $F_p(\cdot)$. Define the characteristic function

$$P(\omega) \equiv \int_{-\infty}^{\infty}\exp[-i\omega x]\,dF_p(x) = \sum_{k=-\infty}^{\infty}\exp[-i\omega kd]p_k$$

where

$$p_k \equiv P(\mathbf{x}_1 = kd)$$

It is trivial to verify that $P(\omega)$ is periodic with period $2\pi/d$. The characteristic function of $\mathbf{S}_n$ is (by the convolution property of Fourier transforms) $P(\omega)^n$. The characteristic function of $\mathbf{S}_n/\sqrt{n}$ is hence $P(\omega/\sqrt{n})^n$, which is periodic with period $2\pi\sqrt{n}/d$. Hence we may expand $P(\omega/\sqrt{n})^n$ in a Fourier series. The constant term in the series representation must correspond to $P(\mathbf{S}_n/\sqrt{n} = 0)$. Hence

$$P(\mathbf{S}_n = 0) = P\left(\frac{\mathbf{S}_n}{\sqrt{n}} = 0\right) = \frac{d}{2\pi\sqrt{n}}\int_{-\pi\sqrt{n}/d}^{\pi\sqrt{n}/d} P\left(\frac{\omega}{\sqrt{n}}\right)^n d\omega$$

Now recall that $P(0) = 1$, $P'(0) = -iE\{x_1\} = 0$, $P''(0) = -\sigma^2, \ldots$. Therefore,

$$P\left(\frac{\omega}{\sqrt{n}}\right)^n = \left[1 - \omega^2\frac{\sigma^2}{2n} + o(n^{-1})\right]^n \approx \exp\left[-\frac{\omega^2\sigma^2}{2}\right]$$

Hence

$$P(\mathbf{S}_n = 0) \approx \frac{d}{2\pi\sqrt{n}}\int_{-\infty}^{\infty}\exp\left[-\omega^2\frac{\sigma^2}{2}\right]d\omega = \frac{d}{\sqrt{2\pi n}\,\sigma}$$

Hence the limit of interest is

$$\lim_{n\to\infty}\sqrt{n}P(\mathbf{S}_n = 0) = \frac{d}{\sqrt{2\pi}\,\sigma}$$

4. The first theorem in section B will have a d multiplying it since

$$\sum_{x_1,\ldots,x_n:S_n=a}\cdots\sum t(x_1, x_2, \ldots, x_n \mid \mathbf{x}_1 = x, \mathbf{x}_n = y) \approx \frac{d}{\sqrt{2\pi n}\,\sigma}$$

which does depend on the lattice spacing (see exercise 3). Hence we have

$$\lim_{n\to\infty} p^{(n)}(y \mid x) P\left(\mathbf{S}_n = a + \frac{s}{n} \,\middle|\, \mathbf{x}_1 = x, \mathbf{x}_n = y\right)[\exp[nI(a)]\sqrt{n}]$$

$$= r_{\theta_a}(x) r^*_{\theta_a}(y) \exp[-\theta_a(s - f(x))] \frac{d}{\lambda(\theta_a)\sqrt{2\pi}\,\sigma}$$

For the remaining expressions, replace s everywhere by sd to find for the analogs to expressions (52)–(53),

$$\sum_{s=0}^{\infty} \exp[-\theta_a ds] \sum_{x_1,\ldots,x_n:S_n=a+sd/n}\cdots\sum t(x_1, x_2, \ldots, x_n \mid \mathbf{x}_1 = x, \mathbf{x}_n = y)$$

$$\approx \frac{d}{\sqrt{2\pi n}\,\sigma}[1 - \exp[-\theta_a d]]$$

This then gives the modification for the first corollary as

$$\lim_{n\to\infty} p^{(n)}(y \mid x) P(\mathbf{S}_n \geqslant a \mid \mathbf{x}_1 = x, \mathbf{x}_n = y) \exp[nI(a)]\sqrt{n}$$

$$= r_{\theta_a}(x) r^*_{\theta_a}(y) \exp[\theta_a f(x)] \frac{d}{\lambda(\theta_a)\sqrt{2\pi}\,\sigma[1 - \exp[-\theta_a d]]}$$

and hence the second corollary becomes

$$P(\mathbf{S}_n \geqslant a) \approx \exp[-nI(a)]\left[\sum_x p(x) \exp[\theta_a f(x)) r_{\theta_a}(x)\right]$$

$$\cdot\left[\sum_y r^*_{\theta_a}(y)\right] \frac{d}{\lambda(\theta_a)\sqrt{2\pi n}\,\sigma[1 - \exp[-\theta_a d]]}$$

Note that in the iid case with $f(x) = x$ we have that $r_{\theta_a}(\cdot)$ and $r^*_{\theta_a}(\cdot)$ are constants not depending on their arguments and normalized so that

$$\sum_x r_{\theta_a}(x) r^*_{\theta_a}(x) = 1$$

Furthermore,

$$\lambda(\theta_a) = E\{\exp[\theta_a \mathbf{x}]\} = \sum_x p(x)\exp[\theta_a x]$$

Hence

$$P(\mathbf{S}_n \geqslant a) \approx \exp[-nI(a)]\,\frac{d}{\sqrt{2\pi n}\,\sigma[1-\exp[-\theta_a d]]}$$

which matches up with the stated result of the Bahadur–Rao theorem of section A in the lattice random variable case. Taking the limits as d approaches zero is trivial in all these expressions (in order to find the nonlattice constants). One merely substitutes θ_a^{-1} for the $d/[1-\exp[-\theta_a d]]$ term.

5. From outputs on the computer program written to find the optimal memoryless nonlinearity, we compute a numerical second derivative as

$$\lambda''(\theta_0) = \frac{\lambda(\theta_0 + h) + \lambda(\theta_0 - h) - 2\lambda(\theta_0)}{h^2} = .90186$$

where $\theta_0 = .4623$ and $h = 10^{-4}$. This implies that

$$\sigma^2 = \frac{\lambda''(\theta_0)}{\lambda(\theta_0)} = \frac{.90186}{.9256} = .97436$$

Now supposing that we start off the chain at point x with probability 1, we have for the type I error expansion

$$P(\mathbf{S}_n \geqslant 0 \mid H_0) = \lambda(\theta_0)^{n-1}\exp[\theta_0 f(x)]r(x)\frac{\Sigma_y r^*_{\theta_0}(y)}{(1-\exp[-\theta_0])\sqrt{2\pi n}\,\sigma}$$

We compute

$$[r[1] \quad r[2] \quad \cdots \quad r[5]] = [.2155 \quad .3168 \quad .2988 \quad .7248 \quad .7112]$$

and

$$[r^*[1] \quad r^*[2] \quad \cdots \quad r^*[5]] = [.5378 \quad .7112 \quad .7248 \quad .2988 \quad .3168]$$

which implies that

$$\sum_{y} r^*_{\theta_0}(y) = 2.5898$$

We may now substitute in any starting value we choose; e.g., if $x = 1$ (and hence $f[x] = f[1] = 0$), we obtain

$$P(\mathbf{S}_n \geqslant 0 \mid H_0) \approx \frac{(.9256)^{n-1}(.2155)(2.5898)}{1 - \exp[-.4623]\sqrt{2\pi n(.97436)}}$$

$$= \frac{(.9256)^{n-1}}{\sqrt{n}[.60935]}$$

CHAPTER VIII

1. We wish to minimize

$$\int \frac{p(x)^2 \eta(x)^2}{q(x)}\, dx$$

subject to $\int q(x)\, dx = 1$ and $q(x) \geqslant 0$. Ignoring the inequality constraint for now we set up the objective as

$$\frac{\delta}{\delta\varepsilon}\left[\int \frac{p(x)^2 \eta(x)^2}{q(x) + \varepsilon m(x)}\, dx + \mu \int [q(x) + \varepsilon m(x)]\, dx\right]\Bigg|_{\varepsilon=0}$$

which formally yields

$$\int m(x)\left[\mu - \frac{(p(x)\eta(x))^2}{q(x)^2}\right] dx$$

which must be true for all perturbations $m(\cdot)$ such that $\int m(x) = 0$. This implies that

$$\mu - \frac{(p(x)\eta(x))^2}{q(x)^2} = 0$$

or

$$q(x) \text{ is proportional to } |p(x)\eta(x)|$$

Note: Holder's inequality may also be used to prove this result.

2. The original transition matrix of interest is

$$\begin{bmatrix} 0 & 1 & 0 & 0 & 0 \\ .7 & 0 & .3 & 0 & 0 \\ 0 & .7 & 0 & .3 & 0 \\ 0 & 0 & .7 & 0 & .3 \\ 0 & 0 & 0 & .7 & .3 \end{bmatrix}$$

and twisted matrix

$$\begin{bmatrix} 0 & 1 & 0 & 0 & 0 \\ .5146 & 0 & .4854 & 0 & 0 \\ 0 & .5049 & 0 & .4951 & 0 \\ 0 & 0 & .4951 & 0 & .5049 \\ 0 & 0 & 0 & .4854 & .5146 \end{bmatrix}$$

Using the Turbo Pascal random number generator, three stimulation runs of 10,000 trials each for $n=100$ gave 2.024E-5, 1.99E-5, and 2.04E-5. The asymptotic value computed in chapter VII is 2.89E-5. Three simulation runs of 1000 trials each for $n=1000$ gave 5.151E-36, 5.14E-36, and 4.85E-36. The asymptotic value is 5.51E-36.

3. For the $n = 2$ case, the asymptotic expressions derived in exercise 8 of chapter II indicate that the probability should lie between .975E-45 and 1.049E-45. A simulation run of 1000 sums gave the value 1.23E-45 (our random number generator is not too good!). For the $n=10$ case the asymptotic expression indicates that the probability is approximately 8.988E-220. A simulation run of 1000 sums gave the value 9.29E-220 (same random number generator). These probabilities are so astronomically small that a direct simulation would be absurd to even attempt.

4. A couple of values were found by computer search. For $a=.75$, $p=.25$, $\theta_0=2.63$, and $\lambda(\theta_0)=.707107$. For $a=.9$, $p=.4$, $\theta_0=2.89$, and $\lambda(\theta_0)=.596908$.

5. Start with (15), the expression that k times the variance of our estimate will be given by

$$\sum_{\Sigma f(x_i)\geq 0}\left[\exp\left[-\theta_0\sum_{i=1}^{n-1} f(x_{i+1})\right] r(x)\frac{\lambda(\theta_0)^{n-1}}{r(x_n)}\right]^2 t(x_1, x_2, \ldots, x_n \,|\, \mathbf{x}_1 = x) - P(A)^2$$

First let us consider the analog to (16):

$$\sum_{s=0}^{\infty} \sum_{\Sigma f(x_i)=sd} \left[\exp\left[-\theta_0 \sum_{i=1}^{n-1} f(x_{i+1})\right] r(x) \frac{\lambda(\theta_0)^{n-1}}{r(x_n)}\right]^2 t(x_1, x_2, \ldots, x_n \mid \mathbf{x}_1 = x)$$

$$= \sum_{s=0}^{\infty} \sum_{\Sigma f(x_i)=sd} \left[\exp\left[-\theta_0 \sum_{i=1}^{n} f(x_i) - f(x)\right] r(x) \frac{\lambda(\theta_0)^{n-1}}{r(x_n)}\right]^2 t(x_1, x_2, \ldots, x_n \mid \mathbf{x}_1 = x)$$

where s is a nonnegative integer. Now conditioning on the event that $\mathbf{x}_n = y$, we may expand the expression above to obtain the analog to (17):

$$\sum_y t^{(n)}(y \mid x) \exp[-2\theta_0(sd - f(x))] \frac{r(x)^2}{r(y)^2} \lambda(\theta_0)^{2(n-1)} \times \left[\sum_{\Sigma f(x_i)=sd} t(x_1, \ldots, x_n \mid \mathbf{x}_1 = x, \mathbf{x}_n = y)\right]$$

which is exactly like (17) in the text except for an additional $\exp[2\theta_0 f(x)]$ term. Hence we must obtain for the analog to (22):

$$\sum_{\Sigma f(x_i) \geqslant 0} \left[\exp\left[-\theta_0 \sum_{i=1}^{n-1} f(x_{i+1})\right] r(x) \frac{\lambda(\theta_0)^{n-1}}{r(x_n)}\right]^2 t(x_1, x_2, \ldots, x_n \mid \mathbf{x}_1 = x)$$
$$\approx r(x)^2 \lambda(\theta_0)^{2(n-1)} \exp[2\theta_0 f(x)] \frac{d}{\sqrt{2\pi n}\,\sigma} \sum \frac{r^*(y)}{r(y)} \Big/ [1 - \exp[-2\theta_0 d]]$$

Now let us consider the other term,

$$P(A) = P(\mathbf{S}_n \geqslant a)$$
$$\approx \exp[-nI(a)] \left(\sum_x p(x) \exp[\theta_a f(x)] r_{\theta_a}(x)\right) \times \left(\sum_y r^*_{\theta_a}(y)\right) \frac{d}{\lambda(\theta_a)(1 - \exp[-\theta_a d]\sqrt{2\pi n}\,\sigma}$$

Now in our setting $a = 0$, $p(\cdot) = \delta(\cdot - x)$, $I(a) = -\log \lambda(\theta_0)$, which gives the analog to (25) as

$$P(A) \cong \lambda(\theta_0)^{n-1} \exp[\theta_0 f(x)] r(x) d \frac{\Sigma_y r^*(y)}{(1 - \exp[-\theta_0 d])\sqrt{2\pi n}\,\sigma}$$

6. In this setting $\zeta_q(\gamma)$ is the largest positive eigenvalue of the operator [with associated eigenvector $e_\gamma(\cdot)$]

$$K_\gamma(g)(y) \equiv \sum_x \exp[\gamma f(x)] \frac{p(x \mid y)^2}{q(x)} g(x)$$

or equivalently, $\zeta_q(\gamma)$ is the largest positive eigenvalue of the transpose operator [with associated eigenvector $e^*_\gamma(\cdot)$]:

$$K^*_\gamma(g)(x) \equiv \frac{\exp[\gamma f(x)]}{q(x)} \sum_y p(x \mid y)^2 g(y)$$

Let us consider a perturbation analysis. Let us define a new perturbed operator

$$K^\varepsilon_\gamma(g)(y) \equiv \sum_x \exp[\gamma f(x)] \frac{p(x \mid y)^2}{q(x) + \varepsilon m(x)} g(x)$$

where $\Sigma_x m(x) = 0$. Now we have that

$$K^\varepsilon_\gamma(g)(y) = \sum_x \exp[\gamma f(x)] \frac{p(x \mid y)^2}{q(x)[1 + \varepsilon m(x)/q(x)]} g(x)$$

$$= K_\gamma(g)(y) - \varepsilon \sum_x m(x) \frac{\exp[\gamma f(x)] p(x \mid y)^2}{q(x)^2} g(x) + O(\varepsilon^2)$$

Now we may write out the first order perturbation expansion of the principal eigenvalue as

$$\zeta_{q+\varepsilon m}(\gamma) = \zeta_q(\gamma) - \varepsilon \sum_x \sum_y m(x) \exp[\gamma f(x)] p(x \mid y)^2 \frac{e_\gamma(x) e^*_\gamma(y)}{q(x)^2}$$

Now note that

$$\sum_y p(x \mid y)^2 e^*_\gamma(y) = \zeta_q(\gamma) \exp[-\gamma f(x)] q(x) e^*_\gamma(x)$$

or

$$\zeta_{q+\varepsilon m}(\gamma) = \zeta_q(\gamma) - \varepsilon \zeta_q(\gamma) \sum_x m(x) \frac{e_\gamma(x) e^*_\gamma(x)}{q(x)}$$

This last expression implies that for $q(\cdot)$ to be a critical point of the perturbation, we must have

$$q(x) = e_\gamma(x) e_\gamma^*(x)$$

This last expression suggests an algorithm (much as in section C of chapter VI).

1. Find the unique solution to the system above.
2. Vary γ to make the principal eigenvalue $\zeta_q(\gamma)$ of this system as small as possible, denoted as $\zeta_q(\gamma_0)$.
3. The solution that goes with this minimizing γ will be the optimal choice for $q(\cdot)$. Furthermore,

$$\lim_{n\to\infty} \frac{1}{n} \log[k \operatorname{var}(\hat{\mathbf{p}}_q)] = \log[\zeta_q(\gamma_0)]$$

CHAPTER IX

1. $$c_n(t) \equiv \left\{x : \frac{1}{n} \log \frac{q(x_1, \ldots, x_n)}{p(x_1, \ldots, x_n)} \leqslant t\right\}$$

$$= \left\{x : \frac{1}{n} \sum_{i=1}^{n-1} \log \frac{q(x_{i+1} \mid x_i)}{p(x_{i+1} \mid x_i)} + \frac{1}{n} \log \frac{q(x_1)}{p(x_1)} \leqslant t\right\}$$

$$K(q, p) \equiv \inf\left\{t : \lim_{n\to\infty} Q_n\{c_n(t)\} = 1\right\}$$

Now under the $Q_n(\cdot)$ distributions we have that

$$\frac{1}{n} \sum_{i=1}^{n-1} \log \frac{q(x_{i+1} \mid x_i)}{p(x_{i+1} \mid x_i)}$$

converges to

$$\sum_x \sum_y q(x \mid y) \log \frac{q(x \mid y)}{p(x \mid y)}$$

Therefore, if and only if $t \geqslant \Sigma_x \Sigma_y\, q(x \mid y) \log[q(x \mid y)/p(x \mid y)]$ will $\lim_{n\to\infty} Q_n\{c_n(t)\} = 1$. Therefore,

$$K(q, p) = \sum_x \sum_y q(x \mid y) \log \frac{q(x \mid y)}{p(x \mid y)}$$

which is the rate function for the two-dimensional empirical distributions of a finite-dimensional Markov chain (see exercise 4 of chapter V).

2. $$K(p(\cdot,\delta), p(\cdot,\gamma)) = \int \exp\left[-\frac{(x-\delta)^2}{2}\right] \log[\exp[-.5[(x-\delta)^2-(x-\gamma)^2]]]\,dx$$
$$= .5(\gamma-\delta)^2$$

Now we have, by definition,

$$b(\varepsilon,\gamma) = \inf\{.5(\gamma-\delta)^2 : \delta \in \{\delta : |\delta-\gamma| > \varepsilon\}\} = .5\varepsilon^2$$

and furthermore,

$$a_n(\varepsilon,\gamma) = \lim_{n\to\infty} \frac{1}{n} \log P\left(\frac{1}{n}\sum_{i=1}^n \mathbf{x}_i \geqslant \gamma+\varepsilon \text{ or } \frac{1}{n}\sum_{i=1}^n \mathbf{x}_i \leqslant \gamma-\varepsilon\right)$$
$$= -\inf_{x\in[\gamma+\varepsilon,\infty)\cup(-\infty,\gamma-\varepsilon)} .5(x-\gamma)^2 = -\frac{\varepsilon^2}{2} = b(\varepsilon,\gamma)$$

so the bound is attained.

3. $$p(x,\gamma) = \prod_{i=1}^n \gamma \exp[-\gamma x_i]\cdot 1_{[0,\infty)}(x_i)$$
$$= \gamma^n \exp\left[-\gamma\sum_{i=1}^n x_i\right]\prod_{i=1}^n 1_{[0,\infty)}(x_i)$$

Therefore, to find the maximum likelihood estimate we need to evaluate

$$\frac{\delta}{\delta\gamma}\log[p(\mathbf{x},\gamma)]\big|_{\gamma=\hat{\gamma}_{\mathrm{ml},n}} = 0$$

or

$$-\sum_{i=1}^n \mathbf{x}_i + \frac{n}{\gamma}\bigg|_{\gamma=\hat{\gamma}_{\mathrm{ml},n}} = 0$$

which implies that

$$\hat{\gamma}_{\mathrm{ml},n} = \left[\frac{1}{n}\sum_{i=1}^n \mathbf{x}_i\right]^{-1}$$

Now

$$P\left(\left|\gamma - \left[\frac{1}{n}\sum_{i=1}^{n} \mathbf{x}_i\right]^{-1}\right| > \varepsilon\right) = P\left(\frac{1}{n}\sum_{i=1}^{n} \mathbf{x}_i > \frac{1}{\gamma-\varepsilon}\right) + P\left(\frac{1}{n}\sum_{i=1}^{n} \mathbf{x}_i < \frac{1}{\gamma+\varepsilon}\right)$$

Therefore, the large deviation asymptotics of the estimate will be controlled by the large deviation asymptotics of a sample average. Let us now compute the rate function for this sum.

$$M(\theta) \equiv E\{\exp[\theta \mathbf{x}]\} = \int_0^\infty \exp[\theta x]\gamma \exp[-\gamma x]\, dx = \frac{\gamma}{\gamma-\theta} \qquad (\text{for } \theta \leqslant \gamma)$$

Hence

$$\begin{aligned} I(x) &= \sup_\theta [\theta x - \log[M(\theta)] = [\theta_{\text{opt}} x - \log[M(\theta_{\text{opt}})] \\ &= \left[\left(\gamma - \frac{1}{x}\right)x - \log\left[\gamma \Big/ \left(\gamma - \left(\gamma - \frac{1}{x}\right)\right)\right]\right] \\ &= \gamma x - 1 - \log[\gamma x] \end{aligned}$$

Hence

$$\lim_{n\to\infty} \frac{1}{n} \log P\left(\left|\gamma - \left[\frac{1}{n}\sum_{i=1}^{n} \mathbf{x}_i\right]^{-1}\right| > \varepsilon\right) = -\min\left[I\left(\frac{1}{\gamma-\varepsilon}\right), I\left(\frac{1}{\gamma+\varepsilon}\right)\right]$$

Using the Taylor series expansions

$$\frac{1}{1-x} \approx 1 + x + x^2$$

and

$$\log[1-x] \approx -x - \frac{x^2}{2}$$

we have

$$I\left(\frac{1}{\gamma-\varepsilon}\right) \approx I\left(\frac{1}{\gamma+\varepsilon}\right) \approx \frac{\varepsilon^2}{2\gamma^2}$$

Hence we have the limit

$$\lim_{\varepsilon \to 0} \lim_{n \to \infty} \frac{1}{\varepsilon^2 n} \log P\left(\left| \gamma - \left[\frac{1}{n} \sum_{i=1}^{n} \mathbf{x}_i \right]^{-1} \right| > \varepsilon \right) = -[2\gamma^2]^{-1}$$

On the other hand,

$$k(\gamma) = E_\gamma \left\{ \left(\frac{\delta}{\delta\gamma} \log p(\mathbf{x}_1, \gamma) \right)^2 \right\} = E_\gamma \left\{ \left(\frac{1}{\gamma} - \mathbf{x}_1 \right)^2 \right\}$$

$$= \frac{1}{\gamma^2} - \frac{2}{\gamma} E_\gamma\{\mathbf{x}\} + E_\gamma\{\mathbf{x}^2\} = \frac{1}{\gamma^2} - \frac{2}{\gamma^2} + \frac{2}{\gamma^2} = \frac{1}{\gamma^2}$$

Therefore,

$$-\frac{k(\gamma)}{2} = -[2\gamma^2]^{-1}$$

and the expressions match up.

4. $\int p(x, \gamma + \varepsilon) \log[p(x, \gamma + \varepsilon)/p(x, \gamma)]\, dx$
$= \int p(x, \gamma + \varepsilon)[\log[p(x, \gamma + \varepsilon)] - \log[p(x, \gamma)]]\, dx.$

We also have the Taylor series expansions

$$\log[p(x, \gamma + \varepsilon)] = \log[p(x, \gamma)] + \varepsilon \frac{d}{d\gamma} \log[p(x, \gamma)] + \frac{\varepsilon^2}{2} \frac{d^2}{d\gamma^2} \log[p(x, \gamma)] + O(\varepsilon^3)$$

and

$$p(x, \gamma + \varepsilon) = p(x, \gamma) + \varepsilon \frac{d}{d\gamma} p(x, \gamma) + O(\varepsilon^2)$$

We will also make use of the fact that (assuming some smoothness conditions)

$$\int p(x, \gamma) \frac{d}{d\gamma} \log[p(x, \gamma)]\, dx = \int \frac{d}{d\gamma} p(x, \gamma)\, dx = \frac{d}{d\gamma} \int p(x, \gamma)\, dx$$

$$= \frac{d}{d\gamma}(1) = 0$$

Therefore,

$$\int p(x, \gamma+\varepsilon)[\log[p(x, \gamma+\varepsilon)] - \log[p(x, \gamma)]]\, dx$$

$$= \int \left[p(x, \gamma) + \varepsilon \frac{d}{d\gamma} p(x, \gamma) + O(\varepsilon^2) \right] \left[\varepsilon \frac{d}{d\gamma} \log[p(x,\gamma] \right.$$

$$\left. + \frac{\varepsilon^2}{2} \frac{d^2}{d\gamma^2} \log[p(x, \gamma)] + O(\varepsilon^3) \right] dx$$

$$= \varepsilon^2 \int \left[\frac{(d/d\gamma)p(x, \gamma)}{p(x, \gamma)} \right]^2 p(x, \gamma)\, dx + \frac{\varepsilon^2}{2} \int \left[\frac{d^2}{d\gamma^2} \log[p(x, \gamma)] \right] p(x, \gamma)\, dx$$

$$= \varepsilon^2 \int \left[\frac{d}{d\gamma} \log[p(x, \gamma)] \right]^2 p(x, \gamma)\, dx + \frac{\varepsilon^2}{2} \int \left[\frac{d^2}{d\gamma^2} \log[p(x, \gamma)] \right] p(x, \gamma)\, dx$$

$$= \frac{\varepsilon^2}{2} \int \left[\frac{d}{d\gamma} \log[p(x, \gamma)] \right]^2 p(x, \gamma)\, dx$$

$$= \frac{\varepsilon^2}{2} k(\gamma)$$

(See appendix F and the Cramér–Rao inequality.)

CHAPTER X

1. We wish to minimize

$$\sum_{i=1}^{d} q_i \log \frac{q_i}{p_i}$$

subject to constraints

$$\sum_{i=1}^{d} q_i \log q_i = R \qquad \text{and} \qquad \sum_{i=1}^{d} q_i = 1$$

Setting up our standard Lagrange multiplier minimization, we have

$$\frac{\delta}{\delta q_j} \left[\sum_{i=1}^{d} q_i \log \frac{q_i}{p_i} + \mu \sum_{i=1}^{d} q_i \log q_i + \nu \sum_{i=1}^{d} q_i \right] = 0$$

for all j. This implies that

$$(1+\mu)\left[\log[q_j]+\frac{q_j}{q_j}\right]-\log[p_j]+\nu=0$$

or

$$q_j = C_1 p_j^{C_2}$$

where C_1 is chosen so that $\Sigma q_j = 1$ and C_2 is chosen so that $\Sigma q_j \log q_j = R$.

2. Note that for $Q \in Q_D$ we have $d(Q) \leqslant D$. Hence

$$\begin{aligned}
&I(Q)-\theta D+f(\theta) \geqslant I(Q)-\theta d(Q)+f(\theta)\\
&=\sum_j\sum_k P(j)Q(k\mid j)\log\left[\frac{Q(k\mid j)\exp[-\theta\rho(j,k)]\,\Sigma_{k'}\exp[\theta\rho(j,k')Q(k')}{Q(k)}\right]\\
&\geqslant \sum_j\sum_k P(j)Q(k\mid j)\left[1-\left[\frac{Q(k\mid j)\exp[-\theta\rho(j,k)]\,\Sigma_{k'}\exp[\theta\rho(j,k')]Q(k')}{Q(k)}\right]^{-1}\right]\\
&=1-\sum_j P(j)\sum_k \exp[\theta\rho(j,k)]Q(k)\left[\sum_{k'}\exp[\theta\rho(j,k')Q(k')\right]^{-1}\\
&=1-\sum_j P(j)\\
&=1-1=0
\end{aligned}$$

Then since

$$R(D) \equiv \inf_{Q \in Q_D} I(Q)$$

we immediately have the lemma statement

$$R(D) \geqslant \sup_{\theta \leqslant 0}\,[\theta D - f(\theta)]$$

This proof is a very slight modification of Theorem 2.53 in Berger [1971].

3. Now

$$\sum_j\sum_k \rho(j,\,k)Q^*(k\mid j)P(j) = D$$

which implies that there exist constants α_j such that

$$\alpha_j \geqslant 0$$

and

$$\sum_{j=1}^{M} \alpha_j = 1$$

such that

$$\sum_k \rho(j,k)Q^*(k\,|\,j) \leqslant \alpha_j \frac{D}{P(j)}$$

To see this, simply set up a simple argument by contradiction.

Consider

$$Q(S_n(x)) \equiv Q\left\{\mathbf{y}: \frac{1}{n}\sum_{i=1}^{n} \rho(x_i, \mathbf{y}_i) < D + \delta\right\}$$

This is an average of independent but not identically distributed random variables. Each output letter from the source can take on only one of M possible values with probability $P(\mathbf{x}_1 = j) = P_j$. For the vector x, let $n_j \equiv$ number of times that x_i, $i = 1, 2, \ldots, n$ equals j. Then

$$Q(S_n(x)) = Q\left[\frac{n_1}{n}\frac{1}{n_1}\sum_{i=1}^{n_1} \rho(1, \mathbf{y}'_{1i}) + \frac{n_2}{n}\frac{1}{n_2}\sum_{i=1}^{n_2} \rho(2, \mathbf{y}'_{2i}) + \cdots \right.$$
$$\left. + \frac{n_M}{n}\frac{1}{n_M}\sum_{i=1}^{n_M} \rho(M, \mathbf{y}'_{Mi}) < D + \delta\right]$$

where the $\{\mathbf{y}'_{ji}\}$ are iid and distributed as Q^*. Hence

$$Q(S_n(x)) \geqslant Q\left(\left(\frac{n_j}{n}\right)\frac{1}{n_j}\sum_{i=1}^{n_j} \rho(j, \mathbf{y}'_{ji}) < \alpha_j(D + \delta), j = 1, 2, \ldots, M\right)$$

Since $n_j/n \rightarrow P(j)$ for all source vectors x, we must have that the above is

$$Q(S_n(x)) \geqslant \prod_{j=1}^{M} Q\left(\frac{1}{n_j}\sum_{i=1}^{n_j} \rho(j, \mathbf{y}'_{ji}) < \alpha_j \frac{D + \delta/2}{P(j)}\right)$$

for n sufficiently large. Therefore,

$$\begin{aligned}\lim_{n\to\infty}\frac{1}{n}\log Q(S_n(x)) &\geq \sum_{j=1}^{M}\lim_{n\to\infty}\frac{1}{n}\log Q\left(\frac{1}{n_j}\sum_{i=1}^{n_j}\rho(j,\mathbf{y}'_{ji})<\alpha_j\frac{D+\delta/2}{P(j)}\right)\\ &=\sum_{j=1}^{M}\lim_{n\to\infty}\frac{n_j}{n}\left(\frac{1}{n_j}\right)\log Q\left(\frac{1}{n_j}\sum_{i=1}^{n_j}\rho(j,\mathbf{y}'_{ji})<\alpha_j\frac{D+\delta/2}{P(j)}\right)\\ &=\sum_{j=1}^{M}P(j)\lim_{n_j\to\infty}\log Q\left(\frac{1}{n_j}\sum_{i=1}^{n_j}\rho(j,\mathbf{y}'_{ji})<\alpha_j\frac{D+\delta/2}{P(j)}\right)\end{aligned}$$

Now by Sanov's theorem and the contraction principle we have

$$\lim_{n_j\to\infty}\log Q\left(\frac{1}{n_j}\sum_{i=1}^{n_j}\rho(j,\mathbf{y}'_{ji})<\alpha_j\frac{D+\delta/2}{P(j)}\right)\geq -\inf_{\nu\in F_j}\sum_k \nu(k)\log\frac{\nu(k)}{Q^*(k)}$$

where

$$F_j\equiv\left\{\nu:\sum_k\rho(j,k)\nu(k)<\alpha_j\frac{D+\delta/2}{P(j)}\right\}$$

Then

$$\lim_{n\to\infty}\frac{1}{n}\log Q(S_n(x))\geq -\sum_{j=1}^{M}P(j)\inf_{\nu\in F_j}\sum_k\nu(k)\log\frac{\nu(k)}{Q^*(k)}$$

Now note that by the defining relations for the α_j we have that $Q^*(\cdot\,|\,j)$ is in F_j for all j. Replacing the infimum by a particular value contained in the set will maintain the overall lower bound. Hence

$$\begin{aligned}\lim_{n\to\infty}\frac{1}{n}\log Q(S_n(x)) &\geq -\sum_{j=1}^{M}P(j)\sum_k Q^*(k\,|\,j)\log\frac{Q^*(k\,|\,j)}{Q^*(k)}\\ &=\sum_j\sum_k P(j)Q^*(k\,|\,j)\log\left[\frac{Q^*(k\,|\,j)P(j)}{Q^*(k)P(j)}\right]=-R(D)\end{aligned}$$

References

V. Anantharam (1990). A large deviations approach to error exponents in source coding and hypothesis testing. *IEEE Trans. Inform. Theory*, Vol. IT-36, No. 4 (in press).

R. Azencott (1980). Grandes déviations et applications. *Ecole d'été de Probabilités de Saint-Flour VIII*. Lecture Notes in Mathematics 774. Springer-Verlag, New York.

R. Azencott (1983). Densité des diffusions en temps petit: développements asymptotiques. *Séminaire Probabilités XVII*. Lecture Notes in Mathematics. 1059, Springer-Verlag, New York.

R. Azencott (1984). Petites perturbations aléatoires des systèmes dynamiques: développements asymptotiques. Preprint, University of Paris–South.

R. Azencott and G. Ruget (1977). Mélanges d'équations différentielles et grand écarts à la loi des grands nombres. *Z. Wahrsch. Verw. Gebiete*, Vol. 38, pp. 1–54.

R. Bahadur (1960). Asymptotic efficiency of tests and estimates. *Sankhya*, Vol. 22, pp. 229–252.

R. Bahadur (1967). Rates of convergence of estimates and test statistics. *Ann. Math. Statist.*, Vol. 38, pp. 303–324.

R. Bahadur (1971). *Some Limit Theorems in Statistics*. Regional Conference Series in Applied Mathematics. SIAM, Philadelphia.

R. Bahadur and R. R. Rao (1960). On deviations of the sample mean. *Ann. Math. Statist.*, Vol. 31, pp. 1015–1027.

R. Bahadur and S. Zabell (1979). Large deviations of the sample mean in general vector spaces. *Ann. Probab.*, Vol. 7, pp. 587–621.

R. Bahadur, S. Zabell, and J. Gupta (1980). Large deviations, tests, and estimates. In *Asymptotic Theory of Statistical Tests and Estimation*, edited by I. M. Chatcravarti, pp. 33–64. Academic Press, New York.

M. Basseville (1981). Edge detection using sequential methods for change in level: II. Sequential detection of change in mean. *IEEE Trans. Acoust. Speech Signal Process.*, Vol. ASSP-29, pp. 32–50.

M. Basseville (1988). Detecting changes in signals and systems:—a survey. *Automatica*, Vol. 24, No. 3, pp. 309–326.

M. Basseville and A. Benvensite (1983). Design and comparative study of some sequential jump detection algorithms for digital signals. *IEEE Trans. Acoust. Speech Signal Process.*, Vol. ASSP-31, June 1983, pp. 521–534.

M. Basseville, B. Espiau, and J. Gasnier (1981). Edge detection using sequential methods for change in level: I. A sequential edge detection algorithm. *IEEE Trans. Acoust. Speech Signal Process.*, Vol. ASSP-29, pp. 16–31.

G. Benitz and J. Bucklew (1987). An asymptotic analysis of nonlinear detectors in Gaussian noise. Presented at *1987 Johns Hopkins Conference on Information Sciences and Systems*; to appear *IEEE Trans. Inform. Theory.*

T. Berger (1971). *Rate Distortion Theory: A Mathematical Basis for Data Compression.* Prentice-Hall, Englewood Cliffs, N.J.

P. Billingsley (1968). *Convergence of Probability Measures.* Wiley, New York.

R. Blahut (1987). *Principles and Practice of Information Theory.* Addison-Wesley, Reading, Mass.

E. Bolthausen (1986). Laplace approximations for sums of independent random vectors. *Probab. Theory Rel. Fields*, Vol. 72, pp. 305–318.

A. Borokov (1967). Boundary value problems for random walks and large deviations in function spaces. *Theory Probab. Appl.*, Vol. 12, pp. 575–595.

L. Boza (1971). Asymptotically optimal tests for finite Markov chains. *Ann. Math. Statist.*, Vol. 42, pp. 1992–2007.

J. Bucklew (1987). The source coding theorem via Sanov's theorem. *IEEE Trans. Inform. Theory*, Vol. IT-33, No. 6, pp. 907–909.

J. Bucklew (1988). A large deviation theory proof of the abstract alphabet source coding theorem. *IEEE Trans. Inform. Theory*, Vol. IT-34, No. 5, pp. 1081–1083.

J. Bucklew and P. Ney (1988). The optimal design of memoryless nonlinearities via large deviation theory techniques. Presented at *1988 Princeton Conference on Information Sciences and Systems.*

J. Bucklew, P. Ney, and J. Sadowsky (1988). Fast simulation techniques. Presented at *1988 Regional Meeting of the Institute of Mathematical Statistics*, Madison, Wis., May 1988; to appear March 1990, *J. Appl. Probab.*

H. Chernoff (1952). A measure of asymptotic efficiency for tests of a hypothesis based on the sum of observations. *Ann. Math. Statist.*, Vol. 23, pp. 493–507.

M. Cottrell, J.-C. Fort, and G. Malgouyres (1983). Large deviations and rare events in the study of stochastic algorithms. *IEEE Trans. Automat. Control*, Vol. AC-28, No. 9, pp. 907–920.

H. Cramér (1938). Sur un nouveaux theorème-limite de la théorie des probabilités. Actualités Scientifiques et Industrielles 736, pp. 5–23. *Colloque consacré à la théorie des probabilités*, October 1937, Vol. 3. Hermann, Paris.

I. Csiszar and G. Longo (1971). *On the Error Exponent for Source Coding and for Testing Simple Statistical Hypotheses.* Hungarian Academy of Sciences, Budapest.

D. Dacunha-Castelle (1979). Formule de Chernoff pour des rapports de vraisemblance. In *Grandes Déviations et Applications Statistiques.* Société Mathématique de France, astérisque 68.

A. de Acosta (1985). Upper bounds for large deviations of dependent random vectors. *Z. Wahrsch. Verw. Gebiete,* Vol. 60, pp. 551–565.

A. de Acosta (1988). Large deviations for vector-valued functionals of a Markov chain: Lower bounds. *Am. Probab.,* Vol. 16, No. 3, pp. 925–960.

J. Deshayes and D. Picard (1982). Tests of disorder of regression: asymptotic comparison. *Theory Probab. Appl.,* Vol. 27, No. 1, pp. 100–115.

J. Deshayes and D. Picard (1986). Off-line statistical analysis of change-point models using nonparametric and likelihood methods. In *Detection of Abrupt Changes in Signals and Dynamical Systems,* edited by M. Basseville and A. Benvensite. Springer-Verlag, New York.

J.-D. Deuschel and D. Strook (1989). *Large Deviations.* Academic Press, New York.

M. Donsker and S. Varadhan (1975a). Asymptotic evaluation of certain Markov process expectations for large time: — I. *Comm. Pure Appl. Math.,* Vol. 28, pp. 1–47.

M. Donsker and S. Varadhan (1975b). Asymptotic evaluation of certain Markov process expectations for large time:—II. *Comm. Pure Appl. Math.,* Vol. 28, pp. 279–301.

M. Donsker and S. Varadhan (1976). Asymptotic evaluation of certain Markov process expectations for large time:—III. *Comm. Pure Appl. Math.,* Vol. 29, pp. 389–461.

P. Dupuis (1988). Large deviations analysis of some recursive algorithms with state dependent noise. *Ann. Probab.,* Vol. 16, No. 4, pp. 1509–1536.

R. Ellis (1984). Large deviations for a general class of random vectors. *Ann. Probab.,* Vol. 12, No. 1, pp. 1–12.

R. Ellis (1985). *Entropy, Large Deviations, and Statistical Mechanics.* Springer-Verlag, New York.

S. Ethier and T. Kurtz (1986). *Markov Processes: Characterization and Convergence.* Wiley, New York.

M. Freidlin and A. Wentzell (1984). *Random Perturbations of Dynamical Systems.* Springer-Verlag, New York.

J. Fu (1975). The rate of convergence of estimates and test statistics. *Ann. Math. Statist.,* Vol. 3, pp. 234–240.

F. Gantmacher (1960). *Matrix Theory* (Part II). Chelsea, New York.

J. Gärtner (1977). On large deviations from the invariant measure. *Theory Probab. Appl.,* Vol. 22, pp. 24–39.

I. Gelfand and S. Fomin (1963). *Calculus of Variations.* Prentice-Hall, Englewood Cliffs, N.J.

R. Gray (1972). On the asymptotic eigenvalue distribution of Toeplitz matrices. *IEEE Trans. Inform. Theory,* Vol. IT-18, pp. 767–800.

U. Grenander and G. Szego (1958). *Toeplitz Forms and Their Applications.* Univ. California Press, Berkeley.

P. Groeneboom, J. Oosterhoff, and F. Ruymgaart (1979). Large deviation theorems for empirical probability measures. *Ann. Probab.*, Vol. 7, No. 4, pp. 553–586.

D. Gustafson, A. Willsky, J.-Y. Wang, M. Lancaster, and J. Triebwasser (1978a). ECG/VCG rhythm diagnosis using statistical signal analysis: I. Identification of persistent rhythms. *IEEE Trans. Biomed. Eng.*, Vol. BME-25, pp. 344–353.

D. Gustafson, A. Willsky, J.-Y. Wang, M. Lancaster, and J. Triebwasser (1978b). ECG/VCG rhythm diagnosis using statistical signal analysis: II. Identification of transient rhythms. *IEEE Trans. Biomed. Eng.*, Vol. BME-25, pp. 353–361.

P. Hahn and M. Jeruchim (1987). Developments in the theory and applications of importance sampling. *IEEE Trans. Commun.*, Vol. COM-35, No. 7, pp. 706–714.

P. Hall and C. Heyde (1980). *Martingale Limit Theory and Its Application.* Academic Press, New York.

D. Halverson and G. Wise (1984). Asymptotic memoryless discrete-time detection of φ-mixing signals in φ-mixing noise. *IEEE Trans. Inform. Theory*, Vol. IT-30, pp. 189–198.

W. Hoeffding (1965). Asymptotically optimal tests for multinomial distributions. *Ann. Math. Statist.*, Vol. 31, pp. 369–408.

P. Hoel, S. Port, and C. Stone (1972). *Introduction to Stochastic Processes.* Waveland Press, Prospect Heights, Ill.

V. Hunkel and J. Bucklew (1988). Fast simulation for functionals of Markov chains. Presented at 1988 Princeton Conference on Information Sciences and Systems.

I. Iscoe, P. Ney, and E. Nummelin (1985). Large deviations of uniformly recurrent Markov additive processes. *Adv. Appl. Math.*, Vol. 6, pp. 373–412.

F. Jelinek (1968). *Probabilistic Information Theory.* McGraw-Hill, New York.

S. Karlin and H. Taylor (1975). *A First Course in Stochastic Processes.* Academic Press, New York.

T. Kato (1976). *Perturbation Theory for Linear Operators*, 2nd ed. Wiley, New York.

A. Kester and W. Kallenberg (1986). Large deviations of estimators. *Ann. Statist.*, Vol. 14, pp. 648–664.

G.-H. Kim and H. David (1979). Large deviations of functions of Markovian transitions and mathematical programming duality. *Ann. Probab.*, Vol. 7, pp. 874–881.

L. Koopmans (1960). Asymptotic rate of discrimination for Markov processes. *Ann. Math. Statist.*, Vol. 31, pp. 982–994.

S. Kullback (1959). *Information Theory and Statistics.* Wiley, New York.

S. Kullback and R. Leibler (1951). On information and sufficiency. *Ann. Math. Statist.*, Vol. 22, pp. 79–86.

T. Kurtz (1971). Limit theorem for sequences of jump Markov processes approximating ordinary differential processes. *Appl. Probab.*, Vol. 8, No. 2, pp. 344–356.

H. Kushner (1984). Robustness and approximation of escape times and large deviation estimates for systems with small noise effects. *SIAM J. Appl. Math.*, Vol. 44, No. 1, pp. 160–182.

H. Miller (1961). A convexity property in the theory of random variables on a finite Markov chain. *Ann. Math. Statist.*, Vol. 32, pp. 1260–1270.

J. Miller and J. Thomas (1972). Detectors for discrete-time signals in non-Gaussian noise. *IEEE Trans. Inform. Theory*, Vol. IT-18, pp. 241–250.

M. Milosavljevic (1988). The modified generalized likelihood ratio algorithm (MGLR) for automatic detection of abrupt changes in stationarity of signals. *Proceedings of 1988 Princeton Conference on Control, Information, and System Science.*

S. Natarajan (1985). Large deviations, hypotheses testing, and source coding for finite Markov chains. *IEEE Trans. Inform. Theory*, Vol. 31, No. 3, pp. 360–365.

P. Ney (1983). Dominating points and the asymptotics of large deviations for random walk on $\mathbb{R}^d$. *Ann. Probab.*, Vol. 11, No. 1, pp. 158–167.

P. Ney (1984). Convexity and large deviations. *Ann. Probab.*, Vol. 12, No. 3, pp. 903–906.

P. Ney and E. Nummelin (1987a). Markov additive processes: I. Eigenvalue properties and limit theorems. *Ann. Probab.*, Vol. 15, No. 2, pp. 561–592.

P. Ney and E. Nummelin (1987b). Markov additive processes: II. Large deviations. *Ann. Probab.*, Vol. 15, No. 2, pp. 593–609.

S. Orey (1983). Large deviations and Shannon–McMillan theorems. University of Minnesota Mathematics Report 83–124.

S. Orey and S. Pelikan (1986). Large deviation principles for stationary processes. Institute for Mathematics and Its Applications Preprint 227. University of Minnesota, Minneapolis.

P. Papantoni-Kazakos (1979). Algorithms for monitoring changes in quality of communications links. *IEEE Trans. Commun.*, Vol. COM-27, pp. 682–693.

A. Papoulis (1984). *Probability, Random Variables, and Stochastic Processes.* McGraw-Hill, New York.

S. Parekh and J. Walrand (1986). Quick simulation method for excessive backlogs in networks of queues. Electronics Research Laboratory Memorandum UCB/ERL M86/74. College of Engineering, University of California, Berkeley.

A. Perez (1971). Asymptotic discernability of two stationary Markov chains. In *2nd International Symposium of Information Theory in USSR*, pp. 365–374.

A. Perez (1974). Generalization of Chernoff's results on the asymptotic discernability of two random processes. In *Progress in Statistics:—European Meeting of Statisticians, Budapest*, Vol. II, pp. 619–632. North-Holland, Amsterdam.

M. Pincus (1968). Gaussian processes and Hammerstein integral equations. *Trans. Amer. Math. Soc.*, Vol. 134, pp. 193–216.

H. Poor (1988). *An Introduction to Signal Detection and Estimation.* Springer-Verlag, New York.

H. Poor and J. Thomas (1979). Memoryless discrete-time detection of a constant signal in *m*-dependent noise. *IEEE Trans. Inform. Theory*, Vol. IT-25, pp. 54–61.

C. Rao (1973). *Linear Statistical Inference and Its Applications*, 2nd ed. Wiley, New York.

F. Riesz and B. Sz.-Nagy (1955). *Functional Analysis*, 2nd ed. Ungar, New York.

B. Ripley (1987). *Stochastic Simulation*. Wiley, New York.

R. Rockafellar (1970). *Convex Analysis*. Princeton University Press, Princeton, N.J.

H. Royden (1968). *Real Analysis*, 2nd ed. Macmillan, Toronto.

W. Rudin (1974). *Real and Complex Analysis*, 2nd ed. McGraw-Hill, New York.

J. Sadowsky (1986). Some large deviations results for robust dependent data memoryless detector performance. Presented at *18th Annual Allerton Conference on Communications, Control, and Computing*, Montecello, Ill.

J. Sadowsky and J. Bucklew (1986). A non-local approach for asymptotic memoryless detection. *IEEE Trans. Inform. Theory*, Vol. IT-32, pp. 115–120.

I. Sanov (1957). On the probability of large deviations of random variables (in Russian), *Mat. Sb.*, Vol. 42, pp. 11–44. (English translation in *Selected Translations in Mathematical Statistics: I*, pp. 213–244, 1961).

M. Schilder (1966). Some asymptotic formulae for Wiener integrals. *Trans. Amer. Math. Soc.*, Vol. 125, pp. 63–85.

S. Shaban (1980). Change-point problem and two-phase regression: an annotated bibliography. *Internat. Statist. Rev.*, Vol. 48, pp. 83–93.

C. Shannon (1948). A mathematical theory of communication. *Bell System Tech. J.*, Vol. 27, pp. 379–423, 623–656.

C. Shannon (1959). Coding theorems for a discrete source with a fidelity criterion. *IRE National Convention Record*, part 4, pp. 142–163.

F. Stremler (1990). *Introduction to Communication Systems*, 3rd ed. Addison-Wesley, Reading, Mass.

D. Strook (1984). *An Introduction to the Theory of Large Deviations*. Springer-Verlag, New York.

J. Van Tiel (1984). *Convex Analysis:—An Introductory Text*. Wiley, New York.

H. Van Trees (1968). *Detection, Estimation, and Modulation Theory* (Part I). Wiley, New York.

S. Varadhan (1984). *Large Deviations and Applications*. SIAM, Philadelphia.

A. Weiss (1987). Large deviation theory and birth–death models with applications. Preprint.

A. Wentzell (1976a). Rough limit theorems on large deviations for Markov stochastic processes:—I. *Theory Probab. Appl.* (*USSR*), Vol. 21, pp. 227–242.

A. Wentzell (1976b). Rough limit theorems on large deviations for Markov stochastic processes:—II. *Theory Probab. Appl.* (*USSR*), Vol. 21, pp. 499–512.

A. Willsky (1976). A survey of several failure detection methods. *Automatica*, November, pp. 601–611.

E. Wong and B. Hajek (1985). *Stochastic Processes in Engineering Systems*. Springer-Verlag, New York.

Index

GUTTMAN, WILKS, and HUNTER • Introductory Engineering Statistics, *Third Edition*
HAHN and SHAPIRO • Statistical Models in Engineering
HALD • Statistical Tables and Formulas
HALD • Statistical Theory with Engineering Applications
HAND • Discrimination and Classification
HEIBERGER • Computation for the Analysis of Designed Experiments
HOAGLIN, MOSTELLER, and TUKEY • Exploring Data Tables, Trends and Shapes
HOAGLIN, MOSTELLER, and TUKEY • Understanding Robust and Exploratory Data Analysis
HOCHBERG and TAMHANE • Multiple Comparison Procedures
HOEL • Elementary Statistics, *Fourth Edition*
HOEL and JESSEN • Basic Statistics for Business and Economics, *Third Edition*
HOGG and KLUGMAN • Loss Distributions
HOLLANDER and WOLFE • Nonparametric Statistical Methods
HOSMER and LEMESHOW • Applied Logistic Regression
IMAN and CONOVER • Modern Business Statistics
JESSEN • Statistical Survey Techniques
JOHNSON • Multivariate Statistical Simulation
JOHNSON and KOTZ • Distributions in Statistics
Discrete Distributions
Continuous Univariate Distributions—1
Continuous Univariate Distributions—2
Continuous Multivariate Distributions
JUDGE, GRIFFITHS, HILL, LÜTKEPOHL, and LEE • The Theory and Practice of Econometrics, *Second Edition*
JUDGE, HILL, GRIFFITHS, LÜTKEPOHL, and LEE • Introduction to the Theory and Practice of Econometrics, *Second Edition*
KALBFLEISCH and PRENTICE • The Statistical Analysis of Failure Time Data
KASPRZYK, DUNCAN, KALTON, and SINGH • Panel Surveys
KAUFMAN and ROUSSEEUW • Finding Groups in Data: An Introduction to Cluster Analysis
KEENEY and RAIFFA • Decisions with Multiple Objectives
KISH • Statistical Design for Research
KISH • Survey Sampling
KUH, NEESE, and HOLLINGER • Structural Sensitivity in Econometric Models
LAWLESS • Statistical Models and Methods for Lifetime Data
LEAMER • Specification Searches: Ad Hoc Inference with Nonexperimental Data
LEBART, MORINEAU, and WARWICK • Multivariate Descriptive Statistical Analysis: Correspondence Analysis and Related Techniques for Large Matrices
LINHART and ZUCCHINI • Model Selection
LITTLE and RUBIN • Statistical Analysis with Missing Data
McNEIL • Interactive Data Analysis
MAGNUS and NEUDECKER • Matrix Differential Calculus with Applications in Statistics and Econometrics
MAINDONALD • Statistical Computation
MALLOWS • Design, Data, and Analysis by Some Friends of Cuthbert Daniel
MANN, SCHAFER, and SINGPURWALLA • Methods for Statistical Analysis of Reliability and Life Data
MARTZ and WALLER • Bayesian Reliability Analysis
MASON, GUNST, and HESS • Statistical Design and Analysis of Experiments with Applications to Engineering and Science
MILLER • Beyond ANOVA, Basics of Applied Statistics
MILLER • Survival Analysis
MILLER, EFRON, BROWN, and MOSES • Biostatistics Casebook